COMMON CORE BASICS

Building Essential Test Readiness Skills

MATHEMATICS

Mc
Graw
Hill
Education

Bothell, WA • Chicago, IL • Columbus, OH • New York, NY

mheonline.com

Send all inquiries to:
McGraw-Hill Education
8787 Orion Place
Columbus, OH 43240

ISBN: 978-0-07-657519-0
MHID: 0-07-657519-5

Printed in the United States of America.

4 5 6 7 8 9 QLM 18 17 16 15 14

Contents

To the Student

Common Core Basics: Building Essential Test Readiness Skills, Mathematics will help you learn or strengthen the skills you need when you take any Common Core State Standards-aligned mathematics test. Regardless of your background, you can use this book to learn and practice the mathematical and problem-solving skills that are most important.

Before beginning the lessons in this book, take the Pretest. This test will help you identify which skill areas you need to concentrate on most. Use the chart at the end of the Pretest to pinpoint the types of questions you have answered incorrectly and to determine which skills you need to work on. You may decide to concentrate on specific areas of study or to work through the entire book. It is highly recommended that you work through the whole book to build a strong foundation in the core areas in which you will be tested.

Common Core Basics: Building Essential Test Readiness Skills, Mathematics is divided into twelve chapters:

- **Chapter 1: Whole Numbers** introduces place value, ordering and comparing numbers, operations on numbers, factoring, rounding, estimation, and solving real-world problems with whole numbers.

- **Chapter 2: Decimals** teaches decimal place value, operations with decimals, and solving real-world problems with decimals.

- **Chapter 3: Fractions** introduces operations with fractions and mixed numbers.

- **Chapter 4: Integers** explains integer concepts, including absolute value, integer operations, and the coordinate grid.

- **Chapter 5: Expressions and Equations** teaches writing and evaluating expressions, solving one- and two-step equations and inequalities, and identifying patterns.

- **Chapter 6: Linear Equations and Functions** introduces linear equations, graphing linear equations, functions, and scatter plots.

- **Chapter 7: Ratios, Proportions, and Percents** introduces ratios and rates, solving proportions, and solving percent problems.

- **Chapter 8: Exponents and Roots** teaches how to evaluate expressions with exponents, roots, and scientific notation.

- **Chapter 9: Data** introduces the measures of central tendency and range, graphs and line plots, and misleading graphs.

- **Chapter 10: Probability** explains counting methods and introduces probability and compound events.

- **Chapter 11: Measurement** teaches using customary units, metric units, and converting among units.

- **Chapter 12: Geometry** explores various geometric figures and solids, scale drawings and measurement, the Pythagorean theorem, and finding perimeter, circumference, area, and the volume of cones and spheres.

In addition, *Common Core Basics: Building Essential Test Readiness Skills, Mathematics* has a number of features designed to familiarize you with and prepare you for mathematics tests.

- The **Chapter Opener** provides an overview of the content and a goal-setting activity.

- **Lesson Objectives** state what you will be able to accomplish after completing the lesson.

- **Skills** list the Core Skills and Core Practice that are taught and applied to the lesson content.

- **Vocabulary** critical for understanding lesson content is listed at the start of every lesson. All bold words in the text can be found in the Glossary.

- The **Key Concept** summarizes the content that is the focus of the lesson.

- **Think About Math** questions check your understanding of the content throughout the lesson as you read.

- **Math Link** activities include tips and strategies related to the mathematics skills and practices.

- End-of-lesson **Vocabulary Review** checks your understanding of important lesson vocabulary, while the **Skill Review** checks your understanding of the content and skills presented in the lesson.

- **Skill Practice** exercises appear at the end of every lesson to help you apply your learning of content and skill fundamentals.

- The **Chapter Review** tests your understanding of the chapter content.

- **Check Your Understanding** charts allow you to check your knowledge of the skills you have practiced.

- The **Answer Key** explains the answers for questions in the book.

- The **Glossary** and **Index** contain lists of key terms found throughout the book and make it easy to review important skills and concepts.

After you have worked through the book, take the Posttest to see how well you have learned the skills presented in this book.

Good luck with your studies! Keep in mind that mathematics and problem-solving skills will help you succeed on any mathematics test and in other future tasks, whether at school, at home, or in the workplace.

Mathematics

Take the Mathematics Pretest before you begin working on any of the chapters in this book. The purpose of the pretest is to help you determine which skills you need to develop to improve your mathematical and problem-solving skills. The test consists of 25 multiple-choice questions that correspond to the twelve chapters of this book.

Directions: Choose the best answer to each question. Answer each question as carefully as possible. If a question seems to be too difficult, do not spend too much time on it. Work ahead and come back to it later when you can think it through carefully. When you have completed the test, check your work with the answers and explanations on pages 6–7.

Use the Evaluation Chart on page 8 to determine which areas you need to study the most. If you missed many of the questions that correspond to a certain mathematical skill, you will want to pay special attention to that skill as you work your way through this book.

Mathematics

Directions: Choose the best answer to each question.

1. After Nikos left his car in the parking lot of a park, he walked a total of 347 paces due east and then 284 paces due south to the information booth. If he entered a park after walking only 59 paces, how many paces did he walk after entering the park and before arriving at the information booth?

   A. 690   C. 572
   B. 631   D. 288

2. Simplify the following decimal expression.
   $4.56 - 0.932 + 11.087$

   A. 10.611
   B. 14.715
   C. 7.11
   D. 15.498

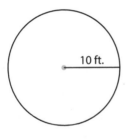

3. A courtyard for an office building is in the shape of a circle, as shown in the diagram above. It has a radius of 10 feet. What is the circumference, in feet, of the courtyard? (Use 3.14 for π.)

   A. 15.7
   B. 31.4
   C. 33.4
   D. 62.8

4. Multiply the following decimal expression.
   5.63×8.72

   A. 49.0936
   B. 40.4536
   C. 490.936
   D. 404.536

5. Last week, Kristin bought $5\frac{3}{4}$ pounds of apples. This week, she bought one-half of that amount. How many pounds of apples did Kristin buy this week?

   A. $2\frac{1}{2}$
   B. $2\frac{7}{8}$
   C. $5\frac{1}{4}$
   D. $11\frac{1}{2}$

6. In the graph below, which point has coordinates $(-1, -3)$?

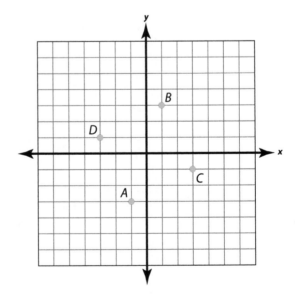

   A. A
   B. B
   C. C
   D. D

Mathematics

7. In the diagram below, which point represents the value of −3 + (−5)?

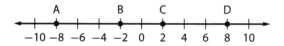

A. A
B. B
C. C
D. D

Use the following table for Questions 8 and 9.

HIGH TEMPERATURES FOR A WEEK

Day	Farenheit Temperature
Sunday	72°
Monday	80°
Tuesday	83°
Wednesday	78°
Thursday	70°
Friday	74°
Saturday	75°

8. What is the median high temperature in °F for the week?

A. 75
B. 76
C. 78
D. 83

9. What is the range of high temperatures in °F for the week?

A. 2
B. 3
C. 7
D. 13

10. Write the number 1,234,567,890 in scientific notation.

A. $1.23456789 \times 10^{10}$
B. 12.3456789×10^{8}
C. 1.23456789×10^{9}
D. 123.456789×10^{7}

11. Liam drank 54 milliliters of water. How many liters of water did Liam drink?

A. 5.4
B. 0.54
C. 0.054
D. 0.0054

12. Which expression shows the probability of getting doubles when you roll two number cubes?

A. $\dfrac{\text{favorable outcomes}}{\text{possible outcomes}} = \dfrac{30}{36} = \dfrac{5}{6}$

B. $\dfrac{\text{favorable outcomes}}{\text{possible outcomes}} = \dfrac{6}{36} = \dfrac{1}{6}$

C. $\dfrac{\text{possible outcomes}}{\text{favorable outcomes}} = \dfrac{36}{6} = 6$

D. $\dfrac{\text{number of dice}}{\text{number of doubles}} = \dfrac{2}{6} = \dfrac{1}{3}$

13. Frozen bread dough is priced at $3.39 for 3 loaves. What is the unit price (the price of one loaf)?

A. $0.88
B. $1.13
C. $2.26
D. $10.17

Mathematics

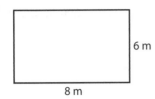

6 m

8 m

14. What is the area, in square meters, of the rectangle above?

   A. 14
   B. 28
   C. 48
   D. 96

15. Which function below represents converting feet (f) to yards (y)?

   A. $y = 3f$
   B. $f = 3y$
   C. $f = y + 3$
   D. $y = f + 3$

16. What is the value of $3x$ if $x = -1$?

   A. -31
   B. -3
   C. 2
   D. 3

17. Percy is 3 years older than his brother. If Percy is 27, how old in years is his brother?

   A. 9
   B. 24
   C. 26
   D. 30

18. Gary is planting trees in a park. He wants to plant at least 5 trees for every 2 picnic tables. If there are 12 picnic tables in the park, what is a possible number of trees Gary could plant?

   A. 40
   B. 25
   C. 10
   D. 5

19. Refer to the circle graph below. The graph shows the results of a survey of the number of bedrooms in homes that were sold in 2012. Which statement is supported by the information in the circle graph?

PERCENTAGE OF HOMES THAT SOLD IN 2012 WITH 1, 2, 3, AND 4 OR MORE BEDROOMS

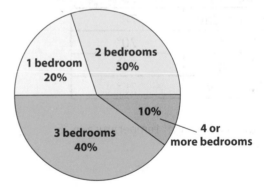

   A. The number of 2-bedroom homes sold is twice the number of 1-bedroom homes sold.
   B. The number of 4 or more bedroom homes sold was equal to the number of 1-bedroom homes sold.
   C. More than half the homes sold had either 1 or 2 bedrooms.
   D. More homes were sold with 3 bedrooms than with 4 or more bedrooms.

Mathematics

20. What is the volume, in cubic inches, of the cube shown below?

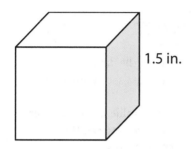

1.5 in.

- **A.** 4.5
- **B.** 3.375
- **C.** 2.25
- **D.** 1.5

21. Company A offers a yearly cable television plan for $100 installation with a $34 monthly fee. Company B offers a yearly plan for $30 installation with a $41 monthly fee. In which month will Company B's plan cost more than Company A's plan?

- **A.** 10 months
- **B.** 9 months
- **C.** 12 months
- **D.** 11 months

22. What is the slope of a line that passes through the points (3, 8) and (–2, 5)?

- **A.** $\frac{5}{3}$
- **B.** $-\frac{5}{3}$
- **C.** $\frac{3}{5}$
- **D.** $-\frac{3}{5}$

23. What is the next number in the pattern?

5, 10, 20, 40

- **A.** 50
- **B.** 60
- **C.** 80
- **D.** 100

24. Which of the statements explains how to find the next number in the following pattern?

33, 30, 27, 24, 21

- **A.** add 3 to 21
- **B.** multiply 3 by 21
- **C.** subtract 3 from 21
- **D.** divide 21 by 3

25. Mindy is writing thank-you notes. She has a mixed pile of cards that contains 2 yellow cards, 3 blue cards, and 1 white card. She randomly chooses one card and writes her note. What is the probability that the card she chose was yellow?

- **A.** $\frac{2}{3}$
- **B.** $\frac{1}{2}$
- **C.** $\frac{1}{6}$
- **D.** $\frac{1}{3}$

Answer Key

1. **A.** added 59 instead of subtracting it
   B. failed to subtract 59 from the sum of 347 and 284
   C. KEY: $(347 - 59) + 284 = 572$
   D. answered the question "How many paces due east did Nikos walk after entering the park and before arriving at the information booth?"

2. **A.** added $0.456 - 0.932 + 11.087$
   B. KEY: $4.56 - 0.932 + 11.087 = 14.715$
   C. added $4.56 - 9.32 + 11.87$
   D. added $4.56 - 0.932 + 11.87$

3. **A.** divided the product πr by 2 instead of multiplying by 2
   B. failed to multiply the product πr by 2
   C. added 2 to the product πr instead of multiplying by 2
   D. KEY: $2\pi r$ is about $2(3.14)(10) = 62.8$ ft.

4. **A.** KEY: $5.63 \times 8.72 = 49.0936$
   B. multiplied 5×8, then $.63 \times .72$, and added together
   C. multiplied correctly, misplaced decimal
   D. multiplied same as (B.), misplaced decimal

5. **A.** multiplied 5 by $\frac{1}{2}$ instead of multiplying $5\frac{3}{4}$ by $\frac{1}{2}$
   B. KEY: $5\frac{3}{4} \times \frac{1}{2} = \frac{23}{4} \times \frac{1}{2} = \frac{23}{8} = 2\frac{7}{8}$
   C. subtracted $\frac{1}{2}$ instead of multiplying by $\frac{1}{2}$
   D. divided by $\frac{1}{2}$ instead of multiplying by $\frac{1}{2}$

6. **A.** KEY: for $(-1, -3)$, locate one unit to the left of the origin, then 3 down
   B. found $(1, 3)$
   C. found $(3, -1)$
   D. found $(-3, 1)$

7. **A.** KEY: $-3 + (-5) = -8$
   B. added $+3$ and -5
   C. added -3 and $+5$
   D. added $+3$ and $+5$

8. **A.** KEY: 70, 72, 74, 75, 78, 80, 83; 75 is middle number
   B. found the mean of the data set
   C. failed to put the data set in order from least to greatest
   D. found the greatest value in the data set

9. **A.** found the difference between the two lowest temperatures
   B. found the difference between the two highest temperatures
   C. found the total number of high temperatures for the week
   D. KEY: $83 - 70 = 13$

10. **A.** correct decimal number, but incorrect exponent
    B. incorrect decimal number, incorrect exponent
    C. KEY: $1{,}234{,}567{,}890 = 1.23456789 \times 10^9$
    D. incorrect decimal number, incorrect exponent

11. **A.** divided by 10 instead of by 1,000
    B. divided by 100 instead of by 1,000
    C. KEY; 1 mL $= 0.001$ L; 54 mL $= 54(0.001)$ or 0.054 L
    D. divided by 10,000 instead of by 1,000

12. **A.** answered the question "Which expression shows the probability of not rolling doubles?"
    B. KEY: $(1, 1), (2, 2), (3, 3), (4, 4), (5, 5), (6, 6)$, so there are 6 ways to get doubles out of a total of 36 possible outcomes. $\frac{6}{36} = \frac{1}{6}$
    C. found the inverse of the probability
    D. answered the question "Which expression shows the ratio of the number of dice to the number of possible doubles?"

Answer Key

13. **A.** divided 3 by 3.39 instead of 3.39 by 3
    B. KEY: $3.39 ÷ 3 = $1.13
    C. answered the question "Which expression shows the price of two loaves?"
    D. multiplied by 3 instead of dividing

14. **A.** added instead of multiplying
    B. found the perimeter
    C. KEY: $6 \times 8 = 48$
    D. found twice the area

15. **A.** KEY: There are three feet in a yard, and so to convert from feet to yards, multiply the number of feet by 3, so $y = 3f$
    B. translated the sentence "1 foot is 3 yards" to an equation
    C. added 3 instead of multiplying 3
    D. added 3 instead of dividing 3

16. **A.** does not understand integer multiplication
    B. KEY: $3 \times (-1) = -3$
    C. added 3 and -1
    D. did not use negative sign for -1 to multiply

17. **A.** divided by 3
    B. KEY: $27 - 3 = 24$
    C. subtracted incorrectly
    D. added 3 to 27

18. **A.** KEY: $12 ÷ 2 = 6$ (There will be at least 5 trees for every 2 tables); Gary will plant at least 5×6 (or 30) trees; 40 is the only answer choice greater than or equal to 30.
    B. found the fewest number of trees if there were 10 picnic tables
    C. found the fewest number of trees if there were 4 picnic tables
    D. found the fewest number of trees if there were 2 picnic tables

19. **A.** did not compare 30% to 20% accurately
    B. did not compare 10% to 20% accurately
    C. did not compare 20% + 30% to 50% accurately
    D. KEY: 40% > 10%

20. **A.** multiplied 1.5 times 3 instead of cubing 1.5
    B. KEY: volume = $s^3 = (1.5)^3 = (1.5)(1.5)(1.5) = 3.375$
    C. found the surface area of one face of the cube
    D. found the length of one side of the cube

21. **A.** found month total costs are equal
    B. found month before total costs are equal
    C. chose the months in a year
    D. KEY: Company A costs $70 more upfront, but $7 less per month. It will take 10 months for the two to agree. During the 11th month, Company B will cost more.

22. **A.** computed slope as change of x divided by change of y, and numbers out of order
    B. computed slope as change of x divided by change of y
    C. KEY: slope is $\frac{(5-8)}{(-2-3)} = \frac{3}{5}$
    D. computed slope with numbers out of order

23. **A, B, D.** did not know that the pattern is to multiply previous number by 2
    C. KEY: $2 \times 40 = 80$

24. **A, B, C.** did not try rule with pattern
    C. KEY: $33 - 3 = 30$, $30 - 3 = 27$, $27 - 3 = 24$, $24 - 3 = 21$, so the rule is to subtract 3

25. **A.** found probability of all cards other than yellow
    B. found probability of a blue card
    C. found probability of a white card
    D. KEY: 2 yellow cards out of 6 cards is $\frac{2}{6} = \frac{1}{3}$.

Evaluation Chart

On the following chart, circle the number of any problem you got wrong. After each problem, you will see the name of the section where you can find the skills you need to solve the problem.

Problem	Unit: Section	Starting Page
	Number Sense and Operations	
1	Add and Subtract Whole Numbers	18
17	Arithmetic Expressions	36
2	Add and Subtract Decimals	54
4	Multiply Decimals	60
5	Multiply and Divide Fractions	88
	Basic Algebra	
7	Add Integers	108
6	The Coordinate Grid	124
16	Expressions	134
23, 24	Identify Patterns	156
22	Linear Equations	166
21	Pairs of Linear Equations	184
15	Functions	200
	More Number Sense and Operations	
13	Ratios and Rates	212
18	Solve Proportions	224
10	Scientific Notation	260
	Data Analysis and Probability	
8, 9	Measures of Central Tendency and Range	270
19	Graphs and Line Plots	274
12, 25	Introduction to Probability	296
	Measurement and Geometry	
11	Metric Units	316
3, 14	Perimeter and Circumference	332
20	Geometric Solids and Volume	360

UNIT 1

Number Sense and Operations

Whole Numbers

People use numbers throughout the day to tell the time or date, to understand how much something costs, to make a phone call, or to find an address. This chapter reviews the basic principles of numbers. How can knowing more about numbers help you in your daily life?

The set of whole numbers is made up of the counting, or natural, numbers 1, 2, 3, 4, 5, 6, 7, 8, 9, ..., and 0. The whole numbers are the basic set of numbers. Addition, subtraction, multiplication, and division are introduced with these numbers.

The Key Concepts you will study include:

Lesson 1.1: Place Value
Represent, compare, and order whole numbers to better understand the meaning and value of whole numbers.

Lesson 1.2: Add and Subtract Whole Numbers
Addition and subtraction are basic operations in mathematics.

Lesson 1.3: Multiply and Divide Whole Numbers
Multiplication is the operation of adding a certain quantity a set number of times. Division is the operation that is used to separate a quantity into parts.

Lesson 1.4: Factoring
Learning how to factor numbers enables you to compare the factors of two numbers and also to find the greatest common factor of two numbers.

Lesson 1.5: Rounding and Estimation
Rounding and estimation are useful when an answer does not need to be exact or when checking an exact answer.

Lesson 1.6: Arithmetic Expressions
Basic mathematical operations must be performed in a specified order to obtain the correct answer.

Lesson 1.7: Problem Solving
Problem solving is an important part of the study of mathematics and an important part of everyday life.

Goal Setting

Before starting this chapter, set goals for your learning. Think about the ways that strengthening your understanding of whole numbers will benefit you.

- How do you use numbers and mathematics in your daily life?

  _____

- What tasks would be easier if you were more confident about your math skills?

  _____

Understanding whole numbers is the foundation for building strong math skills. Like a house that needs a strong, solid foundation to support it, a clear understanding of whole numbers is necessary to help you learn about other mathematical concepts. Check the strength of your understanding by returning to this page after you have completed the chapter. Can you answer each question? If so, you are ready to move to Chapter 2.

- Why is place value an important concept?

  _____

- How are addition and subtraction similar?

  _____

- How are addition and subtraction different?

  _____

- How are multiplication and division related?

  _____

- Why is estimation a key problem-solving skill?

  _____

- How does factoring simplfy problem solving?

  _____

- What is the five-step approach to problem solving?

  _____

Place Value

KEY CONCEPT: Represent, compare, and order whole numbers to better understand the meaning and value of whole numbers.

1. What number is 1 more than 8?

2. What number is 1 less than 73?

3. What number is 10 more than 60?

4. What number is 10 less than 45?

Lesson Objectives

You will be able to
• Use place value to read and write whole numbers
• Compare and order whole numbers

Skills

• **Core Skill:** Apply Number Sense Concepts
• **Core Practice:** Model with Mathematics

Vocabulary

approximate
chart
digit
number line
period
value
whole number

Place Value

Digits are the ten number symbols: 0, 1, 2, 3, 4, 5, 6, 7, 8, 9. A number is an arrangement of digits in a particular order. The numbers beginning with 0, 1, 2, 3, and so on are the set of **whole numbers**. The position of a digit in a number determines its **value**, or how much it represents.

Starting from the ones place, commas are inserted every third number to separate a number into groups of three, called **periods**.

PLACE VALUE CHART

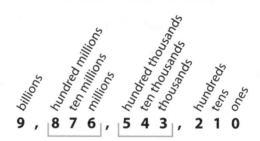

Example 1 Use a Place-Value Chart

In the number 137,258,406, which digit is in the ten millions place?

Step 1 Locate the ten millions place in the place-value chart. A chart is an arrangement of numbers or other information.

Step 2 Find the digit in 137,258,406 that is in that position. The 3 is in the ten millions place.

Example 2 Determine the Value of the Digits

What is the value of each digit in the number 105?

Step 1 1 is in the hundreds place. Its value is 1 hundred or 100.

Step 2 0 is in the tens place. No tens are in the number 105.

Step 3 The 5 is in the ones place. Its value is 5 ones, or 5.

IDENTIFY THE MAIN IDEA

Most of the material you read both at home and in school contains a **main idea**. The main idea tells what the paragraph, article, or lesson is about. The other sentences support the main idea.

The main idea is not always found in the first sentence or even in the first paragraph of a passage. It might be found almost anywhere within a passage. Sometimes the main idea is not even stated directly.

To identify the main idea ask: *What is this passage about?*

Read the following paragraph and identify the main idea.

> (1) For some problems, an exact answer is not needed.
> (2) An estimate (an **approximate**, or "about," answer) will be sufficient. (3) It is also good to estimate an answer, then solve the problem, and finally check the solution by comparing the estimate to the exact answer.

Sentence 1 is a general suggestion that exact answers are not always needed. Sentence 2 explains what an estimate is and states that it may be all that is needed to answer a math problem. Sentence 3 states the usefulness of making an estimate first, finding an exact solution, and then comparing the two. Sentences 1 and 3 support Sentence 2. Sentence 2 states the main idea.

Charts, which are diagrams that show information, can help you understand place value. In the place-value chart on page 12, for example, the place value is shown for each digit in the number, and periods are indicated. In a notebook, write a sentence explaining how the value of each digit in the number 5,555 changes as the digits move from the right of the number to the left.

THINK ABOUT MATH

Directions: Identify the value of the underlined digit.

1. 3,4<u>7</u>8 _____   4. 7,<u>3</u>00,561,892 _____

2. 15,7<u>8</u>9,200 _____   5. <u>8</u>,570,213,000 _____

3. 702,432,51<u>6</u> _____

Read and Write Whole Numbers

In general, we read whole number in words and use the digits 0, 1, 2, 3, 4, 5, 6, 7, 8, and 9 to write them.

Example 3 Read Whole Numbers

Read the number 28,304.

Step 1 Begin at the left of the number. Read the number in each period, and replace the comma with the name of the period.
So, the number 28,304 means 28 thousands, 3 hundreds, 0 tens, 4 ones.

Step 2 Read the number 28,304 as "twenty-eight thousand, three hundred four."

When reading whole numbers, remember to concentrate on each period and the positions of all the digits.

Example 4 Write Whole Numbers

Write the number *six million, two hundred ninety-one thousand, fifty* as a whole number.

Step 1 Six million becomes 6,000,000.
Two hundred ninety-one thousand becomes 291,000.
Fifty becomes 50.

Step 2 Combine the whole number parts.
6,000,000 + 291,000 + 50 = 6,291,050

When you write whole numbers, think about the place-value chart. Remember to insert zeros as needed.

MATH LINK

Remember that zeros hold a position and should not be ignored. When writing numbers, write a zero for each place that is not expressed in words.

THINK ABOUT MATH

Directions: Match the number to its name in words.

_____ 1. fourteen thousand, two hundred sixty

_____ 2. five thousand, eighty-five

_____ 3. seventy-eight

_____ 4. twenty-six million

_____ 5. one hundred eleven thousand

A. 78
B. 111,000
C. 26,000,000
D. 5,085
E. 14,260

Compare and Order Whole Numbers

Compare numbers by using a **number line**. A number line is a list of numbers arranged in order from left to right on a line. The numbers are larger the farther right they are on the number line.

Example 5 Use a Number Line to Compare Numbers

Which is greater, 35 or 45?

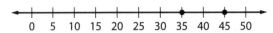

Step 1 Locate each number on the number line.

Step 2 45 is to the right of 35, so 45 is greater than 35. Write this as 45 > 35 (45 is greater than 35) or 35 < 45 (35 is less than 45).

Example 6 Use Place Value to Compare Numbers

Compare. Write < or > in the blank to make a true statement.

12,358 _____ 12,421

Step 1 To compare, align the numbers by place value.

12,358
12,421

Step 2 Start at the left. Compare the digits until they differ.

12,358
12,421

The digits in the hundreds place are different. 3 is less than 4, so 12,358 < 12,421.

Example 7 Order Whole Numbers

Write the set of numbers in order from greatest to least.

4,134,805   5,883,081   4,147,001

Step 1 Align the numbers by place value. Start at the left and compare digits.

4,134,805
5,883,081
4,147,001

Step 2 5 > 4, so 5,883,081 is the greatest number. Continue comparing the other numbers until the digits differ.

4,134,805
4,147,001

4,147,001 > 4,134,805
So, 5,883,081 > 4,147,001 > 4,134,805.

MATH LINK

When using the inequality symbols < and >, remember that the pointed end of the symbol always points to the smaller number.

THINK ABOUT MATH

Directions: Answer the following.

1. How would you use place value to compare 203,478 and 204,210?

   _____

   _____

2. Place the numbers below in order from least to greatest.
   701,286   698,321   698,432

   _____

Vocabulary Review

Directions: Complete the sentences below using one of the following words.

approximate   digits   number line   periods   value   whole numbers

1. The position of a digit in a number determines its _____.

2. The set of numbers beginning with 0, 1, 2, 3, and so on is the set of _____.

3. Starting from the ones place, insert commas every third place to separate a number into groups of three called _____.

4. The number system used today is based on the _____ 0 through 9 arranged in a particular pattern.

5. Antonia did not need an exact answer, so she found an _____ answer.

6. Numbers can be compared by using a _____.

Skill Review

Directions: Write the value of each digit based on the place-value chart.

Millions	Hundred thousands	Ten thousands	Thousands	Hundreds	Tens	Ones
2	4	7	3	0	1	5

1. 2 _____

2. 4 _____

3. 7 _____

4. 3 _____

5. 0 _____

6. 1 _____

7. 5 _____

Skill Review (continued)

Directions: Create a place value chart for each number below.

8. 6,729

9. one million, thirty-five

Directions: Locate each point on a number line. Then compare the numbers.

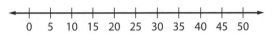

0   5   10   15   20   25   30   35   40   45   50

10. 16; 40

11. 20; 13

12. 5; 27

13. 50; 2

Directions: Identify the main idea of the passage.

14. One way to compare numbers is to line the digits up by place value, then compare the first place where the digits are different. A number line can also be used. You should arrive at the same conclusion with either method. There are several ways to compare numbers correctly.

15. Digits represent numbers. 1, 2, 3, 4, 5, 6, 7, 8, 9, and 0 are the digits we use. The digits can be combined and re-ordered to create infinitely many numbers. Digits have different meanings based on which position they are in.

Skills Practice

Directions: Choose the best answer to each question.

1. Which of the following has the numbers arranged from greatest to least?

A. 45,378; 55,210; 56,345
B. 5,010,000; 5,100,321; 5,002,146
C. 78; 75; 72
D. 379; 389; 398

2. Laquita wrote a check for $241 to pay her heating bill. How did she write this number in words on the check?

A. two hundred forty-one
B. two hundred forty
C. two hundred fourteen
D. twenty-one

3. Ashley priced four cars she was interested in buying. Which car was the most expensive?

A. Car A: $23,456
B. Car B: $22,201
C. Car C: $22,345
D. Car D: $23,712

4. For the number 601,295, what is the place value of 0?

A. hundred thousands
B. ten thousands
C. thousands
D. hundreds

Add and Subtract Whole Numbers

KEY CONCEPT: Addition and subtraction are basic operations in mathematics.

Write each number in words.
1. 37 _____   2. 1,008 _____ 3. 152 _____   4. 32,000 _____

Use a less than (<), greater than (>), or equal to (=) symbol to compare each pair of numbers.
5. 15 □ 39   6. 301 □ 108   7. 222 □ 44   8. 1,234 □ 1,324

Add Whole Numbers

The most basic of all **operations**, or processes, in mathematics is addition. **Addition** is the combining of two or more numbers. Suppose you have two sets of pencils: 4 in one set and 3 in the other. Find the total number of pencils by combining the two sets or adding $4 + 3$.

$$4 \quad + \quad 3 \quad = \quad 7$$

The answer to an addition problem is called the **sum,** or total. So the sum of 4 and 3 is 7.

Example 1 Add Two-Digit Numbers

Find the sum of 32 and 47.

Step 1 To **calculate** means to find the answer using a mathematical process. To calculate the sum of an addition problem, line up the digits with ones under ones, tens under tens, and so on.

$$\begin{array}{r} 3\,2 \\ +\,4\,7 \\ \hline 7\,9 \end{array}$$

Step 2 Add the ones column.

Step 3 Add the tens column.

Example 2 Column Addition

Add: 248 + 36 + 1,987

Step 1 Line up the digits by place value.

Step 2 Add the digits in the ones column and write the sum at the bottom. If the sum has more than one digit, carry the left digit to the next column.

Step 3 Repeat until all columns have been added.

$$\begin{array}{r} {\scriptstyle 1\ 1\ 2} \\ 2\,4\,8 \\ 3\,6 \\ +\,1\,9\,8\,7 \\ \hline 2,2\,7\,1 \end{array}$$

Example 3 Add Whole Numbers on a Calculator

Use a calculator to find the sum of 2,179 + 873.

Press (on)

Press (2)(1)(7)(9)(+)(8)(7)(3)(enter)

The display should read

```
2179 + 873    Math ▲

              3052
```

THINK ABOUT MATH

Directions: Solve each problem.

1. 48 + 31
2. 57 + 28
3. 14 + 165 + 374

Subtract Whole Numbers

Subtraction is deducting, or taking away, an amount from another amount. To find how many objects remain in a set of objects after some of them are removed, use subtraction. Suppose you have 8 pencils in a set and take away 5 of them. Find the number of pencils by performing the operation 8 − 5.

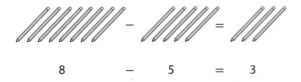

$$8 \quad - \quad 5 \quad = \quad 3$$

The answer to a subtraction problem is called the **difference**. The difference between 5 and 8 is 3. Subtraction is also used to compare one amount to another: for example, the question, "How many more people registered to vote this year than last year?"

Example 4 Subtract Numbers Without Regrouping

Subtract 254 from 497.

Step 1 The sentence translates to 497 − 254. In order to calculate the difference, write the digits in the ones under ones, tens under tens, and so on. Start with subtracting the ones, 7 − 4.

$$\begin{array}{r} 497 \\ -254 \\ \hline 243 \end{array}$$

Step 2 Subtract the digits in the tens and then the hundreds columns.

Example 5 Subtract Numbers with Regrouping

Find 2,754 − 657.

Step 1 Begin in the ones column. Since you cannot subtract 7 from 4, regroup a ten as 10 ones. Subtract.

Step 2 Move to the tens column. Since you cannot subtract 5 from 4, regroup a hundred as 10 tens. Subtract.

$$\begin{array}{r} {}^{6}\,{}^{\cancel{4}}\,{}^{14} \\ 2,\cancel{7}\cancel{5}\cancel{4} \\ -\ 657 \\ \hline 2,097 \end{array}$$

Step 3 Move to the hundreds column. Subtract.

Step 4 Bring down the 2 in the thousands place.

Example 6 Subtract Whole Numbers on a Calculator

Use a calculator to find the difference of 587 − 398.

Press (on)

Press  (enter).

The display should read

| 587 − 398 | Math ▲ |
| | 189 |

THINK ABOUT MATH

Directions: Solve each problem.

1. 73 − 48   2. 2,387 − 455   3. 800 − 171

Vocabulary Review

Directions: Complete each sentence with the correct word.

calculate   difference   operations   sum

1. The result in subtraction is called the _____.

2. To _____ is to find an answer using a mathematical process.

3. The answer to an addition problem is its _____.

4. Addition and subtraction are basic _____ in mathematics.

Skill Review

Directions: Identify the context clue for each word problem. Solve the problem.

1. Last night, the attendance at the Ten Screen Movie Theater was 3,183. Tonight, the attendance increased by 459 people. How many people were at the movie theater tonight?

2. The football stadium store sold 2,498 T-shirts on one Saturday and 3,565 on the next Saturday. How many T-shirts were sold on those two Saturdays combined?

3. Ahmed's gross pay for the week is $525. If his deductions are $138, what is his net pay?

4. Janine bought a car for $8,750. After two years, the car had lost value. It had depreciated $1,900. How much is her car worth now?

5. Jermaine has $576 in his checking account. He uses his ATM card and withdraws $110. How much money is left in his checking account?

6. The bookstore ordered 317 books on Friday. On Monday, 489 books were ordered. How many total books were ordered?

7. The James family's earnings for last year were $63,789. This year, the earnings were $69,123. How much greater were the earnings this year than last year?

Skill Practice

Directions: Choose the best answer to each question.

1. During the week, Bernardo drove these distances: 456 miles, 482 miles, 449 miles, 479 miles, and 468 miles. How many miles did he drive during the week?

   A. 2,000
   B. 2,034
   C. 2,304
   D. 2,334

2. Last year, a food company sold 937,642 packages of carrots. This year, 1,000,000 packages were sold. How many more packages were sold this year?

   A. 62,358
   B. 124,598
   C. 999,098
   D. 1,937,642

3. Warren County collected $23,470 in taxes last year and $31,067 this year. How much more money was collected this year?

   A. 7,597
   B. 8,567
   C. 7,957
   D. 8,957

4. At an insurance company, there were 380 cubicles on the first floor, 407 cubicles on the second floor, 298 cubicles on the third floor, and 321 cubicles on the fourth floor. How many cubicles are there in total at the insurance company?

   A. 986
   B. 1,406
   C. 2,314
   D. 1,043

Multiply and Divide Whole Numbers

KEY CONCEPT: Multiplication is the operation of adding a certain quantity a set number of times. Division is the operation that is used to separate a quantity into parts.

Add.

1. $4 + 8$ **2.** $57 + 13$ **3.** $142 + 89$ **4.** $909 + 111$

Subtract.

5. $86 - 53$ **6.** $718 - 81$ **7.** $100 - 54$ **8.** $21 - 9$

Multiply Whole Numbers

The answer to a **multiplication** problem is called the **product**. The numbers that are multiplied are the **factors**.

$$\text{factor} \times \text{factor} = \text{product}$$

An $\times$ or a dot ($\cdot$) can be used to show multiplication. Here are two ways to write 3 times 9 equals 27.

$$3 \times 9 = 27 \qquad 3 \cdot 9 = 27$$

Example 1 Multiply Two Numbers

Multiply 736×45.

Step 1 Line up the digits you want to multiply with ones under ones, tens under tens, and so on. Put the number with more digits on top. Multiply each digit in 736 by 5 in 45 to find the first partial product.

$$\begin{array}{r} 7\,3\,6 \\ \times\,4\,5 \\ \hline 3{,}6\,8\,0 \end{array}$$

Step 2 Multiply each digit in 736 by the 4 in 45. Start the second partial product under the 8.

Step 3 Add the partial products.

$$\begin{array}{r} 7\,3\,6 \\ \times\,4\,5 \\ \hline 3\,6\,8\,0 \\ 2\,9\,4\,4\,0 \\ \hline 3\,3{,}1\,2\,0 \end{array}$$

MATH LINK

In multiplication, the product of two single-digit numbers can be a double-digit number. In such cases, place the ones digit of the number in the partial product, and carry the tens digit over to the next multiplication. In the first step of Example 1 ($5 \times 6 = 30$), for example, the 0 is placed in the ones place. Then add (or carry) 3 to the next product, $5 \times 3 = 15$, to get 18. **Now place 8 in the tens place, and carry and add 1 to the next product, $5 \times 7 = 35$, to get 36.** The final product is 3,680. In most multiplication problems, some carrying to the next place value is needed.

Example 2 Multiply Whole Numbers on a Calculator

Use a calculator to find the product of 489×15.

Press (on)

Press (4) (8) (9) (×) (1) (5) (enter)

The display should read

489 × 15	Math ▲
	7335

.

THINK ABOUT MATH

Directions: Find the product.

1. 17×4
2. 46×9
3. 390×4
4. 63×311
5. 394×29
6. $38 \cdot 18$
7. $96 \cdot 37$
8. 48×207
9. 100×482
10. $1{,}467 \times 35$

Core Skill
Find Reverse Operations

You learned that addition and subtraction and also multiplication and division are reverse operations. This is not simply a "fun fact." A reverse operation allows you to check an answer that you came up with when solving a problem. Consider this division problem: Divide 25 by 5. You come up with the answer 20. You can check the answer by using the reverse operation. You turn the original divisor and the quotient in your solution into factors, and after multiplying the two numbers, you check the product against the original dividend. As soon as you do so, you see that your original answer was wrong, because $5 \times 20 \neq 25$. You didn't divide; you subtracted when solving the problem.

After you complete the Skill Practice questions on page 25, check your answers by using the reverse operation in each case.

Divide Whole Numbers

The answer to a **division** problem is called the **quotient**. The number that is divided is the **dividend,** and the number that is dividing it is the **divisor**. There are several ways to show division.

$$\text{dividend} \div \text{divisor} = \text{quotient} \qquad \text{divisor} \overline{)\text{dividend}}^{\text{quotient}}$$

$$24 \div 8 \qquad 8\overline{)24} \qquad 24/8 \qquad \frac{24}{8}$$

Example 3 Divide Two Numbers

Divide: $372 \div 6$

Step 1 Find the largest number that you can multiply the divisor by to get a product that is less than or equal to the dividend. Since you cannot divide 6 into 3, start this problem by dividing 6 into 37.

$$6\overline{)372}^{\,6}$$

$$\begin{array}{r} 62 \\ 6\overline{)372} \\ -36 \\ \hline 12 \end{array}$$

Step 2 Multiply $6 \times 6 = 36$, and subtract $37 - 36 = 1$. Continue to multiply, subtract, and bring down the next number. Divide 6 into 12.

$$\begin{array}{r} 62 \\ 6\overline{)372} \\ -36 \\ \hline 12 \end{array}$$

Step 3 Multiply and subtract.

$$\begin{array}{r} 62 \\ 6\overline{)372} \\ -36 \\ \hline 12 \\ -12 \\ \hline 0 \end{array}$$

Context is the setting, events, or ideas surrounding something. In reading, context is the surrounding words and sentences that help explain the meaning of a certain word. In math word problems, context clues can help the reader determine which operation to use in a word problem. Phrases such as *product* and *times* indicate multiplication. Phrases such as *quotient*, *split*, and *divided* mean division.

Sometimes a little more detective work is needed to determine the operation(s) that will solve a problem. Read the following word problems.

Jorgé puts $20 in his savings every week. What is the total he will have saved at the end of 17 weeks?

Keisha has a case of 100 granola bars. She eats five bars every week. How many weeks will the case of granola bars last?

Both problems contain the words *every week*. The first problem gives the number of weeks and is asking for a *total*, so multiplication is indicated. The second problem tells the *total* number of granola bars in a case and is asking for *how many weeks*, so division is the best choice.

In a notebook, make a list of phrases that are clues to using multiplication and another list of clues to using division.

Example 4 Divide

Divide: $4\overline{)2{,}374}$

Step 1 Divide 4 into 23.

Step 2 Multiply, subtract, and bring down the next number.

Step 3 Divide 4 into 37.

Step 4 Multiply, subtract, and bring down the next number.

Step 5 Divide 4 into 14.

Step 6 Multiply and subtract. There are no more numbers to bring down. The number 2 is the remainder.

```
     593 R2
4)2,374
  -20
   37
  -36
   14
  -12
    2
```

Example 5 Divide Whole Numbers on a Calculator

Use a calculator to find the quotient of $611 \div 13$.

Press **on**

Press **6** **1** **1** **÷** **1** **3** **enter**.

The display should read

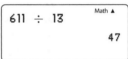

$611 \div 13$ Math ▲
                    47

THINK ABOUT **MATH**

Directions: Find the quotient.

1. $41 \div 3$
2. $186 \div 6$
3. $15\overline{)480}$
4. $16\overline{)3{,}246}$
5. $1{,}200/6$

6. $409 \div 12$
7. $200 \div 10$
8. $16\overline{)248}$
9. $9\overline{)984}$
10. $625/39$

Vocabulary Review

Directions: Write each word next to its meaning.

dividend   division   divisor   factor   multiplication   product   quotient

_____ 1. the number by which the dividend is being divided

_____ 2. one of the numbers in a multiplication problem

_____ 3. the answer to a multiplication problem

_____ 4. the number being divided

_____ **5.** repeating a quantity a set number of times

_____ **6.** the answer to a division problem

_____ **7.** separating a quantity into parts

Skill Review

Directions: Write the words or phrases that give a context clue to the correct operation. Solve the problem.

1. During the year, Juanita spent $480 for electricity. She paid the same amount each month. How much did she pay monthly for electricity? (1 year = 12 months)

2. Bob owns a hardware store. He sells hammers for $15. On Saturday, he sold 36 hammers. How much money did he collect for hammers that day?

3. Harold had 4,866 stamps in his stamp collection that were equally divided into six stamp books. How many stamps are in each of his six stamp books?

4. If Rafael pays $75 monthly for health insurance, how much will he have paid after two years?

5. The ticket office sold 64,750 tickets to the playoff game. If the cost is $27 per ticket, how much money did the ticket sales generate?

6. A city swim team generated $432 in ticket sales. If the tickets are $3 each, how many tickets were purchased?

7. Sun is decorating tables for an event. She needs to place 30 rose petals on each table. There are 17 tables. How many petals does she need?

Skill Practice

Directions: Choose the best answer to each question.

1. The computer repair shop pays each of five employees $589 each week. How much is this total each week?
   - A. $2,505
   - B. $2,914
   - C. $2,945
   - D. $2,954

2. Chantou pays $525 for rent every month. How much rent does she pay in one year?
   - A. $40 R5
   - B. $43 R9
   - C. $1,575
   - D. $6,300

3. There are 320 people expected for an awards dinner. Each table can seat 16 people. How many tables will be needed?
   - A. 10
   - B. 20
   - C. 24
   - D. 32

4. A rock band earns $1,315 from ticket sales at a concert. For each ticket sold, the band gets $5. How many tickets were sold?
   - A. 263
   - B. 343
   - C. 1,163
   - D. 6,575

Factoring

KEY CONCEPT: A whole number is the product of two or more factors, and the greatest common factor is the greatest factor shared by those whole numbers.

Begin at 0 and count forward: 0, 1, 2, 3, 4, and so on. The number 0 and the numbers that follow are whole numbers. A place-value chart is a valuable tool for reading the value of each digit in a whole number. For example, the value of the 2 in the whole number 721,465 is 20,000.

Billions			Millions			Thousands			Ones		
H	T	O	H	T	O	H	T	O	H	T	O
						7	2	1	4	6	5

Lesson Objectives

You will be able to

• Determine the set of factors of a number

• Determine the greatest common factor of two numbers

• Identify and apply patterns

Skills

• **Core Skill:** Apply Number Sense Concepts

• **Core Skill:** Build Solutions Pathways

Vocabulary

Commutative Property of Multiplication
Distributive Property of Multiplication
equation
evaluate
expression
factor
greatest common factor
operation

What Is a Factor?

We can use the following equation, or mathematical statement, to answer the question "What is a factor?"

$$6 \times 4 = 24$$

A factor is a number that is multiplied by another number to give a product, or total. Start with the number 6 in the equation. The example shows that you can multiply 6 by another number to get the product 24. So, the number 6 is a factor of 24.

Look at the equation again. It shows that you can multiply 4 by another number to get the product 24. So, the number 4 is also a factor of 24.

Apply Number Sense Concepts: Finding the Factors of a Number

You know that the numbers 6 and 4 are factors of 24. There are more factors of 24, too. Look at the different factors you can multiply to get 24, starting with 1 and working your way up to 24.

$$1 \times 24 = 24$$
$$2 \times 12 = 24$$
$$3 \times 8 = 24$$
$$4 \times 6 = 24$$
$$6 \times 4 = 24$$
$$8 \times 3 = 24$$
$$12 \times 2 = 24$$
$$24 \times 1 = 24$$

Do you notice any similarities among these equations? Think about the **Commutative Property of Multiplication.** This property states that if you switch the order in which two numbers are multiplied, the product is the same. For example:

$$2 \times 12 = 24$$
$$12 \times 2 = 24$$

Because $12 \times 2 = 24$ and $2 \times 12 = 24$, both of these equations tell us the same thing—that the numbers 2 and 12 are factors of 24. There is no need to write the same factors more than once, so you can cross out one of these equations from the list.

Look for other examples of the Commutative Property of Multiplication in the list. Cross out one equation in each pair.

$$1 \times 24 = 24$$
$$2 \times 12 = 24$$
$$3 \times 8 = 24$$
$$4 \times 6 = 24$$
$$\cancel{6 \times 4 = 24}$$
$$\cancel{8 \times 3 = 24}$$
$$\cancel{12 \times 2 = 24}$$
$$\cancel{24 \times 1 = 24}$$

Now that you have crossed out equations that use identical factors, you can write each factor in the list of equations in numerical order. Starting from the top, the factors are 1, 2, 3, 4, 6, 8, 12, and 24.

You can shorten the set of factors even further. Notice the first multiplication equation in the list:

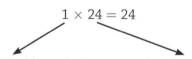

$$1 \times 24 = 24$$

The number 1 is a factor of 24, but 1 is a factor of every number! For this reason, we do not include 1 in any set of factors.

The number 24 is a factor of 24, but every number is a factor of itself! For this reason, we do not include a number in its own set of its factors.

So, the factors of 24 are 2, 3, 4, 6, 8, and 12. You can follow the same procedure to find the set of factors of any number.

You can apply properties of numbers to solve mathematical problems. The Associative Property, for example, applies to grouping. In addition, the rule is $a + (b + c) = (a + b) + c$, so the order of the **addends**, or numbers being added, doesn't matter. The sum remains the same. The property also applies to multiplication. The order of the factors, or numbers being multiplied $a(b \times c) = (a \times b)c$, doesn't matter. In either case, the product, or answer, remains the same.

The Commutative Property also applies to addition and multiplication. In addition, the sum remains the same even when the order of two addends changes: $a + b = b + a$

In multiplication, the product remains the same even when the order of two factors changes: $ab = ba$

You can shorten the amount of time you spend factoring a number by looking for patterns among the equations you write when factoring. For example, factor the number 28.

$$2 \times 14 = 28$$
$$4 \times 7 = 28$$
$$7 \times 4 = 28 \leftarrow$$
$$14 \times 2 = 28$$

Notice that the third equation is the first time you can apply the Commutative Property of Multiplication. It tells you that these factors have been multiplied before: $7 \times 4 = 4 \times 7$. When you move to the fourth equation in the list, you see the same thing: $14 \times 2 = 2 \times 14$. You have found a pattern. You can use this pattern to determine the factors of a number more quickly.

As long as you begin factoring with the lowest factor, you can stop writing equations the first time that you identify an equivalent equation using the Commutative Property of Multiplication. By this time, you will have identified all of the factors of a number. In this example, the only equations you have to look at are

$$2 \times 14 = 28$$
$$4 \times 7 = 28$$

So, the factors of 28 are 2, 4, 7, and 14.

Comparing Sets of Factors

Often in mathematics, it is necessary to compare sets of factors for two different numbers. For example, you can compare the sets of factors for 24 and 100. Recall the factors of 24. They are 2, 3, 4, 6, 8, and 12.

Now you can use the procedure for finding factors to determine the factors of 100.

- List the equations.
- Apply the Commutative Property of Multiplication. Cross out equivalent equations.
- Delete equations containing the factors 1 and the number itself.
- Order the remaining factors.

$$\cancel{1 \times 100 = 100}$$
$$2 \times 50 = 100$$
$$4 \times 25 = 100$$
$$5 \times 20 = 100$$
$$10 \times 10 = 100$$
$$\cancel{20 \times 5 = 100} \longleftarrow$$
$$\cancel{25 \times 4 = 100}$$
$$\cancel{50 \times 2 = 100}$$
$$\cancel{100 \times 1 = 100}$$

This is the first time the Commutative Property of Multiplication applies.

The factors of 100 are 2, 4, 5, 10, 20, 25, and 50.

Now that you know both sets of factors, you can find common factors, or factors the numbers share.

Factors of 24: **2**, 3, **4**, 6, 8, 12
Factors of 100: **2**, **4**, 5, 10, 20, 25, 50

The **greatest common factor** is the largest common factor shared by two numbers. In this example, there are only two shared factors, and the greater factor is 4. So, the greatest common factor of 24 and 100 is 4.

Factoring Mathematical Expressions

In mathematics, an **expression** is a phrase that shows terms separated by **operation** symbols, such as symbols for addition, subtraction, multiplication, and division. You can use the greatest common factor of two numbers to rewrite mathematical expressions to make them easier to **evaluate**, or find the value of. For example, consider the addition expression $36 + 8$.

First, find the greatest common factor of 36 and 8. Begin by factoring each number and identifying common factors.

Factors of 36: **2**, 3, **4**, 6, 8
Factors of 8: **2**, **4**
The common factors are 2 and 4.
The greatest common factor is 4.

Next, use the factors to simplify the expression. Each number in the expression is an addend, or a number that is added to another number. Express the addends in $36 + 8$ as a product of two numbers, one of which is the greatest common factor, 4.

$$36 + 8 = (4 \times 9) + (4 \times 2)$$

Now you can apply the **Distributive Property of Multiplication** to rewrite this expression as:

$$(4 \times 9) + (4 \times 2) = 4 \times (9 + 2)$$
$$36 + 8 = 4 \times (9 + 2)$$

THINK ABOUT **MATH**

Directions: Think about this example of a subtraction expression: $63 - 28$.

The factors of 63 are: _____.

The factors of 28 are: _____.

The only common factor is _____.

So, the greatest common factor is _____.

Now, apply the Distributive Property of Multiplication to rewrite the subtraction expression.

$63 - 28 =$ _____ − _____

_____ − _____ $= 7 \times ($ _____ − _____ $)$

MATH LINK

To *commute* means "to move or change." The Commutative Property of Multiplication states that if you change the order of the numbers you're multiplying, the result will stay the same. In the box below, provide an example of the Commutative Property of Multiplication. Then use the example to explain how applying the Commutative Property of Multiplication shortens the time it takes to factor a number.

```
_____ × _____ = 20
_____ × _____ = 20
_____ × _____ = 20
_____ × _____ = 20
```

Factors of 20: _____

Vocabulary Review

Directions: Match each term to its example.

1. _____ greatest common factor
2. _____ equation
3. _____ Commutative Property of Multiplication
4. _____ Distributive Property of Multiplication
5. _____ evaluate
6. _____ operation
7. _____ expression
8. _____ factor

A. $9 \times 4 = 36$

B. $6 \times 54 = (6 \times 50) + (6 \times 4)$

C. 12: 1, 2, 3, 4, 6, 12
16: 1, 2, 4, 8, 16

D. $9 + 3$

E. $12 \times 2 = 2 \times 12$

F. 3 cases = \$45, so 1.25 cases = ?

G. division

H. the number 9 in 9×28

Skill Review

Directions: Use the factoring procedure to find all factors of the following numbers:

1. 12 _____

2. 32 _____

3. 45 _____

4. 44 _____

5. 88 _____

Directions: Use the factoring procedure to find the greatest common factor of each of the following pairs of numbers:

6. 4 and 20 _____

7. 30 and 42 _____

8. 35 and 49 _____

9. 66 and 88 _____

10. 50 and 100 _____

Directions: Use the Distributive Property to write the following expressions in the form $a \times (b + c)$ or $a \times (b - c)$, where a is the greatest common factor.

11. 24 + 36 _____

12. 45 − 27 _____

13. 20 + 64 _____

14. 48 + 72 _____

15. 66 − 44 _____

Skill Practice

Directions: Use the flow chart to show the stages of factoring. Number each stage.

1.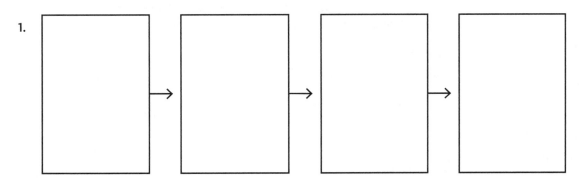

Directions: Choose the best answer.

2. Which of the following sets of numbers represents all of the factors of 52?

   A. 2, 26
   B. 2, 4, 13, 26
   C. 1, 52
   D. 2, 4, 6, 13, 26
   E. 2, 6, 13, 26

3. Which of the following sets of numbers represents all of the factors of 90?

   A. 2, 3, 5, 6, 9, 10
   B. 2, 3, 5, 6, 9, 10, 15, 18, 30, 45
   C. 2, 3, 5, 9, 10, 18, 30, 45
   D. 2, 3, 30, 45
   E. 2, 45

4. What is the greatest common factor of 36 and 60?

   A. 2
   B. 6
   C. 12
   D. 36
   E. 60

5. What is the greatest common factor of 25 and 70?

   A. 2
   B. 5
   C. 15
   D. 25
   E. 70

6. Which of the following expressions is equivalent to $60 + 84$, where the first term is the greatest common factor of $60 + 84$?

   A. $2 \times (30 + 42)$
   B. $3 \times (15 + 21)$
   C. $6 \times (10 + 14)$
   D. $12 \times (5 + 7)$
   E. $24 \times (5 + 7)$

7. Which of the following expressions is equivalent to $96 - 72$, where the first term is the greatest common factor of $96 - 72$?

   A. $2 \times (48 - 36)$
   B. $4 \times (24 - 18)$
   C. $6 \times (16 - 12)$
   D. $12 \times (8 - 6)$
   E. $24 \times (4 - 3)$

Rounding and Estimation

KEY CONCEPT: Rounding and estimation are useful when an answer does not need to be exact or when checking an exact answer.

Solve each problem.

1. $45 + 853$ **3.** $692 - 132$ **5.** $43 - 32$ **7.** $341 + 35$

2. 23×100 **4.** $112 \div 16$ **6.** 18×912 **8.** $1,170 \div 26$

Lesson Objectives

You will be able to

- Identify situations in which rounding or estimating is appropriate

- Round numbers to the nearest specified place value

- Use estimation appropriately

Skills

- **Core Skill:** Paraphrase Data

- **Core Practice:** Use Appropriate Tools Strategically

Vocabulary

compatible numbers
estimate
front-end digits
rounding

Rounding

For some problems, an exact answer is not necessary. An **estimate**, an approximate answer, will be sufficient. It is also good practice to estimate an answer first, solve the problem, and then check your solution by comparing the estimate to the exact answer.

One of the most common estimation strategies is **rounding**. Think of a number as being part of a hilly number line like the one below.

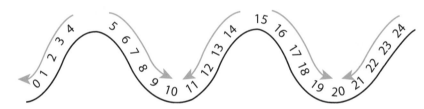

The numbers roll forward or backward to the closest valley. Numbers ending in 0, 1, 2, 3, and 4 roll back to the nearest 0. Numbers ending in 5, 6, 7, 8, or 9 roll ahead to the nearest 0.

Example 1 Round Numbers

Round 53 and 227 to the nearest ten or nearest hundred.

To the nearest ten, 53 rounds backward to 50.

To the nearest hundred, 53 rounds forward to 100.

To the nearest ten, 227 rounds forward to 230.

To the nearest hundred, 227 rounds backward to 200.

THINK ABOUT **MATH**

Directions: Round each number to the nearest ten.

1. 57		**3.** 125
2. 92		**4.** 1,345

DRAW CONCLUSIONS

Drawing conclusions requires you to make decisions about information in the text. It is taking the facts presented by the author and determining their relationship to each other or their most logical outcome.

It is important to base conclusions on only the facts given and not to create meaning that is not presented in the text.

What conclusions can be drawn from the following paragraph?

> In the hot months of summer, some people use air conditioning to cool off. Air conditioners use electricity. Electric bills are based on the amount of electricity used by a household. They vary from month to month.

What conclusion can you draw about electric bills in the summer, based on the passage above?

_____ (A) They rise for all people.

_____ (B) They fall for all people.

_____ (C) They rise for people who use air conditioning.

_____ (D) They rise for people who do not use air conditioning.

The correct answer is C. The passage states that some people use air conditioning, which uses electricity. Using more electricity raises electric bills. The passage does not give any facts about people who do not use air conditioning, so no conclusions about their electric bills can be drawn.

Estimation

Estimation can often save time when an exact answer is not needed or when you want to check whether an exact answer is reasonable.

Front-End Digits

We have already used the strategy of rounding. Another estimation strategy involves using the far-left digits or **front-end digits**.

Example 2 Estimate the Sum of 259, 673, and 110

Step 1 Rewrite each number using its front-end digit followed by zeros to replace the other numbers.

$259 \longrightarrow 200 \qquad 673 \longrightarrow 600 \qquad 110 \longrightarrow 100$

Step 2 Perform the appropriate operation, in this case addition.

$$
\begin{array}{r}
2\,5\,9 \\
6\,7\,3 \\
+1\,1\,0 \\
\end{array}
\quad \text{becomes} \quad
\begin{array}{r}
2\,0\,0 \\
6\,0\,0 \\
+1\,0\,0 \\
\hline
9\,0\,0 \\
\end{array}
$$

The sum of 259, 673, and 110 is about 900.

Compatible Numbers

Sometimes, numbers that are close to the original numbers are used instead of the original numbers to make the solution easier or quicker to achieve. These numbers are called **compatible numbers**. They are also helpful in mental math.

Example 3 Estimate the Quotient of 1,239 Divided by 37

Step 1 Change 1,239 and 37 to numbers that are easier to divide.

$1{,}239 \longrightarrow 1{,}200 \qquad 37 \longrightarrow 40$

Step 2 Divide using the new numbers.

Answer: 1,239 divided by 37 is about 30.

$$
\begin{array}{r}
30 \\
40\overline{)1{,}200} \\
120 \\
\hline
00 \\
\end{array}
$$

Step 3 Check the estimate.

Multiply 30 and 40 (mentally if possible). The product 1,200 is close to 1,239, so the estimate is reasonable.

THINK ABOUT MATH

Directions: Estimate each answer using front-end numbers or compatible numbers.

1. $563 + 215$
2. $2{,}610 \div 485$
3. $251 + 358 + 608$
4. $1{,}115 \div 8$

Directions: Match each word to its definition.

A. compatible numbers   **B.** estimate   **C.** front-end digits   **D.** rounding

_____ **1.** a common estimation strategy in which the number goes up or down to the nearest 0

_____ **2.** an approximate answer

_____ **3.** numbers that are close to the original numbers and make the solutions easier or quicker to achieve

_____ **4.** an estimation strategy involving the use of far-left digits

Skill Review

Directions: State the conclusion you draw about the result of the estimation method. Explain.

1. Jamie needs to estimate the number of buses he will need in order to transport people to the company picnic. Each bus can carry 56 people, and there are 1,344 people going to the picnic. Jamie decides to use the compatible numbers 60 and 1,200. Will he have enough buses for the picnic?

2. Mai has $485 in her checking account. She writes a check for repairs totaling $211 done to her car. She also needs to pay her telephone bill of $232 in the next week. After writing the check for the car repair, she estimates the amount in her checking account and the amount of the car repair check by using front-end estimation. Does she have enough in her checking account to pay her phone bill?

Skill Practice

Directions: Choose the best answer to each question.

The following chart gives the weights of three animals on a particular farm.

Animal	Weight
Clydesdale Horse	2,067 lb
Welsh Pony	478 lb
Arabian Horse	952 lb

1. About how many pounds more does the Clydesdale weigh than the Welsh pony?

    A. 500   **C.** 1,500

    B. 1,000   **D.** 2,500

2. Estimate the quotient of 5,398 ÷ 87 using compatible numbers.

    A. 54   **C.** 60

    B. 55   **D.** 62

3. Mario is buying a pair of pants for $49, a shirt for $25, and socks for $6. Which method of estimation should he use to make sure he has enough money to buy them?

    A. rounding up

    B. rounding down

    C. front-end digits

    D. compatible numbers

4. Julietta is buying six new windows for her home. The windows she wants cost $365 each. How much should she budget to buy the windows?

    A. $60

    B. $400

    C. $1,800

    D. $2,400

Arithmetic Expressions

KEY CONCEPT: Basic mathematical operations must be performed in a specified order to obtain the correct answer.

Solve.

1. $46 + 51$ **3.** 15×172 **5.** $1,145 - 132$ **7.** $1,092 \div 26$

2. $34,762 - 4,875$ **4.** 102×72 **6.** $3,413 \div 17$ **8.** $1,892 + 412$

Lesson Objectives

You will be able to

- Understand that operations must be performed in a specific order

- Solve problems involving the order of operations

- Use mental math to solve problems without paper and pencil

Skill

- **Core Practice:** Make Use of Structure

- **Core Skill:** Solve Real-World Arithmetic Problems

Vocabulary

compensation
mental math
order of operations
strategy

Order of Operations

An **arithmetic expression** is an expression that has a number value. It often includes more than one operation. When finding the value of an arithmetic expression, you need to know the proper **order of operations**, the order in which the operations should be performed. For example, the expression $8 - 4 \times 2$ suggests two possible ways to solve the problem. Working from left to right, subtract first and get $8 - 4 = 4$; $4 \times 2 = 8$, which is an incorrect answer. Multiplying first gives $4 \times 2 = 8$; $8 - 8 = 0$. This is the correct answer.

Use the following set of rules whenever you want to find the value of an arithmetic expression. If an expression includes more than one operation of the same type, work from left to right.

1) Do operations within parentheses.

2) Do multiplication and division.

3) Do addition and subtraction.

Example 1 Use the Order of Operations

Find the value of the expression $5 \times (6 - 2) \div 2 - 2$.

   Step 1 Do operations within parentheses.
         $5 \times 4 \div 2 - 2$

   Step 2 Do multiplication and division. Work from left to right.
         $20 \div 2 - 2$; $10 - 2$

   Step 3 Do addition and subtraction.
         $10 - 2 = 8$; The value of the expression is 8.

MATH LINK

A way to remember the order of operations is PEMDAS, or parentheses, exponents, multiplication, division, addition, subtraction.

Exponents are covered in Chapter 7.

If the expression does not include all the operations listed in the rules, skip the step with that rule and go on to the next step. If an expression includes a set of parentheses inside another set, work from the inside to the outside.

Example 2 Find the Value of an Expression

$100 - (2 \times (9 - 2)) \times 3$

Step 1 Do operations within parentheses.
$100 - (2 \times 7) \times 3$; $100 - 14 \times 3$

Step 2 Do multiplication and division.
$100 - 42$

Step 3 Do addition and subtraction.
$100 - 42 = 58$; The value of the expression is 58.

THINK ABOUT MATH

Directions: Use the order of operations to solve each problem.

1. $(12 + 4) \div (1 + 7)$
2. $((30 - 8) \div 2) + 5$
3. $13 + 6 \times 2 \div 4 - 9$
4. $60 - (25 - (13 - 4))$

Mental Math Strategies

Sometimes, a solution can be determined without using paper, pencil, or a calculator. **Mental math** is done by applying certain strategies to find an answer without writing. A **strategy** is a plan. Knowing and practicing these strategies will save time in situations such as taking timed tests.

Zeros

One strategy involves shortcuts for adding or multiplying when zeros are involved. For example, to multiply 48 and 100, first write 48. Since 100 ends in two zeros, insert two zeros after the 48.

Think: "$48 \times 1\underline{00} = 4,8\underline{00}$."

Another example would be adding 5,000, 14,000, and 6,000 by thinking "$5 + 14 + 6 = 25$ and then attach three zeros."

Example 3 Mentally Multiply 60 and 200

Multiply 6 and 2 and insert 3 zeros (**1** zero in 60 + **2** zeros in 200).

$$6\mathbf{0} \times 2\mathbf{00} = 12,\mathbf{000}$$

$$\mathbf{6} \times \mathbf{2} = \mathbf{12}$$

When you use information you know in a different way or in a new situation, you apply that information.

Many math skills can be applied to everyday problems that are encountered in the context of everyday life. For example, mental math strategies are useful for adding amounts to find a total when shopping.

Some word problems may suggest the kind of situation that you might experience in your own life. For example, *Dae Ho wants to save $400 over the next 4 months. How much should he save every week?* You know there are 4 weeks in a month, so first you multiply 4 weeks times 4 months to get 16 weeks. Then you divide $400 by 4 months to get $25. He must save $25 a week.

In a notebook, write about a time when you made a purchase of some kind. What was the cost of the item? How much money did you hand over when making the purchase? How much change did you receive? Do the calculation.

Break Apart Numbers

Another strategy is known as **breaking apart numbers**. This is thinking of numbers as the sum of two or more smaller numbers. Apply this strategy when it is easier to think of a number as the sum of two numbers. For example, to add $73 + 25$, mentally think of 73 as $70 + 3$ and 25 as $20 + 5$. Then add $70 + 20$ and $3 + 5$ to get an answer of 98.

Example 4 Break Apart Numbers

Jolene bought three cans of fruit juice. If each can costs 58 cents, how much did she spend?

Step 1 Choose an appropriate operation.
Since you are finding the price of several objects when one is given, choose multiplication.

Step 2 Apply the strategy of breaking apart numbers.
Think: $58 = 50 + 8$ so $50 \times 3 = 150$ and $8 \times 3 = 24$.
$150 + 24 = 174$.

Step 3 Write or state the answer to the problem.
174 cents is more commonly referred to as $1.74.
Jolene spent $1.74 for three cans of juice.

The last, most common strategy is referred to as **compensation** (or **substitution**). To apply this strategy, change one number so it is easier to perform the necessary operation and then change the other number by its opposite. For example,
$57 + 34 = (57 + \mathbf{3}) + (34 - \mathbf{3}) = 60 + 31 = 91$.

Example 5 Use Compensation

Miguel drove 101 miles. Connie drove 28 miles. How much farther did Miguel drive than Connie?

Step 1 Choose an appropriate operation.
The key words *farther than* indicate subtraction.

Step 2 Apply the mental math strategy of compensation.
Think: $101 - 28 = 101 - \mathbf{1} - 28 + \mathbf{1} = 100 - 28 + \mathbf{1} = 72 + 1 = 73$

Step 3 Write or state the answer to the problem.
Miguel drove 73 miles farther than Connie drove.

Step 4 Check your answer. Since subtraction and addition are opposite, the answer plus 28 should equal 101.
Add $28 + 73$ mentally.
Think: $20 + 8 + 70 + 3 = 90 + 11 = 101$.
The answer checks out.

The mental math strategies should help save time by allowing shortcuts whenever possible. Remember that you still must know your basic addition and multiplication facts to achieve the correct solution.

THINK ABOUT MATH

Directions: Use a mental math strategy to solve each problem.

1. Meli worked 174 hours last month. She worked 162 hours this month. How many hours did she work in these two months?
2. Chantal has 45 boxes of envelopes. Envelopes are sold in boxes of 100. How many envelopes does she have?

Vocabulary Review

Directions: Fill in the blank with the correct word.

compensation   mental math   order of operations   strategy

1. Use _____ to make numbers in a problem easier to work with mentally.

2. A _____ is a plan for solving.

3. When there are multiple operations in a problem, the _____ tells what order to perform them.

4. _____ is a way of solving problems without paper and pencil or a calculator.

Skill Review

Directions: Solve each problem. Write what information you applied from your own experience.

1. A coffee shop hopes to sell 145 cups of coffee every morning, with an extra 50 each on Saturday and Sunday. How many cups of coffee does it hope to sell in a week?

2. Coretta replaced all of the tires on her car. The tires cost $178 each. How much did she spend on the tires?

3. Sheila's car insurance costs $636 per year. How much does she pay every month?

Skill Practice

Directions: Choose the best answer to each question.

1. What is the value of $(8 + 3) \times 4 - 1$?
   A. 17
   B. 19
   C. 33
   D. 43

2. What is the value of $(44 - (2 \times 14)) \div (3 + 5) + 8$?

   _____

3. Carter is ordering business cards for his coworkers. He needs 100 cards each for 36 people. How many cards should he order?
   A. 36
   B. 360
   C. 3,600
   D. 36,000

Problem Solving

KEY CONCEPT: Problem solving is an important part of the study of mathematics and an important part of everyday life.

Use a calculator to solve each problem.

1. 289×97 **3.** $58,590 \div 62$ **5.** $2,894 + 19,073$ **7.** $96 \div 4$

2. 850×381 **4.** $407 - 388$ **6.** $1,411 \times 3,879$ **8.** $387 + 4,296$

The Five-Step Approach

The following five-step approach can be used to find a **solution**, or an answer, for all types of word problems and help organize thinking.

> **Step 1** Understand the question. After reading and rereading the problem carefully, decide what the problem asks you to find.
>
> **Step 2** Decide what information is needed to solve the problem. Then determine what information is irrelevant to the question.
>
> **Step 3** Choose the most appropriate operation or operations to solve the problem.
>
> **Step 4** Solve the problem. Make sure the solution answers the question asked.
>
> **Step 5** Check your answer by rereading the question to see if the answer is **reasonable**, that it makes sense.

Example 1 Use the Five-Step Approach

Paulo works from 4 p.m. to 7 p.m., Monday through Friday. How many hours does he work each week?

Step 1 Reread the problem for understanding. The problem asks for the number of hours Paulo works each week.

Step 2 Decide what information is needed.
Paulo works from 4 p.m. to 7 p.m. from Monday through Friday.

Step 3 Choose the most appropriate operation or operations.
Multiply 3 hours worked each day by 5 days per week.

Step 4 Solve the problem.
$3 \times 5 = 15$. So, Paulo works 15 hours each week.

Step 5 Check your answer.
If Paulo works three hours for five days a week, he works 15 hours each week. The answer is reasonable.

IDENTIFY IRRELEVANT INFORMATION

Sometimes, a passage will include **irrelevant**, or unnecessary, information. There may be details or other information that can be ignored when reading a passage or problem. Learning to determine which information is irrelevant will help when reading problems.

Not all information given in a problem is necessary, or relevant, to finding the solution. In fact, in real-life situations, the most difficult part of using math is often looking at all the information available and deciding what is actually relevant to the problem. It is important to have a clear sense of what is being asked and which details will help to find the solution.

To find irrelevant information, ask: *How does this detail relate to the question that is being asked?*

Read the following problem, and identify the irrelevant information.

> Sally is taking inventory of the glassware that her store sells. She has 15 cases of green glasses. They come 12 to a case and stand 8 inches tall. She also has 9 cases of blue glasses. They come 20 to a case. They are good for water and juice. If these are the only glasses Sally has in her store, how many glasses does she have in all?

The question asks how many total glasses Sally has. The size, color, and use for them are irrelevant information and can be ignored. The important information is 15 cases with 12 in a case and 9 cases with 20 in a case. Everything else in the problem is irrelevant information.

Precision is important in mathematics, engineering, and sciences. One of the purposes of making multiple measurements or repeating experimental trials is to find precise answers, or answers that repeat the same value.

You may have heard someone say, "Measure twice, cut once," while undertaking a construction project. What is the purpose of measuring twice? And is twice enough?

When builders measure multiple times before they saw a length of wood, for example, they are attending to **precision**. If multiple measures result in the same value, builders can be confident that the value is precise.

In your notebook, describe a time you made multiple measurements or repeated the steps of an experiment to check mathematical values. Explain the value of precision in your work.

Choose an Operation

Many times, the hardest part of solving a problem is deciding whether to add, subtract, multiply, or divide.

One way to determine which operation will be used is to focus on key words. Noticing these words will often provide a clue to determining the appropriate operation. Some key words or phrases are listed below.

Addition	Subtraction	Multiplication	Division
sum	difference	product	quotient
total	more…than	times	split
altogether	less…than	twice ($\times$ 2)	divided by
increased by	minus	when finding several of a different amount	when given amount of many and finding one
when combining different amounts	decreased by	when given part and finding the whole	when sharing, cutting, or splitting
	farther than		
	when comparing one amount to another		

Compare the relevant numbers in the problem to the solution to determine if the correct operation was chosen.

Example 2 Choose an Operation to Solve a Problem

Lamar rode his bicycle for 17 miles on Saturday. On Sunday, he rode 25 more miles. How many miles did Lamar ride altogether?

Step 1 Reread the problem to understand the question.
The problem asks for the number of miles Lamar rode his bicycle on Saturday and Sunday combined.

Step 2 Decide what information is needed.
You need to find the total number of miles Lamar biked over the weekend.

Step 3 Choose the most appropriate operation.
The key word *altogether* gives the clue to add 17 and 25.

Step 4 Solve the problem.
17 + 25 = 42
Lamar rode a total of 42 miles over the weekend.

Step 5 Check your answer. To check an addition problem, subtract one of the numbers in the problem from the answer. In this case, subtract 25 from 42. The answer is 17, the other number in the problem. The answer checks out.

THINK ABOUT MATH

Directions: Solve the following word problem by using the five-step approach. Tell which operation you used.

Marcella bought shampoo for $3, hair rinse for $2, and a brush for $2. She gave the clerk a 10-dollar bill. How much did the items cost altogether?

Problem-Solving Strategies

Draw a Picture

To solve a problem, you can sometimes draw a sketch or diagram to help you understand what is being asked.

Example 3 Draw a Picture to Solve a Problem

An empty water container is 10 meters tall. As it is being filled, the water level goes up 3 meters during the day, but because of a leak, it goes down 1 meter each night. How many days will it be before the water reaches the top of the container?

Step 1 Understand the question. You want to figure out how many days it will take the container to fill.

Step 2 Find the necessary information. Just writing the three numbers that are in the problem (10, 3, and 1) and using division will result in the wrong answer. Drawing a picture should show that subtraction should be used as well.

Step 3 Choose an arithmetic operation or operations. Look at the picture and notice the lines go up 3 meters and then down 1 meter. This indicates that subtraction should be used to find the actual distance at which the water is going up each day and night. Then use division to find the number of days.

Step 4 Solve the problem. Make sure to answer the question asked. One way (but not the only way) is to subtract 1 from 3 and then divide 10 by the result. So, $10 \div (3 - 1) = 5$.

It will take five days to fill the water container.

Step 5 Check the answer. Divide 10 by 5 and get 2, which is the distance the water goes up each day and night. The answer is reasonable.

MATH LINK

You do not have to be an artist to practice the *draw a picture* strategy. If you know what you are drawing and why, this method can be very useful.

THINK ABOUT MATH

Directions: Explain which strategy you would use to solve the following word problems. Then solve the problem.

1. Susan has 32 onions she wants to sell. Each bag she has holds 7 onions. How many onions will Susan have left over after she sells the ones that have been placed in bags?

2. Five basketball players are in a circle. Each player throws the ball to every other player once. How many times is the ball thrown?

Guess and Check

Another popular problem-solving strategy is called guess and check. This is a method people often use in everyday situations.

Example 4 Guess and Check to Solve a Problem

Which two 2-digit numbers made from the digits 1, 2, 3, 4, and 5 give the largest sum when added together? No single digit may be used more than once.

Step 1 Understand the question.
It asks which two 2-digit numbers give the largest sum when added together.

Step 2 Find the necessary information.
Use only the digits 1, 2, 3, 4, and 5. No digit may be used more than once.

Step 3 Choose an arithmetic operation.
Because the question asks for a sum, add the two 2-digit numbers.

Step 4 Solve the problem.

First guess: Second guess:

$54 + 32 = 86$ $53 + 42 = 95$

Answer: $53 + 42 = 95$

Step 5 Check the answer.
Test the first guess. Ask: Is 86 a reasonable sum? No, adding 5 and 3 in the tens place gives 80. By changing 3 to 4 and then adding 5 and 4 in the tens place, you get 90. The number 90 is greater than 80.

Vocabulary Review

Directions: Fill in the blank with the correct word.

irrelevant   reasonable   solution

1. To find the _____ to a word problem is to find the answer.

2. When using addition to solve a word problem, check that you have found a _____ sum.

3. Knowing what a question is asking makes it easy to find _____ information.

Directions: Find the irrelevant information in the following problems. Solve the problem.

1. Manny spent $3.15 for a 96-inch board of wood. How many 8-inch long pieces can he cut from the board?

2. Mrs. O'Rourke worked 28 hours this week. Mr. Martinez worked 32 hours, and Ms. Wong worked 41 hours. How many hours more did Ms. Wong work than Mrs. O'Rourke?

3. Carlos purchased 3 pounds of potatoes, 2 pounds of bananas, and 5 pounds of apples. He handed the clerk $12.43. How many pounds total do his items weigh?

4. At a garage sale, Chaske sold 12 DVDs for $3 each. He also sold his DVD player for $20. How much money did he receive for his DVDs?

Skill Practice

Directions: Choose the best answer to each question.

1. Which two 2-digit numbers made with the digits 5, 6, 7, and 8 give the largest sum when added together? No digit may be used more than once.

   A. 87 + 65
   B. 86 + 75
   C. 88 + 77
   D. 85 + 67

2. In May, Jack's rent will increase $30. If he pays $415 now, how much will his new rent cost each month?

3. Which operation is best to use to solve the following problem: Last year, 64,441,087 passengers flew through a major airport. The year before, there were 51,943,567 passengers. How many more passengers flew through it last year than the year before?

   A. addition
   B. subtraction
   C. division
   D. multiplication

4. Garbage service in the town of Kankakee costs $156 for one year. What is the monthly charge for garbage collection?

Directions: Choose the best answer to each question.

1. What is the value of the underlined digit?

   6,1<u>3</u>5,012

   A. 3
   B. 30
   C. 30,000
   D. 35,012

2. What is the greatest common factor of 36, 48, and 60?

   A. 2
   B. 6
   C. 12
   D. 36

3. Which operation should be performed first in the expression
   $7 + (12 - 4) \times 15 \div 3$?

   A. add 7 and 12
   B. subtract 4 from 12
   C. multiply 4 times 15
   D. divide 15 by 3

4. There are 124 employees in Yilin's office. On a certain day, 14 people are out of the office for illness, and 12 people are out on vacation. How many people are working in the office on that day?

   A. 98
   B. 122
   C. 126
   D. 150

5. What is 4,572,013 rounded to the nearest thousand?

   A. 4,600,000
   B. 4,570,000
   C. 4,572,000
   D. 4,573,000

6. Sandrine went to the grocery store to buy more onions for the soup. Two onions cost $2.79. She gave the grocer $5.00 to pay for the onions. Use estimation to find out how much change Sandrine should receive.

   A. About $3.00
   B. About $1.75
   C. About $2.50
   D. About $2.00

7. What is the greatest common factor of 14, 21, and 42?

   A. 4
   B. 7
   C. 17
   D. 49

8. A stadium has 11,260 seats. If the stadiums sells out for 5 home games in a row, approximately how many people attended those games?

   A. 2,252
   B. 11,000
   C. 55,000
   D. 60,000

9. Winona's mechanic charges $65 dollars per hour for labor plus the cost of parts. How much will she pay if her car needs a new part for $215 and 3 hours of labor?

   A. $280
   B. $410
   C. $710
   D. $840

10. Carl solved $7,820 \div 68 = 115$. How can he check his answer?

    A. add 115 plus 68
    B. subtract 115 from 7,820
    C. multiply 115 times 68
    D. divide 115 by 68

11. Three friends went bowling. The scores for their first game are shown in the chart.

Uppinder	James	Marietta
248	114	187

Which list shows the friends in order of greatest score to least score?

A. James, Marietta, Uppinder
B. Marietta, Uppinder, James
C. James, Uppinder, Marietta
D. Uppinder, Marietta, James

12. What is the value of 15 + (26 − 6) ÷ (7 − (8 ÷ 4))?

A. 19   C. 114
B. 34   D. 147

13. What is 842 ÷ 27?

A. 11 R50   C. 30 R2
B. 21 R5   D. 31 R5

14. Mrs. Cortez is mailing postcards for a fund-raiser for the senior center to which she belongs. She has 2,000 cards. She needs to send 582 cards to people who live near the center, 491 cards to local businesses, and 361 cards to other people on her mailing list. She also wants to have 500 cards to give out at a local shopping mall. How many postcards will she have left over?

A. 66
B. 557
C. 1,934
D. She does not have enough.

15. Quietta has $12,398 in a savings account. She withdraws $762 to pay her car insurance. How much money is left in her account?

A. $4,778
B. $11,636
C. $12,436
D. $13,160

Check Your Understanding

On the following chart, circle the number of any item you answered incorrectly. Under each lesson title, you will see the pages you can review to learn the content covered in the question. Pay particular attention to reviewing skill areas in which you missed half or more of the questions.

Chapter 1: Whole Numbers	Procedural	Conceptual	Application/ Modeling/ Problem Solving
Place Value pp. 12–17	1		11
Add and Subtract Whole Numbers pp. 18–21			4, 14
Multiply and Divide Whole Numbers pp. 22–25	13		8
Factoring pp. 26–31	2, 7		
Rounding and Estimation pp. 32–35	5		6
Arithmetic Expressions pp. 36–39	12	3	9
Problem Solving pp. 40–45		10	15

Decimals

Even if you are unfamiliar with the term *decimals*, every time you purchase something, you are using them. Decimals are used to show the value of amounts that are less than one or the values of amounts that are not whole numbers. If you pay $0.75 for a can of soda, you have paid less than one dollar. Decimals are also used in measuring distances. If you get directions from a website that tells you how to get from your home to the nearest post office, you will see notes such as "Walk 0.4 miles to Washington Street." How far is 0.4 miles? The decimal tells you that it is less than one mile. When you study Chapter 2, you will learn how to determine the distance or value of 0.4 miles, $0.75, and other decimal expressions.

The Key Concepts you will study include:

Lesson 2.1: Introduction to Decimals
Decimals represent a part of a number. They are an extension of the place-value system.

Lesson 2.2: Add and Subtract Decimals
Decimals are added and subtracted by using place value much like whole numbers are added and subtracted.

Lesson 2.3: Multiply Decimals
Multiplying decimals is a similar process to multiplying whole numbers.

Lesson 2.4: Divide Decimals
Dividing decimals is similar to dividing whole numbers. The key difference is the placement of the decimal point in the quotient, or answer.

Goal Setting

Before starting this chapter, set goals for your learning. Think about the ways that strengthening your understanding of decimals will benefit you.

Look for decimals in your life over the course of one week. Copy and complete the chart below.

Decimal	Where You Saw It or Used It	How It Was Used

Introduction to Decimals

KEY CONCEPT: Decimals represent a part of a number. They are an extension of the place-value system.

Identify the value of the underlined digit.

1. 1,<u>7</u>34
2. 2<u>0</u>,015
3. 1<u>8</u>
4. <u>2</u>,986,123

Round each number to the nearest ten.

5. 57
6. 189
7. 296
8. 6,382

Understand Decimals

The U.S. monetary system is based on the **decimal** system. Decimals are based on a whole being split into ten equal parts one or more times. Start by thinking of dollars as whole numbers. For decimals less than 1, think about money values less than a dollar.

1 dollar = 10 dimes = 100 pennies or **cents**, so 1 dime = 1 tenth of a dollar and 1 cent = 1 hundredth of a dollar.

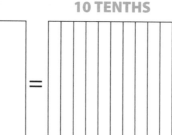

1 WHOLE = **10 TENTHS** = **100 HUNDREDTHS**

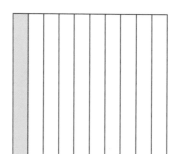
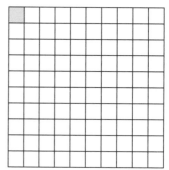

1 **tenth** = $\frac{1}{10}$ of a whole and 1 **hundredth** = $\frac{1}{100}$ of a whole.

Place Value in Decimals

The whole number place-value chart can be extended to include decimal values. The **decimal point** is read as *and*.

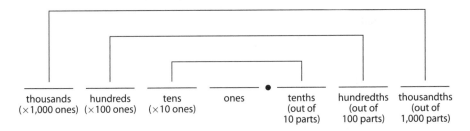

| thousands
(×1,000 ones) | hundreds
(×100 ones) | tens
(×10 ones) | ones | • | tenths
(out of
10 parts) | hundredths
(out of
100 parts) | thousandths
(out of
1,000 parts) |

Remember that each decimal part stands for part of a whole.

Example	Value	Meaning
0.1	1 tenth	1 out of 10 parts
0.01	1 hundredth	1 out of 100 parts
0.001	1 thousandth	1 out of 1,000 parts
0.0001	1 ten-thousandth	1 out of 10,000 parts
0.00001	1 hundred-thousandth	1 out of 100,000 parts
0.000001	1 millionth	1 out of 1,000,000 parts

Compare Decimals

When you **compare** decimals, you decide which has the greater value. Comparing decimals is similar to comparing whole numbers. First, line the decimals up by their decimal points. Start from the left, and compare each place value until there is one that is different.

Example 1 Compare 1.145 and 1.17

Step 1 Align the numbers by decimal point.

$$1.145$$
$$1.17$$

Add zeros to the end of a decimal if needed.

$$1.145$$
$$1.1170$$

Step 2 Starting from the left, look at each place value until you find one that is different.

$$1.1\boxed{4}5$$
$$1.1\boxed{7}0$$

Step 3 Compare the digits.
7 > 4, so 1.17 > 1.145

THINK ABOUT MATH

Directions: Write <, >, or = to compare the numbers.

1. 1.45 ___ 1.045

2. 4.52 ___ 4.273

3. 2.75 ___ 2.750

4. 2.81 ___ 6.81

5. 0.23 ___ 0.2300

Round Decimals

In most cases, decimals are rounded just as whole numbers are rounded. When an amount of money is rounded, the value is usually rounded up to the next nearest cent.

Example 2 Round Decimals

Round 1.537 to the nearest whole number, tenth, and hundredth.

Step 1 Identify the place-value digit to be rounded.

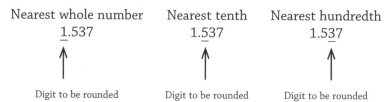

Step 2 Look at the digit immediately to the right of the value to be rounded. If this digit is 5 or greater, round up to the next higher digit. If this digit is less than 5, round down and keep the digit the same. The digits to the right of the rounded digit become 0 (if they are part of the whole number) or are eliminated (if they are part of the decimal).

Example 3 Round Money

Eric computed the sales tax on the items he bought. His calculator displayed the figure 2.0860. How much tax did he pay?

When rounding with money, round to the hundredth (or cent). Look at the first digit to the right of the hundredths place. If it is greater than five, round the decimal to the next higher hundredth (or cent).

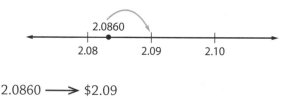

2.0860 ⟶ $2.09

THINK ABOUT **MATH**

Directions: Round each number to the specified place.

1. 6.145, tenth
2. 2.952, tenth
3. 15.876, whole number
4. 5.009, hundredth
5. 4.235, hundredth
6. 12.366, hundredth

Directions: Fill in each blank with the correct word.

cent   decimal point   decimals   hundredths   tenth

1. _____ is/are based on a whole being split into ten equal parts one or more times.

2. The first number to the right of the decimal point is a _____.

3. The _____ separates the whole numbers from the decimal numbers.

4. In 2.34, the 4 is in the _____ place.

5. One hundredth of a dollar is one _____ or one penny.

Skill Review

Directions: Study the chart. Write the value of each digit.

Thousands	Hundreds	Tens	Ones	.	Tenths	Hundredths
2	4	6	1	.	8	5

1. 2 _____   3. 6 _____   5. 8 _____

2. 4 _____   4. 1 _____   6. 5 _____

Directions: Create a place-value chart for each number.

7. 1.45

8. 32.091

9. twenty-four and 31 hundredths

10. one hundred and two hundredths

Skill Practice

Directions: Choose the best answer to each question.

1. Which number rounds 5412.8367 to the nearest thousandth?

   A. 5000
   B. 5400
   C. 5412.836
   D. 5412.837

2. Which drink costs the most?

   HOT DRINKS MENU
Latte	$2.45
Espresso	$2.39
Mocha	$2.99
Black Coffee	$1.89

   A. black coffee   C. mocha
   B. latte   D. espresso

Add and Subtract Decimals

Lesson Objectives

You will be able to
• Add decimals
• Subtract decimals

Skills

• **Core Skill:** Perform Operations

• **Core Practice:** Attend to Precision

Vocabulary

align
annexed
organize
place value
topic
vertically

Core Skill
Perform Operations

Evariste and Sophie are counting money they made by doing chores. Evariste has 2 quarters, 7 dimes, and 12 pennies. Sophie has 3 quarters, 5 dimes, and 7 pennies. Evariste believes he has more money than Sophie because he has more coins. In a notebook, determine how much each person made. Is Evariste correct? If not, what was wrong with his thinking?

KEY CONCEPT: Decimals are added and subtracted by using place value much as whole numbers are added and subtracted.

Add or subtract.
1. $41 + 29$ 2. $325 - 149$ 3. $6,009 + 932$ 4. $6,108 - 42$

Compare each pair of numbers. Write an expression using $<$, $>$, or $=$ when comparing.
5. $7.5 \square 2.19$ 6. $9.88 \square 19.1$ 7. $3.1 \square 0.85$ 8. $17.9 \square 17.90$

Add Decimals

Add decimals the same way you add whole numbers. Look at the following examples.

Example 1 Add Money Amounts

Add three quarters and six dimes.

Step 1 Change each amount of money to cents.

$$3 \text{ quarters} \longrightarrow 3 \times 25 \longrightarrow 75 \text{ cents or } \$0.75$$
$$6 \text{ dimes} \longrightarrow 6 \times 10 \longrightarrow 60 \text{ cents or } \$0.60$$

Step 2 Add the cents.

$$
\begin{array}{r}
7\,5 \text{ cents} \\
+6\,0 \text{ cents} \\
\hline
1\,3\,5 \text{ cents}
\end{array}
\quad \text{or} \quad
\begin{array}{r}
^{1} \\
\$0.7\,5 \\
+\$0.6\,0 \\
\hline
\$1.3\,5
\end{array}
$$

Example 2 Column Addition

Add $3.40 + 17.062 + 0.85$.

Step 1 Write the numbers vertically, aligning the decimal points and each **place value** (tens, ones, tenths, hundredths, thousandths).

Step 2 Starting at the right, add as you do whole numbers. Insert zeros to help you align places as necessary. Bring the decimal point straight down.

tens	ones	tenths	hundredths	thousandths
	3	4	0	
1	7	0	6	2
	0	8	5	

$$
\begin{array}{r}
^{1\ 1\ \ 1} \\
3.4\mathbf{00} \\
17.062 \\
+\ 0.85\mathbf{0} \\
\hline
21.312
\end{array}
$$

Move the decimal point straight down.

IDENTIFY TOPIC SENTENCES

The **topic** sentence is usually one sentence in a well-organized paragraph; it states the main idea of the paragraph. Although the topic sentence may appear anywhere in a paragraph, it can usually be found at the beginning. All other sentences in a paragraph are related to the topic sentence. Their purpose is to explain or support the main idea.

To find the topic sentence, ask: *Which sentence tells what this paragraph is about?*

Read the following paragraph, and underline the topic sentence.

(1) Adding decimals is very much like adding whole numbers. (2) For example, when adding 2.3 and 1.5, you first write the numbers vertically, aligning by place value. (3) You also align the decimal points. (4) Start at the right. (5) Add 3 and 5 to get 8. (6) Then add 2 and 1 to get 3. (7) The decimal point is brought straight down to get the answer 3.8.

Sentence 1 makes a general statement about how adding decimals is like adding whole numbers. Sentences 2 through 7 provide an example of adding two decimals, showing how the steps are very much the same as for adding whole numbers. Only sentence 1 is general enough to include the ideas of the other sentences. Sentence 1 is the topic sentence.

Core Practice
Attend to Precision

In most ways, calculating with decimals resembles performing operations with whole numbers— for example, the digits must be **aligned** (lined up) **vertically** (up and down) by place value, the decimal point must be correctly positioned, and the answer must extend the correct number of decimal places (whole numbers need zero decimal places). However, there are differences. Zeroes can be written at the far right of a number, or **annexed**, as needed for decimal numbers, but not for whole numbers.

As shown below, vertical lines can be drawn through the decimal points and the place values of the numbers so that the digits are **organized**, or placed in order, and ready to add or subtract. In a notebook, write down how many decimal places will be needed for the final answer before solving the problem. Then, find the sum.

```
   3.|4| |
  17.|0|6|2
+  0.|8|5|
```

Just as some math
problems must be
completed in a series of
steps that lead to a final
solution, some assigned
school projects must
likewise be completed
in steps. It is important,
therefore, to set up a
schedule before you
begin your work. In this
way, you can budget
your time. You will
complete the project
by the due date if you
finish each step on time.

Make a plan for what
you will do between
the end of the school
day today and the time
you go to bed. Choose
no more than three or
four activities, including
dinner and homework.
Make a schedule. At the
end of the evening, see
if you were successful
in meeting all of your
goals. If not, consider
what you could have
done differently to
achieve what you set
out to do.

MATH LINK

If your display shows
a sum or difference as
a fraction, press **2nd**
table to have it show
the sum or difference
as a decimal.

Example 3 Add Decimals on a Calculator

Use a calculator to find the sum of $1.79 + 8.03$.

Press **on**

Press (1) (.) (7) (9) (8) (.) (0) (3) (enter)

The display will read | $\frac{491}{50}$ | .

Press **2nd** **table** to have it show the sum as a decimal.

The display will read | 9.82 | .

THINK ABOUT MATH

Directions: Solve each problem.

1. Three dimes and six nickels is the same as $ _____.
2. $2.8 + 5.1 =$ _____   3. $1.54 + 0.165 + 0.3 =$ _____

Subtract Decimals

Subtract decimals the same way you subtract whole numbers. Look at the following examples.

Example 4 Subtract Decimals

Subtract 2.13 from 12.6.

Step 1 Write the subtraction problem vertically.

$$\begin{array}{r} 1\,2.6 \\ -\quad 2.1\,3 \end{array}$$ Remember to align the digits and decimal points.

Step 2 Insert zeros, if necessary, at the far right of a number. Subtract. Bring the decimal point straight down.

$$\begin{array}{r} 1\,2.6\,\mathbf{0} \\ -\quad 2.1\,3 \end{array} \longrightarrow \begin{array}{r} \overset{5\ 10}{1\,2.\cancel{6}\,\cancel{0}} \\ -\quad 2.1\,3 \\ \hline 1\,0.4\,7 \end{array}$$ Think of 6 tenths as 5 tenths and 10 hundredths.

Example 5 Subtract Decimals from Whole Numbers

Subtract 3.87 from 10.

Step 1 Write the subtraction problem vertically.

$$\begin{array}{r} 1\,0 \\ -\quad 3.8\,7 \end{array}$$ Remember to align the digits by place value.

Step 2 Insert a decimal point and zeros, if necessary, after a whole number. Subtract.

$$
\begin{array}{r}
1\,0.0\,0 \\
-\;3.8\,7 \\
\end{array}
\longrightarrow
\begin{array}{r}
\overset{9}{\cancel{1}}\overset{9}{0}.\overset{9}{\cancel{0}}\overset{10}{\cancel{0}} \\
-\;3.8\,7 \\
\hline
6.1\,3 \\
\end{array}
$$
Think of 10 ones as 9 ones, 9 tenths, and 10 hundredths.

Example 6 Subtract Whole Numbers from Decimals

Subtract 3 from 5.36.

Step 1 Write the subtraction problem vertically. Remember to align the digits and decimal point appropriately.

$$
\begin{array}{r}
5.\cancel{3}\cancel{6} \\
-\;3.0\,0 \\
\hline
2.3\,6 \\
\end{array}
$$

Step 2 Solve the subtraction problem. Note that because 3 had no decimals, the decimal portion of 5.36 was left unchanged.

Add and Subtract Decimals Summary

Write the numbers to be subtracted or added vertically. Align by place values and by decimal points. Insert any necessary zeros or missing decimal points. Subtract or add the digits from right to left as you would when subtracting or adding whole numbers.

THINK ABOUT **MATH**

Directions: Solve each problem.

1. $5.6 - 2.3$ 2. $12 - 3.47$ 3. $2.165 - 0.18$

Vocabulary Review

Directions: Fill in each blank with the correct word.

align   annexed   organize   place value   vertically

1. The digits in the number 4.29 each have their own _____.

2. When numbers are written one under the other, they are arranged _____.

3. Before adding two decimals, be sure to _____ the decimal points.

4. To _____ an addition or subtraction problem means to write it in a way that makes calculating its sum or difference easier.

5. Zeros can be _____ to the right of the digits in a decimal.

Directions: Draw vertical lines through each problem to show that the digits have been aligned vertically, that decimal points and zeros have been added, and that the problem is organized in a way that is useful for finding the sum or difference.

1.
```
  1 7. 3 5 0
+ 5 0. 9 2 7
```

2.
```
  4 2. 0 0 0
- 3 6. 4 9 8
```

3.
```
  3. 8 9 0
- 1. 4 2 6
```

4.
```
  0. 1 8 0
  8. 9 2 1
+ 3 9. 6 0 0
```

Directions: Rewrite the problems below so that they are organized in a way that is useful for finding the sum or difference.

5. $1.563 + 8.03$

6. $29 - 0.25$

7. $7.5 - 1.004$

8. $0.23 + 1.006 + 80$

Directions: Identify which person is incorrect and why.

9. Marco and Olivia are working together to solve a subtraction problem. When Marco subtracted .57 from 4.28, he got 4.31, because $8 - 7 = 1$ and $5 - 2 = 3$. Olivia subtracted .57 from 4.28 and got 3.71, because $8 - 7 = 1$ and $12 - 7 = 5$.

10. While training for a race, Charlie and Lucy ran 3.56 miles one day and 5.87 miles the next day. Charlie claims that they ran a total of 9.43 miles, while Lucy claims they ran 8.33 miles.

Skill Practice

Directions: Choose the best answer to each question.

1. Calvino drew line segments to show the length of each section of a sidewalk he will be pouring. The lengths of each section were: 1.5 m, 1.8 m, 2.75 m, and 2.9 m. How many meters long is the sidewalk?

    A. 3.37 m
    B. 6.2 m
    C. 6.95 m
    D. 8.95 m

2. Fatima subtracted 0.35 from 2.4 and got the correct answer. Which could have been Fatima's calculations?

    A.
    $$\begin{array}{r} 0.3\ 5 \\ -\ \ 2.4 \\ \hline 0.1\ 1 \end{array}$$

    B.
    $$\begin{array}{r} {\scriptstyle 1\ \ 14\ 10} \\ 2.\cancel{4}\ \cancel{0} \\ -\ 0.3\ 5 \\ \hline 2.1\ 5 \end{array}$$

    C.
    $$\begin{array}{r} {\scriptstyle 3\ \ 10} \\ 2.\cancel{4}\ 0 \\ -\ 0.3\ 5 \\ \hline 2.0\ 5 \end{array}$$

    D.
    $$\begin{array}{r} 2.4\ 0 \\ +\ 0.3\ 5 \\ \hline 2.7\ 5 \end{array}$$

3. The unemployment rate for single men who are 16 years of age is 13.1 percent in Chi's hometown this month. If the unemployment rate for single women is 2.4 percent lower, what is that rate?

    A. 13.6 percent
    B. 15.5 percent
    C. 10.7 percent
    D. 18.4 percent

4. Mia has two dogs. Brutus weighs 12 kg and Buckeye weighs 15.25 kg. How many kilograms do Brutus and Buckeye weigh together?

    A. 27.25 kg
    B. 3.25 kg
    C. 15.37 kg
    D. 16.45 kg

Multiply Decimals

Lesson Objective

You will be able to
- Multiply decimals

Skills

- **Core Skill:** Apply Number Sense Concepts
- **Core Skill:** Represent Real-World Problems

Vocabulary

factor
multiplication
product

KEY CONCEPT: Multiplying decimals is a process that is similar to multiplying whole numbers.

Multiply.

1. 4×9 **2.** 32×5 **3.** 19×24 **4.** 628×317

Add.

5. $1.2 + 1.2 + 1.2 + 1.2 + 1.2$ **6.** $3.15 + 3.15 + 3.15 + 3.15$

Multiply Decimals

Multiplying decimals is done the same way you multiply whole numbers. You must be careful to put the decimal point in the correct place in the **product**— the answer in a multiplication problem. To do this, add the number of decimal places in each number being multiplied. Start at the far right of the product, count that number of decimal places to the left, and insert the decimal point. Look at the following examples.

Example 1 Multiply Decimals

Multiply 2.3 and 1.2.

Step 1 Count the number of decimal places in the original numbers.
2.3 has **1** digit to the right of the decimal point.
1.2 has **1** digit to the right of the decimal point.
The product should have $1 + 1 = $ **2** decimal places.

$$\begin{array}{r} 2.3 \\ \times\, 1.2 \\ \hline 4\,6 \\ 2\,3 \\ \hline 2.7\,6 \end{array}$$

Step 2 Write the problem and multiply as whole numbers.

Step 3 Start at the far right of the product. Move 2 places to the left. Insert the decimal point between the 2 and the 7.

Step 4 Check the answer by estimating the product.
2.3 is about 2.   1.2 is about 1.
$2 \times 1 = 2$
The product of 2.3 and 1.2 should be slightly greater than 2, so 2.76 seems reasonable.

MATH LINK

Apply the same process for multiplying decimals to multiplying **monetary units** (currency) or money amounts by a whole number. For example, to find the value of 13 quarters, multiply $0.25 by 13.

$$\begin{array}{r} 0.25 \\ \times\, 13 \\ \hline 75 \\ +2\,5 \\ \hline 3.25 \end{array}$$

The value of 13 quarters is $3.25.

Example 2 Write Zeros in the Product

Multiply 3.9 and 0.025.

Step 1 Count the number of decimal places in the original numbers. 3.9 has **1** digit to the right of the decimal point. 0.025 has **3** digits to the right of the decimal point. The product should have $1 + 3 = 4$ decimal places.

$$\begin{array}{r} \overset{4}{3.9} \\ \times\,0.025 \\ \hline 195 \\ 78 \\ \hline 0.0975 \end{array}$$

Step 2 Write the problem and multiply as whole numbers.

Step 3 Start at the far right of the product. Move **4** places to the left. A zero will need to be written to the left of 9 in the product.

$$3.9 \times 0.025 = 0.0975 \text{ or } .0975$$

Step 4 Check your answer by estimating the product. 3.9 is about 4. Since **multiplication** is repeated addition, add 0.025 four times. Since 0.100 (the sum) is close to 0.0975 (the product), the product seems reasonable.

$$\begin{array}{r} 0.025 \\ 0.025 \\ 0.025 \\ +\,0.025 \\ \hline 0.100 \end{array}$$

Example 3 Multiply Decimals on a Calculator

Use a calculator to find the product of 0.985×2.1.

Press (on)

Press (0)(.)(9)(8)(5)(×)(2)(.)(1)(enter)

The display will show 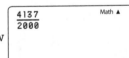.

Press (2nd).

The display will read .

Multiplication with decimals, like multiplication with whole numbers, is repeated addition.

Nina's math teacher announces a pop quiz. "There will be eight questions," he says, "and each one is worth 2.5 points. I'll give extra credit to anyone who can show me on the board two ways to determine how many points you'll get if you answer every question correctly." Nina is called on, and first she writes the number 2.5 eight times and adds the column of numbers. Then she multiplies 2.5 × 8. Each time, her answer is 20. "Twenty points," she says.

With a partner, write the product 2.3 × 3.2 as a repeated addition problem. (Hint: You can write a number as the sum of two other numbers, so the second **factor**, or number that gets multiplied, in this problem, 3.2, can be changed to 3 + 0.2.) When you finish, check your answer using the method you learned for multiplying decimals to other decimals.

THINK ABOUT **MATH**

Directions: Apply the process for multiplying decimals to find the value of the following amounts.

1. 23 dimes 2. 40 quarters 3. 75 nickels 4. 350 pennies

Directions: Find each product.

5. 0.5×0.3 6. 3×6.8 7. 0.025×1.3 8. 7.4×0.31

A zero written before a decimal point can be dropped as long as there is no number to the left of it (0.025 can be written simply as .025, but 10.3 cannot be simplified).

Core Skill
Represent Real-World Problems

There are many real-world scenarios that would require you to multiply decimals. That's because many real-world measurements aren't exactly equal to a whole number. For example, think of the last time you bought something by weight, such as fruit at the grocery store.

Consider the following problem. Jaqueline has a rectangular garden that needs to be covered for the winter to protect it from the weather. The length of the garden is 4.7 feet and the width is 3.56 feet. The amount of cloth she would need would be the area of the garden, which would be a multiplication of decimals problem. In a notebook, determine how much cloth Jaqueline would need to purchase to cover her garden.

Example 4 Multiplying Decimals with Extra Zeroes

Find the product of 1.350 and 6.9.

Step 1 Write the multiplication problem by aligning everything to the right-most space, ignoring place value.

Step 2 Perform the multiplication as you would using whole numbers. Determine the number of places needed after the decimal point. (4)

$$
\begin{array}{r}
1.3\ 5\ 0 \\
\times\ 6.9 \\
\hline
1\ 2\ 1\ 5\ 0 \\
8\ 1\ 0\ 0 \\
\hline
9.3\ 1\ 5\ 0
\end{array}
$$

Looking at the furthest most digit in the multiplication problem, you may have noticed that it is a zero and is not needed. Therefore, the problem only needed 3 decimal places. Why is this? The number of decimal places needed for the problem was the total of decimal places in each factor. After closer inspection, you should notice that one of the factors, 1.350, wasn't written using the least amount of decimal places. It could have been simplified to 1.35 and therefore only 3 decimal places would have been determined.

Multiply Decimals Summary

Count the total number of decimal places in the original problem. Multiply as you would with whole numbers. Starting at the far right of the product, count the required number of places to the left and insert the decimal point. Insert zeros to the left as needed.

Vocabulary Review

Directions: Fill in the blank with the word that makes the sentence true.

factor   multiplication   product

1. The number 0.4 is a _____ in the problem 0.4 × 0.2 = 0.08.

2. The _____ of 4 × 0.6 is 2.4.

3. Repeated addition is more commonly known as _____.

Directions: Write the number of decimal places each product will have.

1. $\begin{array}{r} 1\,7.3 \\ \times\,5.9\,2 \\ \hline \end{array}$
2. $\begin{array}{r} 4\,2.556 \\ \times\,6.293 \\ \hline \end{array}$
3. $\begin{array}{r} 3.8\,9 \\ \times\,1.4 \\ \hline \end{array}$
4. $\begin{array}{r} 3.18 \\ \times\,92 \\ \hline \end{array}$

Directions: Find the product of the whole numbers. Then use that product to find the product of the decimal numbers.

5. 15×12 1.5×1.2 7. 7×9 0.7×9

6. 123×8 1.23×0.8 8. 47×31 0.47×0.031

Skill Practice

Directions: You may use a calculator with these questions. Choose the best answer to each question.

1. Ralf knows the product of 12 and 3 is 36. Which best explains why Ralf knows for sure that 1.2×0.03 equals 0.036 and not 0.36?

   A. 1.2 has one decimal place, so the product should have one zero.
   B. There is one zero in .03, so the product should have only one zero.
   C. The product has three decimal places. A zero needs to be inserted to place the decimal point three places to the left of 6 in the product.
   D. 1.2 has one decimal place, and .03 has two decimal places. $2 - 1 = 1$, so the product should have one zero.

2. Jana wants to purchase 6 new towels. Each towel costs $5.39. She multiplies 539 by 6. Where should she place the decimal point?

   A. two places from the far right of the product
   B. three places from the far right of the product
   C. four places from the far right of the product
   D. six places from the far right of the product

3. Chen wants to buy 2.5 pounds of ground meat. If the cost is $2.38 per pound, what amount will he pay for 2.5 pounds?

   A. $4.88
   B. $5.95
   C. $14.75
   D. $11.90

Divide Decimals

Lesson Objective

You will be able to
• Divide decimals

Skills

• **Core Skill:** Apply Number Sense Concepts

• **Core Skill:** Evaluate Reasoning

Vocabulary

dividend
divisor
evaluate
quotient
reasoning
summarize

KEY CONCEPT: Dividing decimals is similar to dividing whole numbers. The key difference is the placement of the decimal point in the quotient, or answer.

Find each quotient.

1. $5\overline{)835}$ 3. $59,344 \div 16$ 5. $5,000 \div 1,000$ 7. $5,000 \div 10$

2. $30\overline{)510}$ 4. $16,000 \div 40$ 6. $5,000 \div 100$ 8. $5,000 \div 1$

Divide Decimals

Divide decimals the same way you divide whole numbers. When the **divisor** (the number being divided into another number) is a decimal, however, move the decimal point to the right, and write the divisor as a whole number. Next, move the decimal point in the **dividend** (the number being divided) the same number of places as in the divisor. Put a decimal point in the **quotient** (the answer) directly above the dividend's decimal point, and divide. Look at the following examples.

When dividing decimals, it is important to keep track of the number of places the decimal point was moved in the divisor and move the same number of places in the dividend. If not, the answer you get will be off by a power of ten.

If you and a friend earned $27.36 raking leaves, how much would each of you receive if you split the money evenly?

Example 1 Divide a Decimal by a Whole Number

Divide 23.7 by 3.

$$3\overline{)23.7}$$

Step 1 The divisor, 3, is already a whole number, so place the decimal point in the quotient above the decimal point in the dividend.

$$\begin{array}{r} 7.9 \\ 3\overline{)23.7} \\ \underline{21}\downarrow \\ 27 \\ \underline{27} \\ 0 \end{array}$$

Step 2 Divide as you would with whole numbers.

Example 2 Divide a Whole Number by a Decimal

Divide 18 by 2.4.

$$2.4\overline{)18}$$

Step 1 The divisor, 2.4, is not a whole number, so move the decimal point in both the divisor and dividend to the right 1 place. Place the decimal point in the quotient above the new dividend.

$$\begin{array}{r} 7.5 \\ 24\overline{)180.} \\ \underline{168}\downarrow \\ 120 \\ \underline{120} \\ 0 \end{array}$$

Step 2 Divide as you would with whole numbers.

Example 3 Divide a Decimal by a Decimal

Divide 17.5 by 1.25.

$$1.25\overline{)1750\mathbf{0}}$$

Step 1 Move the decimal point in the divisor in order to make it a whole number. In this case, move the decimal point 2 places to the right.

$$\begin{array}{r} 14. \\ 125\overline{)1750.} \\ \underline{125} \\ 500 \\ \underline{500} \\ 0 \end{array}$$

Step 2 Move the decimal point in the dividend the same number of places as in Step 1. Insert zeros if necessary.

Step 3 Insert a decimal point in the quotient directly above the decimal point of the dividend.

Step 4 Divide as with whole numbers.

Step 5 Check your answer by multiplying the quotient by the original divisor. The product should be the dividend.

$$\begin{array}{r} 14 \leftarrow \text{quotient} \\ \times\ 1.25 \leftarrow \text{original divisor} \\ \hline 17.50 \leftarrow \text{original dividend} \end{array}$$

Core Skill
Evaluate Reasoning

Sometimes, when learning procedures in mathematics, you might make an error in **reasoning**, or the process of thinking and finding a solution. So it is important to **evaluate**, or make a judgment about, your reasoning to see if you have introduced any errors into a procedure.

Suppose a student quickly looks at Example 3 and draws a conclusion about how to move the decimal point in all division problems involving decimals. In the problem below, the student took a second look at what he had done and realized the error in reasoning. He used a mathematical skill called *checking the answer.*

$$1.2\overline{)4.92} \longrightarrow \begin{array}{r} 41. \\ \times\ 1.2 \\ \hline 8\,2 \\ 41 \\ \hline 49.2 \end{array}$$

$4.92 \neq 49.2$

In a notebook, explain the error in reasoning made in the problem above. Then show how to find the quotient of $4.92 \div 1.2$.

SUMMARIZE IDEAS

When you **summarize** a passage of text, you read the entire passage, separate the most important from the less important information, and then restate that important information in your own words.

One way to summarize a passage is to follow these three steps: Identify the main ideas, note some of the details about those main ideas, and then write your summary using those main ideas.

Read the following paragraph, and then summarize the ideas.

> There is a close relationship between division of whole numbers and division of decimals. When dividing decimals, count the number of decimal places in the divisor that the decimal point would have to be moved to the right to make the divisor a whole number. Move the decimal points in the divisor and in the dividend that number of places to the right. Then place the decimal point into the quotient straight up above the decimal point in the dividend. For example, $0.2\overline{)4.28}$ becomes $2\overline{)42.8}$.

Main Ideas: dividing whole numbers; dividing decimals; moving the decimal point in the divisor, dividend, and quotient

Summary: To divide a decimal by a decimal, move the decimal points in the divisor and in the dividend, and then place a decimal in the quotient.

You have probably heard
the term *brainstorming*.
Brainstorming is a way
of addressing a problem
that involves a group
of people sharing all
ideas that come to them
as they try to find a
solution to the problem.
Studies have found that
groups can stumble upon
creative solutions when
there is a free exchange
of ideas. One person
may contribute an idea
that sparks other ideas,
and eventually, the
group effort leads to a
solution that no single
individual may have
come up with on his or
her own.

While working in
groups, one person can
draw pictures to help
illustrate the problem,
one person could write
out expressions that
could represent the
problem, while another
could try to help the
entire group find a
solution.

Working with a partner,
brainstorm and try to
find another solution
for Example 4, one that
involves a different
combination of paper
currency and coins.

Example 4 Divide with Money

Yumi and 2 of her friends earned $22.95 one Saturday washing cars.
If they split the money equally, how much did each person receive?

Step 1 Divide the dollars equally. Each friend receives $7 with $1 plus
the original 95 cents left over.

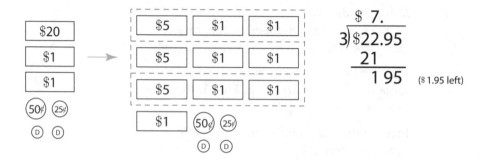

Step 2 The $1.95 left over can be changed into 19 dimes and 1 nickel.
Now divide 19 by 3. Each friend receives 6 dimes, or 60 cents,
with 1 dime and the nickel left over.

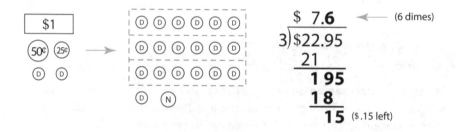

Step 3 Change the dime and nickel to 15 pennies and divide by 3. Each
friend receives 5 pennies plus the 7 dollars and 6 dimes they
received earlier. Each friend receives $7.65.

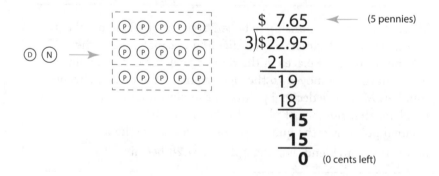

Step 4 Check the answer by finding an estimate using comparable
numbers.

Since $24 divided by 3 is $8, the three friends should each
receive about $8. The quotient, $7.65, rounded to the nearest
dollar is $8, so the answer is reasonable.

Example 5 Divide Decimals on a Calculator

Use a calculator to find the quotient of $9.72 \div 1.2$.

Press (on)

Press (9) (.) (7) (2) (÷) (1) (.) (2) (enter).

The display will show

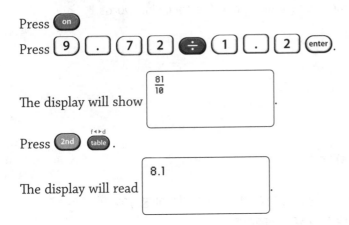

$$\frac{81}{10}$$

.

Press (2nd) (table).

The display will read

8.1

.

MATH LINK

The division symbol ÷ is not the same as the (2nd) (x^2) key, which is used when finding the square root of a number.

Divide Decimals Summary

When dividing decimals, if the divisor is a decimal, follow these steps: Move the decimal point to the right to rewrite the divisor as a whole number, and move the decimal point in the dividend the same number of places. Add zeros as needed. Then insert a decimal point in the quotient directly above the new decimal point in the dividend. Finally, divide as with whole numbers.

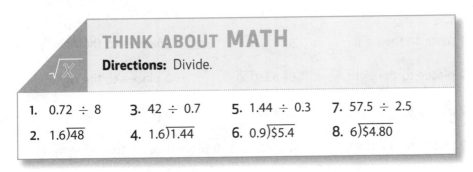

THINK ABOUT MATH

Directions: Divide.

1. $0.72 \div 8$
2. $1.6\overline{)48}$
3. $42 \div 0.7$
4. $1.6\overline{)1.44}$
5. $1.44 \div 0.3$
6. $0.9\overline{)\$5.4}$
7. $57.5 \div 2.5$
8. $6\overline{)\$4.80}$

Directions: Complete each of the sentences below using one the following words.

dividend   divisor   evaluate   quotient   reasoning

1. The _____ in the problem $5\overline{)7.5}$ is 7.5. (quotient 1.5 shown above)

2. The _____ in the problem $5\overline{)7.5}$ is 1.5.

3. The _____ in the problem $5\overline{)7.5}$ is 5.

4. If you place the decimal point in the wrong place when you are dividing, you will have to _____ the steps you took to find your answer.

5. You can use math _____ to determine that a $24 lunch can be divided evenly among 5 friends.

Directions: Next to each problem is the number of places the decimal point should be moved in both the dividend and the divisor. Evaluate whether this is correct or if it shows an error in reasoning. Then tell the correct number of places to move the decimal point.

1. $8.9\overline{)36.223}$ 1 place to the right

2. $0.35\overline{)1.127}$ 2 places to the right

3. $26.35 \div 0.5$ 2 places to the right

4. $18 \div 0.02$ 2 places to the right

5. $5\overline{)21.15}$ 2 places to the right

6. $53.9\overline{)60}$ 0 places to the right

7. $3.92 \div 7$ 0 places to the right

8. $146.72 \div 0.16$ 1 place to the right

Directions: Write a summary of the ideas in each passage.

9. It is easy to understand how to divide a decimal by a whole number. Since the divisor is already a whole number, there is no need to move the decimal point(s). You do, however, need to write the decimal in the quotient directly above the decimal point in the dividend. Then divide as you would with whole numbers.

10. Just as subtraction is the opposite of addition, division is the opposite of multiplication. Thus, checking a division problem is a matter of solving a related multiplication problem. First, turn the division problem into a multiplication problem. Check that the quotient $\times$ the divisor = the dividend. For example, to check that $3.4 \div 0.2 = 17$, show that $17 \times 0.2 = 3.4$.

Skill Practice

Directions: You may use a calculator with these questions. Choose the best answer to each question.

1. During each of the 8 days Zuri was camping, it rained. If the total rainfall was 5.44 inches, what was the average daily rainfall in inches?

   A. 0.068
   B. 0.68
   C. 1.47
   D. 6.8

2. Malik divided 16.56 by 0.9 and got the correct answer. Which calculation could have been Malik's?

   A. $9\overline{)165.6}$ with quotient 18.4
   B. $.9\overline{)16.56}$ with quotient 1.84
   C. $9\overline{)1.656}$ with quotient .184
   D. $9\overline{)1656.}$ with quotient 184.

3. Elvio paid $2.34 for 6 pounds of bananas. What was his cost, in dollars, for each pound?

   A. $0.39
   B. $2.56
   C. $2.28
   D. $3.66

4. Lomasi took 4.5 hours to travel 227.25 miles. How many miles per hour did she average?

   A. 50.5
   B. 5.05
   C. 3.79
   D. 45

Directions: You may use a calculator with these questions. Choose the best answer to each question.

1. In the number 2.707, what is the difference in value between the underlined digits?

   A. 0.007 is 100 times greater than 0.7.
   B. 0.7 is 10 times greater than 0.007.
   C. 0.7 is 100 times greater than 0.007.
   D. 0.7 is 70 more than 0.007.

2. What is 5.43 ÷ 1.2?

   A. 0.4525
   B. 4.525
   C. 45.25
   D. 452.5

3. Ashaki bought 3 pounds of bananas for $0.69 per pound. She also bought 1.2 pounds of cherries for $3.95 per pound and 2.5 pounds of grapes for $4.50 per pound. How much did she spend altogether?

   A. $15.84
   B. $17.96
   C. $18.06
   D. $22.80

4. Two baseball players had batting averages at the end of the season of .206 and .315. What was the difference between them?

   A. .521
   B. .111
   C. .109
   D. .019

5. What is the value of the expression 2.5 + (0.1 × 56) ÷ (3 + 5)?

   A. 0.9
   B. 3.2
   C. 8.7
   D. 27.267

6. What is the sum of 1.3 + 12.502 + 0.045?

   A. 3.0002
   B. 12.56
   C. 13.252
   D. 13.847

7. What is one reason zeros might be added to the end of a decimal?

   A. Zeros are added to increase its value.
   B. Zeros are added to make it easier to align addition and subtraction.
   C. Zeros are added to give the quotient the correct number of decimal places.
   D. Zeros should never be added to the end of a decimal.

8. What is the quotient?

   $0.42\overline{)4.41}$

   A. 1.5
   B. 1.8522
   C. 3.99
   D. 10.5

9. What is the value of the digit 7 in 12.372?

   A. 7 tenths
   B. 7 hundredths
   C. 7 hundreds
   D. 7 tens

10. Rami multiplied 4.52 times 0.95 and got 429.4. What mistake did he make?

    A. He did not align the decimal places when he multiplied.
    B. He did not annex zeros to the end of the decimal.
    C. He did not count the decimal places in both factors.
    D. He did not move the decimal before multiplying.

Review

Directions: Questions 11 and 13 refer to the following chart.

Felipe's Electric Bill	
Month	Amount ($)
April	65.97
May	64.54
June	71.90
July	90.15

11. Which month did Felipe pay the least amount for electricity?

   A. April
   C. May
   B. June
   D. July

12. What is 67.142 rounded to the nearest tenth?

   A. 70
   C. 67.1
   B. 67
   D. 67.14

13. How much more did Felipe pay in July than in April?

   A. $26.61
   C. $24.18
   B. $18.25
   D. $5.93

14. Aponee gets paid by the hour for freelance work. She charges $22.00 per hour. On Friday, she worked 7.25 hours. How much should she charge for the work?

   A. $29.25
   C. $159.50
   B. $308
   D. $1,595

Check Your Understanding

On the following chart, circle the number of any item you answered incorrectly. Near each lesson title, you will see the pages you can review to learn the content covered in the question. Pay particular attention to reviewing these lessons in which you missed half or more of the questions.

Chapter 2: Decimals	Procedural	Conceptual	Application/ Modeling/ Problem Solving
Introduction to Decimals pp. 50–53		1, 9	11, 12
Add and Subtract Decimals pp. 54–59	6	7	4, 13
Multiply Decimals pp. 60–63	10		3, 14
Divide Decimals pp. 64–69	2, 5, 8		

Fractions

Fractions, like decimals, are used to express numbers that are less than one and numbers that are not whole numbers. People often use fractions in conversations that are not directly related to math. "I spent *half* my day trying to get my computer started." "A *third* of the country is expecting snowstorms today."

Fractions are important both at home and at work. Following a recipe will almost always involve fractions. If you use fractions when you buy produce or meat at the grocery store, you can get the exact amount that you need: $1\frac{1}{2}$ pounds of ground beef or $\frac{3}{4}$ pound of mushrooms. Woodworking, plumbing, and other construction-related jobs also require an understanding of fractions: $12\frac{1}{2}$ feet of wood trim or a $\frac{3}{4}$-inch pipe. Even scheduling a meeting or an interview requires fractions: a $\frac{1}{2}$-hour meeting or an interview at a *quarter* past the hour. As you study the rules and operations of fractions, think about the ways fractions are part of your daily life.

The Key Concepts you will study include:

Lesson 3.1: Introduction to Fractions
Represent, compare, and order fractions to understand and develop the meaning and value of fractions.

Lesson 3.2: Add and Subtract Fractions
Understand and apply strategies for finding the sums and differences of fractions that have like or unlike denominators.

Lesson 3.3: Multiply and Divide Fractions
Extend and develop ideas about multiplication and division to include multiplying and dividing fractions.

Lesson 3.4: Mixed Numbers
Understand mixed numbers, and perform the basic operations of addition, subtraction, multiplication, and division with mixed numbers.

Goal Setting

Before starting this chapter, set goals for your learning. Think about the ways that strengthening your understanding of fractions will benefit you.

Use the Action and Fraction table below to think about fractions with every meal. Write down the fractions that you use or hear. You can write words (*half* or *one quarter*), write numerals ($\frac{1}{2}$ or $\frac{1}{4}$), or a combination of both. Add more actions to the first column for additional activities in your day that involve using fractions.

Action	Fraction	How It Was Used
Ordering food		
Buying food		
Cooking food		
Sharing food		

Introduction to Fractions

KEY CONCEPT: Represent, compare, and order fractions to understand and develop the meaning and value of fractions.

Write the decimal shown in each diagram.

1. 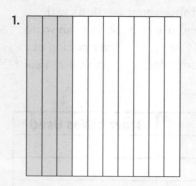　　2.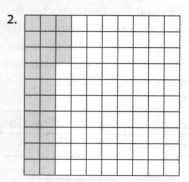

Compare each pair of decimals, using <, >, or =.

3. 0.2 _____ 0.8　　4. 0.67 _____ 0.55　　5. 0.7 _____ 0.70

Understand Fractions

Decimals involve any number of parts out of a whole made up of 10, 100, 1,000, and so on. **Fractions** are another way of showing parts of a whole. A **fraction** is made up of two numbers: the numerator and the denominator. For example, the fraction $\frac{3}{4}$ has a numerator of 3 and a denominator of 4. The **numerator** indicates the number of parts. The **denominator** refers to the number of parts in the whole.

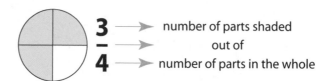

$\frac{3}{4}$ → number of parts shaded
→ out of
→ number of parts in the whole

Example 1　Write the Fraction Shown in a Diagram

What fraction is shown in the diagram?

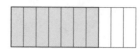

Step 1　Count the total number of shaded parts in the diagram. This is the numerator of the fraction. There are 7 shaded parts, so the numerator is 7.

Step 2　Count the total number of parts in the whole. This is the denominator of the fraction. There are 10 parts in the whole, so the denominator is 10.

Step 3 Write the fraction with the numerator on the top and the denominator on the bottom: $\frac{7}{10}$.

The numerator of a fraction is the top number. It names the number of parts. The denominator of a fraction is the bottom number. It names the number of parts in the whole.

RECOGNIZE DETAILS

In addition to main ideas, passages usually contain details. These are usually specific pieces of information that relate to the main idea. Details can describe, quantify, or support the main idea.

Recognizing details is a skill that will help you to understand what you read.

Read the passage below for details.

> (1) A **fraction** is a way to represent parts of a whole. (2) The **denominator** is the number on the bottom. (3) It shows how many equal parts the whole has been broken into. (4) The **numerator** is the number on top. (5) It tells how many of the equal parts are being counted.

What are the details in the passage above?

_____

_____

_____

The main idea of the passage is stated in sentence 1: A fraction represents parts of a whole. The details that support and expand on this idea are in sentences 2, 3, 4, and 5. Sentences 2 and 4 describe what the different numbers in a fraction represent. Sentences 3 and 5 explain how the numbers relate to parts of a whole.

Core Skill
Interpret Data
Displays

A **diagram** is an illustration or picture that shows mathematical or other types of information. Learning how to interpret the information in diagrams is thus an important skill. The diagrams that appear in this lesson are tools that can be used to represent fractions visually. Each whole diagram is divided into parts, and some of the parts are shaded. The shaded parts of the diagram **represent**, or stand for, the numerator. The total number of parts in the whole represents the denominator.

In a notebook, draw a diagram to represent the fraction $\frac{2}{5}$. Explain how the diagram shows this fraction.

MATH LINK

When you draw a diagram to represent a fraction, it does not matter what kind of shape you draw as long as it can be divided easily into the correct number of equal pieces.

Because students bring different strengths to the classroom, it is often helpful to work together. Each individual can contribute something that will lead to the successful completion of an assignment or project. Math problems may sometimes seem really overwhelming when you try to solve them on your own. But when you partner with a classmate or friend or when you join a group, you can accomplish more together than you could if you were to work alone.

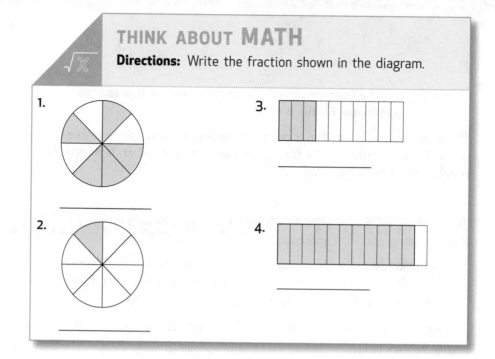

THINK ABOUT MATH

Directions: Write the fraction shown in the diagram.

1.

2.

3.

4.

Name Equivalent Fractions

Equivalent fractions are fractions that have the same value. Any fraction multiplied by some form of 1 will yield an equivalent fraction. Forms of 1 are any fraction in which the numerator is the same as the denominator, such as $\frac{1}{1}$, $\frac{2}{2}$, $\frac{6}{6}$, or $\frac{102}{102}$. Study the diagram below.

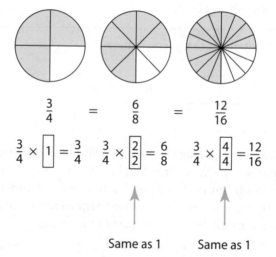

$$\frac{3}{4} \quad = \quad \frac{6}{8} \quad = \quad \frac{12}{16}$$

$$\frac{3}{4} \times \boxed{1} = \frac{3}{4} \qquad \frac{3}{4} \times \boxed{\frac{2}{2}} = \frac{6}{8} \qquad \frac{3}{4} \times \boxed{\frac{4}{4}} = \frac{12}{16}$$

Same as 1   Same as 1

A fraction is in **lowest terms** if the numerator and denominator cannot be divided evenly (remainder of zero) by any whole number other than 1. Since fractions can be written as many equivalent fractions, writing a fraction in lowest terms makes comparing values easier.

Example 2 Name Equivalent Fractions

Name three fractions equivalent to $\frac{4}{5}$.

Step 1 Multiply $\frac{4}{5}$ by three different forms of 1.

$$\frac{4}{5} \times \frac{4}{4} = \frac{16}{20}; \qquad \frac{4}{5} \times \frac{20}{20} = \frac{80}{100}; \qquad \frac{4}{5} \times \frac{50}{50} = \frac{200}{250}$$

Step 2 Write the three equivalent fractions: $\frac{16}{20}$, $\frac{80}{100}$, $\frac{200}{250}$.

Example 3 Find a Specific Equivalent Fraction

Write a fraction with a denominator of 12 that is equivalent to $\frac{21}{36}$.

Step 1 Set up the problem: $\frac{21}{36} = \frac{?}{12}$.

Step 2 Divide by a form of 1.
Since $36 \div 3 = 12$, divide the numerator and denominator by $3\left(\frac{3}{3} = 1\right)$.
$$\frac{21}{36} \div \frac{3}{3} = \frac{7}{12}$$

Step 3 Check your answer by multiplying $\frac{7}{12}$ by $\frac{3}{3}$.
$$\frac{7}{12} \times \frac{3}{3} = \frac{21}{36}$$
The answer is correct. $\frac{7}{12}$ is equivalent to $\frac{21}{36}$.

Example 4 Rewrite a Fraction in Lowest Terms

Rewrite $\frac{20}{24}$ as a fraction in lowest terms.

Step 1 Divide the numerator and denominator by a number or numbers until both numerator and denominator can no longer be divided.
$$\frac{20}{24} = \frac{20}{24} \div \frac{2}{2} = \frac{10}{12} \div \frac{2}{2} = \frac{5}{6} \text{ or}$$
$$\frac{20}{24} = \frac{20}{24} \div \frac{4}{4} = \frac{5}{6}$$

Step 2 Write $\frac{20}{24}$ in lowest terms: $\frac{5}{6}$.

Example 5 Use a Calculator to Reduce a Fraction to Lowest Terms

Use a calculator to reduce $\frac{28}{32}$ to lowest terms.

Press .

Press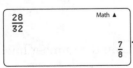

Press (enter)

The display should read

$\frac{28}{32}$	Math ▲
	$\frac{7}{8}$

The fraction $\frac{28}{32}$ reduced to lowest terms is $\frac{7}{8}$.

In summary, to find equivalent fractions, multiply the fraction by forms of 1, such as $\frac{2}{2}$, $\frac{10}{10}$, or $\frac{50}{50}$. To find specific equivalent fractions, multiply or divide by forms of 1.

To rewrite fractions in lowest terms, divide the numerator and denominator by the same number or numbers until the numerator and denominator can no longer be divided, or use a calculator.

MATH LINK

When using the inequality symbols < and >, remember that the open end of the symbol opens to the greater number.

THINK ABOUT MATH

Directions: Match the letter of each fraction or number to the appropriate fraction or whole number.

_____ 1. an equivalent fraction for $\frac{2}{3}$

_____ 2. the missing numerator in the equation $\frac{15}{25} = \frac{?}{5}$

_____ 3. the missing numerator in the equation $\frac{7}{9} = \frac{?}{27}$

_____ 4. $\frac{16}{18}$ written in lowest terms

_____ 5. $\frac{14}{42}$ written in lowest terms

A. 3

B. $\frac{1}{3}$

C. $\frac{10}{15}$

D. $\frac{8}{9}$

E. 21

Compare and Order Fractions

There is often a need to compare fractions. When there are more than two fractions involved, the fractions are usually listed in order from least to greatest or greatest to least. Using a number line is one way to compare fractions. Another way is to find a common denominator for both fractions and then compare the numerators.

Example 6 Use a Number Line to Compare Fractions

Which is greater, $\frac{7}{10}$ or $\frac{3}{5}$?

Step 1 Draw a number line with both fifths and tenths.

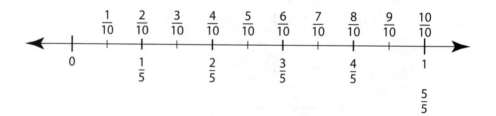

Step 2 Locate both fractions on the number line.

Since $\frac{7}{10}$ is to the right of $\frac{3}{5}$, $\frac{7}{10}$ is greater than $\frac{3}{5}$. Write $\frac{7}{10} > \frac{3}{5}$.

Example 7 Find a Common Denominator to Compare Fractions

Compare $\frac{5}{6}$ and $\frac{7}{8}$.

Step 1 Identify a common denominator for the two fractions. List the multiples of 6 and 8 until a **common multiple**, or a number that is multiple of both denominators, is identified.

Multiples of 6: **6** (6 × 1), **12** (6 × 2), **18** (6 × 3), <u>**24**</u> (6 × 4)

Multiples of 8: **8** (8 × 1), **16** (8 × 2), <u>**24**</u> (8 × 3)

Step 2 Rewrite $\frac{5}{6}$ and $\frac{7}{8}$ as fractions having denominators of 24 (the common multiple).

$$\frac{5}{6} = \frac{?}{24} \longrightarrow \frac{5 \times 4}{6 \times 4} = \frac{20}{24} \qquad \frac{7}{8} = \frac{?}{24} \longrightarrow \frac{7 \times 3}{8 \times 3} = \frac{21}{24}$$

Step 3 Compare the numerators.

Since $20 < 21$, $\frac{20}{24} < \frac{21}{24}$ and $\frac{5}{6} < \frac{7}{8}$.

Example 8 Order Fractions

Write the set of fractions in order from least to greatest.

$$\frac{3}{4}, \frac{2}{3}, \frac{5}{6}$$

Step 1 Rewrite the fractions using a common denominator.
Since 12 is a common multiple of 3, 4, and 6, rewrite the fractions using 12 as the common denominator.

$$\frac{3}{4} = \frac{3 \times 3}{4 \times 3} = \frac{9}{12} \qquad \frac{2}{3} = \frac{2 \times 4}{3 \times 4} = \frac{8}{12} \qquad \frac{5}{6} = \frac{5 \times 2}{6 \times 2} = \frac{10}{12}$$

Step 2 Compare the numerators.

Since $8 < 9 < 10$, $\frac{8}{12} < \frac{9}{12} < \frac{10}{12}$ and $\frac{2}{3} < \frac{3}{4} < \frac{5}{6}$.

THINK ABOUT MATH

Directions: Answer the following.

1. How would you use a number line to compare the fractions $\frac{3}{4}$ and $\frac{7}{8}$?

   _____

2. How would you use a common denominator to compare $\frac{1}{4}$ and $\frac{1}{6}$?

   _____

3. Place the numbers below in order from greatest to least.

   $$\frac{7}{9}, \frac{2}{3}, \frac{5}{6}$$

   _____

Equivalent fractions are fractions that represent the same number, such as $\frac{1}{2}$ and $\frac{2}{4}$ or 3 and $\frac{3}{1}$. Using equivalent fractions can help solve problems found in everyday life. You can use equivalent fractions to determine the number of $51 \times 44 = 204$ or 20 quarters.

Consider the following problem. Mrs. Lenhart made 3 apple pies for her 5th grade class. She cut each pie into 8 slices for her students to share. How many slices of pie were shared among Mrs. Lenhart's students? In a notebook, write the appropriate equivalent fraction by multiplying the correct form of 1.

Before Cut

After Cut

Directions: Complete the sentences below using one of the following words.

common multiple   denominator   equivalent fractions
fraction   lowest terms   numerator

1. A _____ is a made up of two parts. It is a way of showing parts of a whole.

2. The bottom number of a fraction is the _____. It tells the number of parts in the whole.

3. A fraction is in _____ when the numerator and denominator can no longer be divided by a whole number other than 1.

4. The _____ of a fraction is the top number. It tells the number of parts.

5. A common denominator of two or more fractions can be found by finding a _____ of the fractions.

6. Fractions that have the same value are _____.

Skill Review

Directions: Underline and number the details in each passage. Then explain how the details support the main idea.

1. Equivalent fractions have the same value. One way to find equivalent fractions is to multiply each fraction by a form of 1. A form of 1 is any fraction in which the numerator and denominator are the same, such as $\frac{5}{5}$. Another way to find an equivalent fraction is to divide by a form of 1.

_____

_____

_____

2. There are several ways you can compare two fractions. One way is by finding a common denominator. List the multiples of each denominator. The first common multiple is the least common denominator of the two fractions. Rewrite the fractions with the common denominator. Then compare the numerators of the fractions to determine the lesser or greater fraction.

_____

_____

Directions: Use a diagram to represent each fraction below. Explain how the diagram represents the fraction.

3. $\frac{6}{7}$ 4. $\frac{5}{6}$ 5. $\frac{3}{8}$ 6. $\frac{4}{9}$

Directions: Use diagrams to show the equivalent fractions below.

7. the number of twelfths equivalent to $\frac{3}{4}$ 8. the number of fifths equivalent to $\frac{8}{10}$

Skill Practice

Directions: You may use a calculator for these questions. Choose the best answer to each question.

1. Which of the following has the fractions arranged from least to greatest?

   A. $\frac{2}{5}, \frac{1}{2}, \frac{5}{8}$

   B. $\frac{2}{3}, \frac{5}{6}, \frac{5}{8}$

   C. $\frac{3}{4}, \frac{4}{5}, \frac{7}{10}$

   D. $\frac{1}{2}, \frac{3}{5}, \frac{5}{9}$

2. Which fraction represents the shaded part of the diagram?

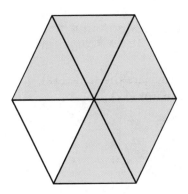

   A. $\frac{1}{8}$ C. $\frac{5}{6}$

   B. $\frac{1}{6}$ D. $\frac{7}{8}$

3. A community center is sponsoring mixed softball teams this summer. Out of the people who signed up for the teams, $\frac{4}{7}$ are female. If 42 people signed up, how many are female?

   A. 28

   B. 24

   C. 20

   D. 16

4. Which fraction is written in lowest terms?

   A. $\frac{7}{35}$

   B. $\frac{4}{18}$

   C. $\frac{18}{21}$

   D. $\frac{14}{25}$

Add and Subtract Fractions

Lesson Objectives

You will be able to

- Add and subtract fractions with like denominators

- Add and subtract fractions with unlike denominators

Skills

- **Core Skill:** Perform Operations

- **Core Skill:** Apply Number Sense

Match each fraction with an equivalent fraction.

_____ 1. $\frac{6}{10}$ _____ 2. $\frac{5}{14}$ _____ 3. $\frac{9}{15}$

A. $\frac{27}{45}$ B. $\frac{25}{70}$ C. $\frac{12}{20}$

Write each fraction in lowest terms.

4. $\frac{12}{15}$ _____   5. $\frac{18}{42}$ _____   6. $\frac{26}{50}$ _____

Vocabulary

common denominator
like denominators
simplify
unlike denominators

Add and Subtract Fractions with Like Denominators

When adding or subtracting fractions, look to see if the denominators are **like denominators** or **unlike denominators**. In other words, do the fractions have denominators that are the same or different?

Example 1 Add Fractions with Like Denominators

Add $\frac{5}{12}$ and $\frac{7}{12}$.

Step 1 Look at the denominators. If they are the same (12), the sum will have this denominator (12).

$$\frac{5}{12} + \frac{7}{12} = \frac{\square}{12}$$

Step 2 Add the numerators.

$$\frac{5}{12} + \frac{7}{12} = \frac{12}{12}$$

Step 3 **Simplify** the answer, or reduce it to lowest terms.

$$\frac{12}{12} = 1, \text{ so } \frac{5}{12} + \frac{7}{12} = \frac{12}{12} = 1$$

MATH LINK

Another way to look at Step 2 in Example 1 is to think: "How many twelfths result from combining the parts?" *12 twelfths*

Example 2 Subtract Fractions with Like Denominators

Subtract $\frac{1}{10}$ from $\frac{7}{10}$.

Step 1 Look at the denominators. They are the same (10), so the difference will have the same denominator (10).

$$\frac{7}{10} - \frac{1}{10} = \frac{\square}{10}$$

Step 2 Subtract the numerators.

$$\frac{7}{10} - \frac{1}{10} = \frac{6}{10}$$

Step 3 Simplify the answer.

$$\frac{6}{10} \div \frac{2}{2} = \frac{6 \div 2}{10 \div 2} = \frac{3}{5}, \text{ so } \frac{7}{10} - \frac{1}{10} = \frac{6}{10} = \frac{3}{5}$$

When reviewing the previous two examples for adding or subtracting fractions with like denominators, remember the sequence of steps. First, add or subtract the numerators. The denominators will be the same as the original ones. Then, if necessary, write the sum or difference in lowest terms.

THINK ABOUT MATH

Directions: Find each sum or difference. Reduce the answer to lowest terms.

1. $\frac{5}{12} + \frac{3}{12}$ _____

2. $\frac{31}{36} - \frac{5}{36}$ _____

3. $\frac{7}{18} - \frac{1}{18}$ _____

4. $\frac{11}{20} + \frac{7}{20}$ _____

Add and Subtract Fractions with Unlike Denominators

When adding or subtracting fractions with denominators that are not the same, first find equivalent fractions that have a **common denominator**—a common multiple of the denominators of the two fractions. Then add or subtract the fractions.

MATH LINK

When adding or subtracting fractions with like denominators, make sure the denominators remain the same. However, if the answer can be simplified (reduced to lowest terms), the final denominator will change.

MATH LINK

One way to find a common denominator when adding or subtracting fractions with unlike denominators is to multiply the original denominators by each other. Remember to simplify or reduce the answer to lowest terms.

Real-world word problems involving fractions will sometimes talk about things like one half of a pizza pie. It is easy to add or subtract fractions when you can visualize individual slices of a pizza that represent part of a whole pie. A pizza has 8 slices, and you and your sister eat 4 of them. How much of the pizza remains? It is easy to see that half of the pie is left. Once you become familiar with fractions, of course, you no longer need to visualize actual parts of a whole in order to add and subtract. You will be able to perform operations involving fractions on paper or with a calculator.

Example 3 Add Fractions with Unlike Denominators

Add $\frac{7}{12}$ and $\frac{1}{3}$.

Step 1 Look at the denominators. They are different (12 and 3). Find a common denominator of these two numbers: 12.

Step 2 Find an equivalent fraction for $\frac{1}{3}$ using a common denominator of 12.

$$\frac{1}{3} = \frac{\square}{12} \qquad \text{Think: } 3 \times ? = 12$$

$$\frac{1}{3} \times \frac{4}{4} = \frac{4}{12} \qquad \text{Multiply by } \frac{4}{4} \text{ or 1.}$$

Step 3 Add using twelfths.

$$\begin{array}{rl} \frac{7}{12} = & \frac{7}{12} \\ + \frac{1}{3} = & \frac{4}{12} \\ \hline & \frac{11}{12} \end{array}$$

Add the new numerators.

The new denominator is 12.

The answer is already in lowest terms.

Example 4 Subtract Fractions with Unlike Denominators

Subtract $\frac{1}{5}$ from $\frac{3}{4}$.

Step 1 Look at the denominators. They are different (5 and 4). One common denominator is 20, since $5 \times 4 = 20$.

$$\begin{array}{rl} \frac{3}{4} = & \frac{\square}{20} \\ - \frac{1}{5} = & \frac{\square}{20} \\ \hline \end{array}$$

Step 2 Find equivalent fractions for $\frac{1}{5}$ and $\frac{3}{4}$ with 20 as the denominators.

$$\frac{3}{4} = \frac{\square}{20} \qquad \text{Think: } 4 \times ? = 20$$

$$\frac{3}{4} \times \frac{5}{5} = \frac{15}{20} \qquad \text{Multiply by } \frac{5}{5} \text{ or 1.}$$

$$\frac{3}{4} = \frac{15}{20}$$

$$\frac{1}{5} = \frac{\square}{20} \qquad \text{Think: } 5 \times ? = 20$$

$$\frac{1}{5} \times \frac{4}{4} = \frac{4}{20} \qquad \text{Multiply by } \frac{4}{4} \text{ or 1}$$

$$\frac{1}{5} = \frac{4}{20}$$

Step 3 Subtract using twentieths.

$$\begin{array}{rl} \frac{3}{4} = & \frac{15}{20} \\ - \frac{1}{5} = & \frac{4}{20} \\ \hline & \frac{11}{20} \end{array}$$

Subtract the new numerators.

The new denominator is 20.

The answer is already in lowest terms.

Example 5 Use a Calculator to Add or Subtract Fractions.

Solve $\frac{9}{10} - \frac{1}{15}$.

Press (on).

Press (n/d) (9) ▼ (1) (0) (◄►) (−) (n/d) (1) ▼ (1) (5).

Press (enter).

The display should read

$\frac{9}{10} - \frac{1}{15}$	Math ▲
	$\frac{5}{6}$

$\frac{9}{10} - \frac{1}{15} = \frac{5}{6}$

Example 6 Solve a Problem with Unlike Denominators

Ray spent $\frac{1}{12}$ hour stretching before he began his morning run. He spent $\frac{2}{3}$ hour running. What part of an hour did Ray spend stretching and running?

Step 1 Set up the problem.

$$\frac{1}{12} = \frac{\square}{12}$$
$$+ \frac{1}{3} = \frac{\square}{12}$$

Step 2 Add.

$$\frac{1}{12} = \frac{1}{12}$$
$$+ \frac{2}{3} = \frac{8}{12}$$
$$\frac{9}{12}$$

Step 3 Simplify. $\frac{9}{12} = \frac{3}{4}$

The sequence for adding and subtracting fractions with unlike denominators is to first find a common denominator. Second, write the equivalent fractions with the new denominator. Third, add or subtract the numerators. Fourth, if necessary, write the answer in lowest terms.

Core Skill
Apply Number Sense

The ability to look at the solution of a problem and determine its inaccuracy is a skill that can strengthen your mathematical ability. For instance, Michael claims that since he received a 70% on his first exam, and an 80% on his second exam, he now has a 150% in the course. This is obviously not the case, since doing better than 100% is rare, depending on the course.

Consider the following problem. Gina works $\frac{3}{5}$ of an hour on her homework on Monday, $\frac{7}{6}$ of an hour on Tuesday, and $\frac{4}{3}$ of an hour on Thursday. What fraction of an hour did Gina work on her homework? Without finding the answer, could Gina have spent more than 6 hours on her homework? In a notebook, solve the equation needed to determine how much time Gina spent.

THINK ABOUT MATH

Directions: Fill in the blanks.

1. To find $\frac{3}{8} + \frac{1}{3}$, you need to find a common denominator for _____ and _____. One common denominator is _____. The sum in lowest terms is _____.

2. To find $\frac{5}{6} - \frac{1}{3}$, you need to find a common denominator for _____ and _____. One common denominator is _____. The difference in lowest terms is _____.

Vocabulary Review

Directions: Match each word to one of the statements below.

_____ 1. common denominator   _____ 3. simplify

_____ 2. like denominators   _____ 4. unlike denominators

A. It is what you do to reduce a fraction, such as $\frac{2}{10}$, to lowest terms.

B. It describes fractions such as $\frac{11}{12}$ and $\frac{2}{15}$, because the denominators of the fractions are different.

C. An example would be 12 because it is a multiple of the denominators of $\frac{2}{3}$ and $\frac{3}{4}$.

D. The fractions $\frac{3}{10}$ and $\frac{7}{10}$ are examples, since the denominators are the same.

Skill Review

Directions: Write the sequence of steps you would use to find the sum or difference of the following fractions. Then find the sum or difference.

1. $\frac{3}{10} + \frac{2}{5}$ 2. $\frac{13}{15} - \frac{4}{15}$

Directions: Describe the steps you would take to find the difference of $\frac{7}{8}$ and $\frac{3}{10}$ on your calculator. Then find the difference.

3. _____

_____

_____

_____

_____

_____

Skill Practice

Directions: Choose the best answer to each question.

1. The sum of $\frac{1}{6}$ and another fraction is $\frac{7}{12}$. What is the other fraction?

   A. $\frac{5}{12}$

   B. $\frac{6}{12}$

   C. $\frac{2}{3}$

   D. $\frac{3}{4}$

2. A chart shows the type of nail to use for different thicknesses of plywood. The thickness of the plywood ranges from $\frac{1}{4}$ inch to $\frac{3}{4}$ inch. In inches, what is the difference in the thicknesses of the plywood?

   A. $\frac{1}{4}$

   B. $\frac{1}{2}$

   C. $\frac{3}{4}$

   D. 1

Directions: Reduce fractions to their lowest terms.

3. Jorge spent $\frac{1}{2}$ hour pulling weeds in the backyard. He spent $\frac{3}{12}$ hour pulling weeds in the front yard. How much longer, in a fraction of an hour, did it take to pull weeds in the back?

   _____

4. Maya bought $\frac{3}{4}$ yard of fabric to cover a footstool. When she tried to cover the footstool, she was short $\frac{1}{8}$ yard. How many total yards of fabric did Maya need to cover the footstool?

   _____

Multiply and Divide Fractions

KEY CONCEPT: Extend and develop ideas about multiplication and division to include multiplying and dividing fractions.

Write each fraction in lowest terms.

1. $\frac{16}{28}$ _____
2. $\frac{45}{80}$ _____
3. $\frac{88}{96}$ _____
4. $\frac{12}{16}$ _____

Multiply Fractions

Two fractions can be multiplied by multiplying the numerators and denominators and simplifying the product.

Example 1 Multiply a Fraction by a Fraction

Multiply $\frac{5}{8}$ and $\frac{3}{4}$.

Step 1 Multiply the numerators. Then multiply the denominators.

$$\frac{5}{8} \times \frac{3}{4} = \frac{5 \times 3}{8 \times 4} = \frac{15}{32}$$

Step 2 Rewrite the answer in lowest terms, if necessary. In this case, the answer is already in lowest terms.

Example 2 Use a Calculator to Multiply Fractions

Find $\frac{8}{15} \times \frac{5}{8}$.

Press (on)

Press (n/d) 8 ▼ 15 ▶ (×) (n/d) 5 ▼ 8.

Press (enter).

The display should read

$\frac{8}{15}$ × $\frac{5}{8}$	Math ▲
	$\frac{1}{3}$

$\frac{8}{15} \times \frac{5}{8} = \frac{1}{3}$

THINK ABOUT MATH

Directions: Match each multiplication problem to its product. All products are in lowest terms.

_____ 1. $\frac{1}{5} \times \frac{3}{8}$ _____ 3. $\frac{9}{10} \times \frac{5}{6}$

_____ 2. $\frac{7}{12} \times \frac{2}{3}$ _____ 4. $\frac{11}{25} \times \frac{9}{11}$

   A. $\frac{3}{4}$ B. $\frac{9}{25}$ C. $\frac{3}{40}$ D. $\frac{7}{18}$

Juan owns a snow shoveling business. Of the money he makes, $\frac{1}{2}$ of it goes to paying his employees (the other half for business costs). Juan has 5 employees (including himself) and pays everyone an equal amount. What fraction of the total money made does Juan pay to a single employee? Use the following fraction bar, if needed, to determine the correct fraction.

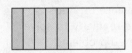

Divide Fractions

Multiplication and division are inverse operations. To divide by a fraction, you must multiply the dividend by the **reciprocal** (the number that when multiplied by the divisor, has a product of 1) of the divisor. To find the reciprocal of any number, **invert**, or turn over, the fraction or change the places of the numerator and denominator. For example, the inverted form of $\frac{3}{7}$ is $\frac{7}{3}$. In this case, $\frac{7}{3}$ is the reciprocal of $\frac{3}{7}$ because $\frac{7}{3} \times \frac{3}{7} = \frac{21}{21} = 1$.

After changing the division symbol to multiplication and the divisor to its reciprocal, multiply the fractions.

Example 3 Divide by a Fraction

Find $3 \div \frac{1}{16}$.

Step 1 Change the division symbol to multiplication and invert the divisor (multiply by the reciprocal of the divisor).

$$3 \div \frac{1}{16} = \frac{3}{1} \times \frac{\mathbf{16}}{\mathbf{1}} \qquad (3 = \frac{3}{1})$$

Step 2 The problem is now a multiplication problem. Multiply the numerators and multiply the denominators.

$$\frac{3}{1} \times \frac{16}{1} = \frac{3 \times 16}{1 \times 1} = \frac{48}{1} = 48$$

Step 3 Simplify, if necessary, by writing the answer in lowest terms. In this case, the answer is in lowest terms.

Example 4 Use a Calculator to Divide Fractions

Find $\frac{5}{9} \div \frac{2}{3}$.

Press (on).

Press

The display should read 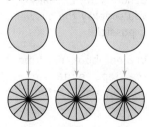.

Another way to look at Example 3 is to ask yourself: *How many sixteenths are in 3 wholes?*

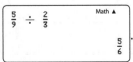

1 whole is the same as 16 (1 × 16) sixteenths.

2 wholes is the same as 32 (2 × 16) sixteenths.

3 wholes is the same as 48 (3 × 16) sixteenths.

Therefore, $3 \div \frac{1}{16}$ is the same as 3 × 16.

$\frac{5}{9} \div \frac{2}{3} = \frac{5}{6}$

MATH LINK

The reciprocal is also known as the **multiplicative inverse**.

Core Skill
Perform Operations

Math relies on **regularity**, or repeating patterns. You may have noticed that a step-by-step approach to problem solving is often used in this book. You may have also noticed that a specific sequence of steps is used regularly to solve related problems.

For example, when you divide a fraction by another fraction, you repeat the same steps in the same order. First, you change the division sign to a multiplication sign. Then you invert the divisor. Next, you multiply the numerators. Then you multiply the denominators. Finally, you simplify the answer, if possible.

In a notebook, describe the steps you take to multiply two fractions.

THINK ABOUT MATH

Directions: Solve the problems below.

1. Write a math problem that shows how to find the number of thirds in 9. Then find the number of thirds. Show your work.

_____

_____

_____

2. A pica is a measure of $\frac{1}{6}$ inch. Picas are most commonly used to measure text. If a line of text is 3 inches long, how many picas is the line of text?

_____

Vocabulary Review

Directions: Fill in the blanks with one of the words below.

invert   multiplicative inverse   reciprocal

1. The reciprocal is the same as the _____.

2. The _____ of 18 is $\frac{1}{8}$.

3. When you divide fractions, you need to _____ the divisor.

Skill Review

Directions: Describe the steps you would take to divide any two fractions using a calculator.

1. _____

_____

_____

_____

_____

2. Fahrenheit and Celsius are temperature scales. You can convert temperatures from one scale to another. For example, to convert from 104 degrees Fahrenheit to degrees Celsius, subtract 32 from 104. Then multiply by $\frac{5}{9}$. Given what you know about inverse relationships, how would you convert 40 degrees Celsius to degrees Fahrenheit? Show calculations to support your reasoning.

_____

_____

_____

_____

_____

Skill Practice

Directions: Choose the best answer to each question.

1. Which of the following is the quotient of $\frac{2}{3}$ divided by $\frac{4}{5}$?

   A. $\frac{8}{15}$

   B. $\frac{5}{6}$

   C. $\frac{6}{5}$

   D. $\frac{15}{8}$

2. How many $\frac{1}{2}$-inch columns can Anna fit in a table that is 10 inches long?

   A. 5
   B. 10
   C. 15
   D. 20

3. A crew is clearing a 15-mile trail after a storm. If the crew clears $\frac{3}{5}$ mile every day, how many days will it take to clear the trail?

   A. 25
   B. 15
   C. 9
   D. 3

4. Tyrone says the quotient of $\frac{1}{4} \div \frac{5}{8}$ is $\frac{5}{32}$. Which of the following best describes Tyrone's statement.

   A. He is correct.
   B. He forgot to invert the divisor before multiplying.
   C. He forgot to invert the dividend before multiplying.
   D. He forgot to invert both fractions before multiplying.

Mixed Numbers

KEY CONCEPT: Understand mixed numbers and perform the basic operations of addition, subtraction, multiplication, and division with mixed numbers.

Find each sum or difference. Simplify the answer, if necessary.

1. $\frac{3}{10} + \frac{3}{5}$ **2.** $\frac{7}{8} - \frac{1}{3}$

Find each product or quotient. Simplify the answer, if necessary.

3. $\frac{5}{9} \times \frac{3}{7}$ **4.** $\frac{2}{3} \div \frac{5}{6}$

Lesson Objectives

You will be able to

- Add and subtract mixed numbers

- Multiply and divide mixed numbers

Skills

- **Core Skill:** Represent Real-World Problems

- **Reading Skill:** Evaluate Arguments

Vocabulary

detail
improper fraction
mixed number
proper fraction
reduce
rename

Add and Subtract Mixed Numbers

The sum of a whole number and a fraction is a **mixed number**. For example, $2\frac{3}{5}$ is the same as $1 + 1 + \frac{3}{5}$, or $\frac{5}{5} + \frac{5}{5} + \frac{3}{5}$. The diagram shows that $2\frac{3}{5}$ is equal to $\frac{13}{5}$.

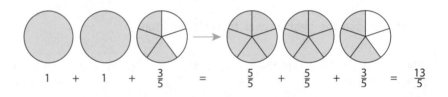

$$1 \quad + \quad 1 \quad + \quad \frac{3}{5} \quad = \quad \frac{5}{5} \quad + \quad \frac{5}{5} \quad + \quad \frac{3}{5} \quad = \quad \frac{13}{5}$$

A fraction with a numerator greater than or equal to its denominator is an **improper fraction**. A fraction with a numerator that is less than the denominator is sometimes called a **proper fraction**. Remember that an improper fraction with a numerator equal to the denominator, such as $\frac{9}{9}$, $\frac{16}{16}$, or $\frac{100}{100}$, is equivalent to 1.

Example 1 Add Mixed Numbers

Add $2\frac{4}{5}$ and $1\frac{1}{3}$.

> **Step 1** Look at the denominators. They are different (5 and 3).
>
> **Step 2** **Rename** the fractions, or find equivalent fractions with a common denominator of 15.
>
> **Step 3** Add the whole numbers.
>
> **Step 4** Add the renamed fractions.
>
> **Step 5** **Reduce**, or simplify, the answer by writing it in lowest terms. Since $\frac{17}{15}$ is an improper fraction, think:
>
> $$3\frac{17}{15} = 3 + \frac{15}{15} + \frac{2}{15} = 4\frac{2}{15}$$

$$2\frac{4}{5} = 2\frac{12}{15}$$
$$+1\frac{1}{3} = 1\frac{5}{15}$$
$$\overline{3\frac{17}{15}}$$

MATH LINK

Follow these steps to change an improper fraction to a whole number or a mixed number:

1) Divide the numerator of the improper fraction by the denominator.

2) If the quotient has a remainder, the improper fraction is written as a mixed number with the remainder as the numerator and the same denominator as the original improper fraction.

SUMMARIZE SUPPORTING DETAILS

A passage will often contain many supporting **details**—information that supports the main idea. Sometimes, trying to remember every detail can be overwhelming. Summarizing details is a way to think about information in a general way. A summary does not include every piece of information from a selection. Only the most important details are presented, and they are presented in a clear manner.

When you summarize, details that are similar can be grouped together. For example, "Javier read 5 books in June, 6 books in July, and 4 books in August" can be summarized by the statement "Javier read 15 books over the summer."

Some details can be omitted from your summary. "Maggie's insurance plan covers health, vision, and dental. She has optional life insurance. Deductibles are low. Her insurance card has blue clouds on it." If the point of this passage is to describe Maggie's insurance, the detail about the clouds can be safely omitted.

Read the passage below. As you read, identify the most important details.

> Isra is shopping for a cable and internet package. The plan she likes is called the Silver Plan. It has 5 broadcast channels, 30 basic cable channels, and 40 specialized cable channels. It also includes 1 premium movie channel. She can add a digital video recording (DVR) device for a small fee. It comes with a remote that takes 2 AA batteries. Internet connection is also included. She thinks it is a good choice because she loves to watch movies. The other plans offered by the same company are not as appealing to her.

A good summary will include the features of the plan: 75 channels, 1 movie channel, option for DVR, internet included. Details such as what kind of batteries the remote uses can be left out of the summary.

Core Skill
Represent Real-World Problems

Because real-life scenarios don't always use whole numbers, it is more likely that mixed fractions, or improper fractions, will appear during calculations. Consider the following problem.

Claire is training for a marathon. She runs $8\frac{1}{3}$ miles on Monday, $9\frac{4}{5}$ miles on Tuesday, $11\frac{5}{6}$ miles on Wednesday, 12 miles on Thursday, and $14\frac{7}{8}$ miles on Friday. How many miles did Claire run during the week? If Claire wanted to run 100 miles during the entire week, including the weekend, how many more miles will she have to run on Saturday and Sunday to achieve her goal? In a notebook, write down the correct equation to determine how many miles Claire has run. Then answer the questions.

Example 2 Subtract Mixed Numbers

Subtract $1\frac{3}{4}$ from $3\frac{1}{6}$.

Step 1 Look at the denominators. They are different (4 and 6).

Step 2 Rename the fractions using 12 as the denominator.

$$3\frac{1}{6} = 3\frac{2}{12}$$
$$-1\frac{3}{4} = 1\frac{9}{12}$$

Step 3 Subtract the fractional parts and whole numbers. If this is not possible, rename the first mixed number by removing 1 from the whole number, converting it to a fraction, and adding it to the fraction. So, $3\frac{2}{12} = 2 + 1 + \frac{2}{12} = 2 + \frac{12}{12} + \frac{2}{12} = 2\frac{14}{12}$.

Step 4 Subtract the renamed fractions and the new whole numbers.

$$3\frac{1}{6} = 3\frac{2}{12} = 2\frac{14}{12}$$
$$-1\frac{3}{4} = 1\frac{9}{12} = 1\frac{9}{12}$$
$$\overline{\qquad\qquad\qquad 1\frac{5}{12}}$$

Step 5 Simplify the answer if necessary. In this case, $1\frac{5}{12}$ is in lowest terms

Multiply and Divide Mixed Numbers

When multiplying and dividing mixed numbers, you must first change the mixed numbers to improper fractions. Then follow the steps for multiplying and dividing fractions.

Example 3 Multiply Mixed Numbers

Multiply $1\frac{2}{3}$ and $2\frac{1}{5}$.

Step 1 Change the mixed numbers to improper fractions.

$$1\frac{2}{3} = \frac{(1 \times 3) + 2}{3} = \frac{5}{3}; \; 2\frac{1}{5} = \frac{(2 \times 5) + 1}{5} = \frac{11}{5}$$

Step 2 Multiply the improper fractions by multiplying the numerators and multiplying the denominators.

$$1\frac{2}{3} \times 2\frac{1}{5} = \frac{5}{3} \times \frac{11}{5} = \frac{55}{15}$$

Step 3 Change the improper fraction to a mixed number, and simplify the fraction.

$$\frac{55}{15} = 3\frac{10}{15} = 3\frac{2}{3}$$

Example 4 Divide Mixed Numbers

Find $1\frac{3}{8} \div \frac{1}{2}$.

Step 1 Change any mixed numbers to improper fractions.

$$1\frac{3}{8} \div \frac{1}{2} = \frac{11}{8} \div \frac{1}{2}$$

Step 2 Change the division symbol to multiplication, and invert the divisor.

$$\frac{11}{8} \div \frac{1}{2} = \frac{11}{8} \times \frac{2}{1}$$

Step 3 The problem is now a multiplication problem. Complete the problem by multiplying the numerators and denominators and simplifying the result, if necessary.

$$\frac{11}{8} \times \frac{2}{1} = \frac{22}{8} = \frac{11}{4} = 2\frac{3}{4}$$

Example 5 Use a Calculator with Mixed Numbers

Solve $2\frac{1}{2} \div 3\frac{1}{2}$.

Press (on).

Press (2) (2nd) (n/d) (1) ▼ (2) (÷) (3) (2nd) (n/d) (1) ▼ (2) (enter).

The display should read

$2\frac{1}{2} \div 3\frac{1}{2}$	Math ▲
	$\frac{5}{7}$

$2\frac{1}{2} \div 3\frac{1}{2} = \frac{5}{7}$

THINK ABOUT MATH

Directions: Indicate the operation you will use to solve the problem. Then solve the problem.

1. One horsepower is defined to be the power needed to lift 33,000 pounds a distance of one foot in one minute. If this is about $1\frac{1}{2}$ times the power an average horse can exert, how much weight would an average horse be expected to lift a distance of one foot in one minute?

_____

Directions: Choose the correct term in parentheses to complete each sentence.

1. If a fraction is (proper, improper), then the numerator is equal to or greater than the denominator.

2. (Mixed numbers, Improper fractions) have a fractional part and a whole number part.

3. When adding or subtracting mixed numbers, you might need to (reduce, rename) the fractions using a common denominator.

4. The fractional part of a mixed number is a(n) (proper fraction, improper fraction).

5. You (reduce, rename) a fraction when you write it in lowest terms.

Skill Review

Directions: Summarize the details in each passage. Explain why any details are left out.

1. To divide fractions and mixed numbers, you need to change any mixed numbers to improper fractions. An improper fraction has a numerator that is equal to or greater than the denominator. For example, if the divisor of a division problem is $3\frac{1}{8}$, you should change it to the improper fraction $\frac{25}{8}$. Next, you need to invert the divisor. The divisor is the number that you divide by. The dividend is the number that is divided. Then you multiply the numerators and the denominators. Sometimes, the result is not in lowest terms. If that is the case, simplify the result. If the simplified result is an improper fraction, rewrite the improper fraction as a mixed number or whole number.

_____

Directions: Use the connections you made in your notebook to describe or explain the following.

2. Describe the similarities either between adding fractions and adding mixed numbers or between subtracting fractions and subtracting mixed numbers.

_____

3. Explain how the steps you used for multiplying or dividing fractions can be applied to multiplying or dividing mixed numbers.

_____

Skill Practice

Directions: Choose the best answer to each question.

1. Kamil used $2\frac{2}{3}$ cups of flour to make one batch of lemon bars for the company picnic. If he wants to make $1\frac{1}{2}$ batches of lemon bars, how many cups of flour should he use?

   A. $1\frac{1}{6}$

   B. $1\frac{7}{9}$

   C. 4

   D. $4\frac{1}{6}$

3. Which is the first step in solving the problem below?
   $4\frac{7}{12} \div 3\frac{1}{10}$

   A. Find a common denominator for $\frac{7}{12}$ and $\frac{1}{10}$.

   B. Invert the divisor and multiply.

   C. Multiply the numerators and denominators of the fractional parts.

   D. Change the mixed numbers to improper fractions.

2. Solve $10\frac{3}{4} - 6\frac{4}{5}$.

   A. $3\frac{19}{20}$

   B. $4\frac{1}{10}$

   C. $4\frac{9}{20}$

   D. $4\frac{19}{20}$

4. A restaurant has $6\frac{1}{2}$ gallons of a secret sauce. A recipe for barbecue pork calls for $\frac{3}{4}$ gallon of the secret sauce. How many *full* recipes can be made with this sauce?

   A. 9   C. 7

   B. 8   D. 5

Directions: Choose the best answer to each question.

1. Which fraction is written in lowest terms?

   A. $\frac{4}{8}$

   B. $\frac{4}{14}$

   C. $\frac{4}{15}$

   D. $\frac{4}{16}$

5. What is the product of factors that are both positive proper fractions?

   A. a fraction whose value is less than either factor

   B. a fraction whose value is greater than either factor

   C. a mixed number

   D. a number whose value is equal to the greater factor

2. What is the first step for finding the sum of $\frac{2}{4}$ and $\frac{1}{3}$?

   A. Add the numerators.

   B. Rename the fractions.

   C. Add the denominators.

   D. Find a common denominator.

6. What is the next step to finding $4\frac{2}{7} \div 2\frac{1}{7}$?

   A. $\frac{7}{30} \times \frac{15}{7}$

   B. $\frac{30}{7} \div \frac{15}{7}$

   C. $2\frac{1}{7} \div 4\frac{2}{7}$

   D. $\frac{7}{30} \times \frac{15}{7}$

3. Yamina worked $4\frac{2}{3}$ hours on Thursday morning and $2\frac{7}{8}$ hours on Thursday afternoon. How many hours did she work in total on Thursday?

   A. $7\frac{13}{24}$ C. $6\frac{13}{24}$

   B. $7\frac{5}{24}$ D. $6\frac{9}{24}$

7. A recipe calls for 1 cup of shortening, $2\frac{2}{3}$ cups of flour, $\frac{1}{2}$ cup of sugar, and 1 cup of raisins. If Kalie doubles the recipe, how many cups of flour should she use?

   A. $6\frac{1}{3}$ C. $4\frac{1}{3}$

   B. $5\frac{1}{3}$ D. $3\frac{1}{6}$

4. A writer wrote $2\frac{3}{8}$ pages each hour for $7\frac{1}{2}$ hours. How many pages did he write in all?

   A. $5\frac{12}{17}$ C. $16\frac{13}{17}$

   B. $14\frac{1}{8}$ D. $17\frac{13}{16}$

8. What is the first step to finding $2\frac{1}{5} - 1\frac{3}{5}$?

   A. Find a common denominator for the fractions.

   B. Rename the first mixed number.

   C. Rename both mixed numbers.

   D. Subtract the numerators of the fractions.

Review

9. Nichelle needs 6 boards that are each $3\frac{1}{3}$ feet long to make bookshelves. She wants to cut the shelves from one long board with no board left over. How many feet long does the original board need to be?

   A. 15   C. $19\frac{1}{2}$

   B. 18   D. 20

10. Jake had $7\frac{3}{5}$ gallons of wood stain. After staining a set of stairs, he had $5\frac{2}{3}$ gallons of wood stain left. How many gallons of wood stain did Jake use on the stairs?

    A. $2\frac{14}{15}$ C. $1\frac{1}{2}$

    B. $1\frac{14}{15}$ D. $1\frac{1}{15}$

11. Ayca subtracted $10\frac{5}{7} - 8\frac{2}{9}$ and got $2\frac{3}{63}$. What mistake did she make?

    A. She did not find a common denominator.
    B. She did not subtract the whole numbers correctly.
    C. She did not rename the fractions correctly.
    D. She added instead of subtracting.

12. What improper fraction in simplest form is equal to $4\frac{2}{7}$?

    A. $\frac{15}{7}$

    B. $\frac{30}{7}$

    C. $\frac{13}{7}$

    D, $\frac{28}{7}$

13. Which set of fractions is in order from greatest to least?

    A. $\frac{4}{5}, \frac{7}{9}, \frac{3}{4}, \frac{4}{6}$

    B. $\frac{3}{4}, \frac{4}{6}, \frac{7}{9}, \frac{4}{5}$

    C. $\frac{4}{5}, \frac{4}{6}, \frac{7}{9}, \frac{3}{4}$

    D. $\frac{3}{4}, \frac{4}{6}, \frac{4}{5}, \frac{7}{9}$

14. Which is the first step in a sequence of steps for dividing mixed numbers?

    A. Change the division symbol to multiplication.
    B. Find a common denominator for the fraction part of the mixed numbers.
    C. Change any mixed numbers to improper fractions.
    D. Multiply the improper fractions.

CHAPTER 3

Review

Check Your Understanding

On the following chart, circle the number of any item you answered incorrectly. Near each lesson title, you will see the pages you can review to learn the content covered in the question. Pay particular attention to reviewing skill areas in which you missed half or more of the questions.

Chapter 3: Fractions	Procedural	Conceptual	Application/ Modeling/ Problem Solving
Introduction to Fractions pp. 74–81	1	13	
Add and Subtract Fractions pp. 82–87		2	
Multiply and Divide Fractions pp. 88–91		5	
Mixed Numbers pp. 92–97	6, 8, 12	11, 14	3, 4, 7, 9, 10

UNIT 2

Basic Algebra

Integers

Integers include whole numbers and the negatives of the non-zero whole numbers. Integers can be compared, ordered, added, subtracted, multiplied, and divided. The concept of *absolute value*—the distance a number is from 0 on the number line—is essential to understanding integers and the operations involving integers. Integers are positive and negative numbers.

Integers are used in daily life. You can track gains (+) or losses (−) in the stock market as trading increases or decreases. If you balance your checkbook or manage your bank account online, you may notice times when your balance goes below zero. If you live in a cold climate, you have probably experienced negative temperatures. Any time you hear or read about positive or negative numbers, you are learning about integers.

The Key Concepts you will study include:

Lesson 4.1: Introduction to Integers and Absolute Value
Identify, compare, and order integers, as well as find their absolute value in order to better understand the meaning and value of integers.

Lesson 4.2: Add Integers
Two ways to find the sum of two integers include using a number line and using a sequence of rules.

Lesson 4.3: Subtract Integers
Subtract two integers by adding the opposite of the integer that is being subtracted.

Lesson 4.4: Multiply and Divide Integers
Use rules to find products and quotients of integers.

Lesson 4.5: The Coordinate Grid
Coordinate grids are a method of locating points in the plane by means of directions and numbers.

Goal Setting

Before starting this chapter, set goals for your learning. Think about the ways that strengthening your understanding of integers will benefit you.

- What do you hope to learn from the lessons in this chapter about integers?

  _____

  _____

  _____

  _____

  _____

- In what ways do you expect integers to be similar to whole numbers?

  _____

  _____

  _____

  _____

Introduction to Integers and Absolute Value

KEY CONCEPT: Identify, compare, and order integers, as well as find their absolute value, in order to better understand the meaning and value of integers.

Locate each number on the number line below.

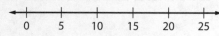

1. 1   **2.** 23   **3.** 7   **4.** 4

Use a less than (<), greater than (>), or equal to (=) symbol to compare each pair of numbers.

5. 5 ☐ 9   **6.** 31 ☐ 0   **7.** 62 ☐ 14   **8.** 18 ☐ 18

Understand Integers

The positive and negative whole numbers, ..., –3, –2, –1, 0, 1, 2, 3, ..., make up the set of **integers** and can be represented on a number line. The set of integers is **infinite**, that is, it extends without end to the left and to the right of 0 on the number line. An integer has both a distance from 0 and a direction (positive or negative). For example, the integer +5 has a positive direction and a distance of 5 from 0. The integer –5 has a negative direction and a distance of 5 from 0. Because +5 and –5 have the same distance from 0 but have different directions, they are called **opposite** numbers.

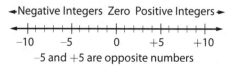

–5 and +5 are opposite numbers

Example 1 Find Opposites of Integers

Use the number line on the previous page to find the opposite of –8.

Step 1 Count how far the given number is from 0.

–8 is 8 units to the left of 0.

Step 2 Count the same distance from 0 in the opposite direction. The number located at this distance is the opposite of the given number. The number 8 units to the right of 0 is +8.

The opposite of –8 is +8.

Example 2 Use a Number Line to Compare Integers

Compare –13 and +7.

Step 1 Locate both integers on a number line.

$$-13 \qquad +7$$
$$-20\ -16\ -12\ \ -8\ \ -4\ \ \ 0\ +4\ \ +8\ +12+16+20$$

Step 2 The numbers increase in value as you move from left to right on the number line. Since –13 is to the left of +7, –13 is less than (<) +7. Since +7 is to the right of –13, +7 is greater than (>) –13. Both statements are accurate comparisons of –13 and +7.
$$-13 < +7 \quad \text{or} \quad +7 > -13$$

Example 3 Use a Number Line to Order Integers

Order the integers +13, –5, +7, and –17 from least to greatest.

Step 1 Locate each integer on a number line.

$$-17 \qquad -5 \qquad +7\ \ +13$$
$$-20\ -16\ -12\ \ -8\ \ -4\ \ \ 0\ +4\ \ +8\ +12+16+20$$

Step 2 Write the numbers from left to right as they appear on the number line. –17, –5, +7, +13

Core Skill
Apply Number Sense Concepts

You already have a solid foundation in number concepts as a result of your study of whole numbers, positive numbers, and negative numbers. You can apply this knowledge of numbers in general as you begin your study of absolute value. You can also rely on the number lines in this lesson when learning about absolute value. They are a great visual tool that can help you grasp and explain this new concept.

After you have read the lesson, write in your notebook and explain the concept of absolute value in your own words.

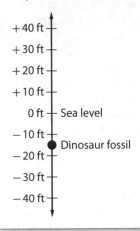

An integer with no + or − sign is understood to be a positive integer.

Core Skill
Represent Real-World Problems

It may seem that negative numbers are not used in the real world, because it's not possible to drive −100 miles in a car. The same may seem true for absolute values, but this is not the case for either topic. Negative numbers are used in everyday life, but you may not know it. This is because absolute values have taken their place!

Consider the following example. Suppose you open a credit card with a local bank. Since you have yet to purchase anything with the card, your balance it $0.00. After making some purchases, you receive a monthly statement saying your balance is $125.00. Does this mean that you have $125.00 in the bank that you can withdraw? It doesn't. Because you made purchases totaling $125.00, it means that your balance is −$125.00, since you owe the bank that money. The number given to you as your balance is the total amount of money from owing nothing (away from zero). This is just absolute value! In a notebook, write down what it would mean if your credit card balance was negative.

THINK ABOUT MATH

Directions: Identify whether each number is an integer. If it is an integer, give its opposite.

1. +4   **2.** −19   **3.** 4.5   **4.** 3   **5.** $\frac{2}{3}$

Directions: Compare. Write > or < for each □.

6. +1 □ −5   **7.** −1 □ +8   **8.** 23 □ −9   **9.** 0 □ +12

Absolute Value

Because opposite integers have the same distance from zero, they have the same **absolute value**. Absolute value is the distance between a number and 0. Distance is never negative, so absolute value is always a positive number. Vertical bars are used to show absolute value. For example, the absolute value of +2 is written $|+2|$. Since +2 is 2 units from 0, $|+2| = 2$.

Example 4 Absolute Value

Find the values of $|+4|$ and $|−4|$.

  Step 1 The integer +4 is 4 units from 0 on a number line. $|+4| = 4$.

  Step 2 The integer −4 is 4 units from 0 on a number line. $|−4| = 4$.

Notice that $|+4| = |−4|$.

THINK ABOUT MATH

Directions: Find each absolute value.

1. $|+9|$ **2.** $|−12|$ **3.** $|+13|$ **4.** $|−25|$

Vocabulary Review

Directions: Complete each sentence with the correct word.

absolute value   infinite   integer   opposite

1. The _____ of a positive integer is a negative integer.

2. Zero is neither a positive nor a negative _____.

3. When a set of numbers is _____, the numbers continue without an end.

4. The _____ of a number is always a positive value.

Directions: Use the number line below to help you determine which number is greater. Indicate your answer by writing *greater than* or *less than* in the blank.

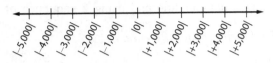

1. −4,351 is _____ +10.

2. +380 is _____ −200.

3. −4,999 is _____ −5,000.

4. 0 is _____ +800.

Directions: Solve each problem. Use a number line to justify your answer.

5. The average low temperatures this week were 20°F, 15°F, −2°F, 0°F, 18°F, −6°F, and 4°F. Order the temperatures from coldest to warmest.

_____

6. A river near Maria's home is 194 feet above sea level. A river near Sheila's home is 600 feet above sea level. Compare the heights of the two rivers using integers and > or < symbols.

_____

Skill Practice

Directions: Choose the best answer to each question.

1. When the furnace broke last winter, the temperature in the auditorium dropped to −3°C. What is the opposite of −3?

   A. −6
   B. −3
   C. +3
   D. +6

2. Which shows −32, +24, −10, 0, and +316 in order from least to greatest?

   A. 0, −10, +24, −32, +316
   B. +316, +24, 0, −10, −32
   C. 0, −32, −10, +24, +316
   D. −32, −10, 0, +24, +316

3. While scuba diving, Ariana dove to 89 feet below sea level. Which integer expresses her position?

   A. 89
   B. −89
   C. +89
   D. |89|

4. Which explains one way to find the absolute value of any integer?

   A. Write the opposite of the integer that is inside the vertical bars.
   B. Compare the integer to its opposite.
   C. Count how far the integer is from 0 on a number line.
   D. Count how far the integer is from its opposite on a number line.

Add Integers

KEY CONCEPT: Two ways to find the sum of two integers include using a number line and using a sequence of rules.

Lesson Objectives

You will be able to

- Use a number line to find the sum of two integers
- Use integer addition rules to find the sum of two integers

Skills

- **Core Practice:** Reason Abstractly
- **Core Skill:** Perform Operations

Vocabulary

addend
illustrate
negative
positive
sequence
sign

Find each sum or difference.

1. $4 + 5$ **2.** $8 + 2$ **3.** $9 - 7$ **4.** $10 - 4$

Use a less than (<), greater than (>), or equal to (=) symbol to compare each pair of numbers.

5. $|+5|\ \square\ |-9|$ **6.** $|-3|\ \square\ |0|$ **7.** $|+8|\ \square\ |+4|$ **8.** $|-81|\ \square\ |+14|$

Use a Number Line to Add Integers

You can use a number line to add two integers. Use the **sign** (whether the integer is positive [+] or negative [–]) of the **addends**, or numbers that are added together, to tell which way to move on the number line.

Example 1 Add a Negative Integer

Add $+4 + (-6)$.

Step 1 Locate the first addend, +4, on the number line.

Step 2 Move 6 units to the left, since you are adding a **negative** (–), less than zero, number whose absolute value is 6.

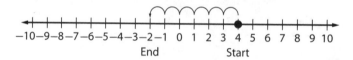

Step 3 The sum is the number of the point where you end. In this case, it is –2.

$$+4 + (-6) = -2$$

Example 2 Adding a Positive Integer to a Negative Integer

Michael accidently overdraws his bank account by $15. To avoid being charged a fee, he immediately deposits $100 into his account. How much money does he now have in his bank account?

Step 1 Determine the first addend. Because he overdrew his account by $15, Michael has –$15 in his account.

Step 2 Write down the addition problem. Michael added $100 to his account, so the sum is $-15 + 100$.

Step 3 Using a number line, start at –15 and then move 100 units to the right.

Step 4 The sum is the number of the point where you end. In this case, it is 85. Therefore, Michael has $85 in his bank account.

FOLLOW A SEQUENCE OF STEPS

Some passages describe actions that happen in a certain order, called a **sequence**. Words, such as *first*, *then*, *next*, *after*, and *finally*, signal the order in which the actions happened.

Sequences explain the order of an individual's activity. *Sabina went to the store, then she walked to the post office and bought stamps. After that, she mailed her letters. Finally, she went home.*

Actions might not be written in the order in which they happened. *Tahir finished reading the last sixty pages of a novel. Before that, he had done the dishes. When he was done with the novel, he made coffee and poured it into the mug he had washed that morning.* Four actions are described, but they occurred in a different order than they are written. Tahir washed a mug, read sixty pages to finish a novel, made coffee, and poured coffee into the clean mug. The phrases *before that, when he was done,* and *had washed that morning* provide clues about the sequence.

Sometimes, a sequence of events can be given as a set of instructions. *To follow the order of operations, first do operations in parentheses, then perform multiplication and division from left to right. Finally, do addition and subtraction from left to right.*

Read the passage below. Put the sentences in the order in which the actions occurred. Write the clue words that helped you order them.

---

(1) Nayo measured the temperature in the evening as 25°F using a thermometer. (2) She earlier found the afternoon temperature of 57°F using the same method. (3) Nayo had wanted to find the difference in temperature from midday to evening. (4) She then subtracted 25°F from 57°F and got 32°F. (5) She finally knew the temperature dropped 32°F from midday to evening.

---

3 (had wanted), 2 (earlier), 1 (in the evening), 4 (then), 5 (finally)

Core Practice
Reason Abstractly

You may sometimes hear people talking about multitasking, which refers to doing more than one job at the same time. Numbers also multitask. They do the "work" of positive numbers and the work of negative numbers. Understanding what a negative number represents allows you to work with such numbers when performing simple operations, such as addition and subtraction. In these kinds of simple calculations, you are reasoning abstractly— that is, you are working directly with numbers and not with things you can touch and count.

For example, suppose you want to know how far you walk every day in a two-week period. Reason abstractly to find the answer. In other words, write the problem out mathematically, leaving out any words.

MATH LINK

Addends are the numbers being added.

Number lines are a great tool for adding a positive and negative integer, because they allow you to **illustrate**, or use graphic detail, to explain operations with integers. Using number lines, you can actually demonstrate the results of each operation you perform. To add a positive integer, you move right on the number line. To add a negative integer, you move left. When you add two positive integers, you always get a positive answer. Two negative numbers that are added together result in a negative answer. When you add a positive and negative number, first you move in one direction, and then you move in the opposite direction. Your answer will be positive if the absolute value of the negative number was the smaller number. Otherwise, your answer will be zero or negative.

In your notebook, explain why you might get a zero for your answer when adding a positive and negative number. You can use a number line to demonstrate cases in which the two numbers that you picked add up to zero.

Example 3 Add a Positive Integer

Add $+2 + (+6)$.

Step 1 Locate the first addend, $+2$, on the number line.

Step 2 Move 6 units to the right, since you are adding a **positive** (+) (greater than zero) number whose absolute value is 6.

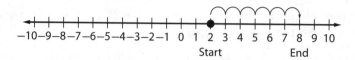

$+2 + (+6) = +8$

Step 3 The sum is the number at the point where you end. In this case, it is $+8$.

In summary, to add integers using a number line, first find the first addend. Then move right to add a positive number or move left to add a negative number.

THINK ABOUT MATH

Directions: Match each problem with its sum. Use a number line to help you.

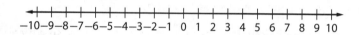

_____ 1. $+10 + (-3)$	**A.** -1
_____ 2. $-2 + (+1)$	**B.** -7
_____ 3. $+3 + (+2)$	**C.** -5
_____ 4. $-5 + (-2)$	**D.** $+7$
_____ 5. $-10 + (+5)$	**E.** $+5$
_____ 6. $-2 + (+3)$	**F.** $+1$

Use Rules to Add Integers

Adding integers with the same sign is similar to adding whole numbers. The sign of the sum depends on the sign of the integers.

Example 4 Add Two Positive Integers

Find $+5 + (+7)$.

Step 1 Find the absolute values.

Step 2 Add the absolute values. Then use the sign of the integers to determine the sign of the answer. If the addends are positive, the answer will be positive.

$+5 + (+7)$
$|+5| = 5 \ |+7| = 7$

$5 + 7 = 12$
$+5 + (+7) = +12$

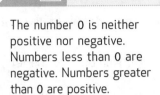

MATH LINK

The number 0 is neither positive nor negative. Numbers less than 0 are negative. Numbers greater than 0 are positive.

Example 5 Add Two Negative Integers

Find $-5 + (-7)$.

Step 1 Find the absolute values.

Step 2 Add the absolute values. Then use the sign of the integers to determine the sign of the answer. If the addends are negative, the answer will be negative.

$+5 + (+7)$
$|+5| = 5 \ |-7| = 7$

$5 + 7 = 12$
$+5 + (-7) = -12$

Adding integers with different signs is similar to subtracting whole numbers. The sign of the sum depends on the sign of the integer with the greater absolute value.

Example 6 Add Integers with Opposite Signs

Find $+5 + (-7)$.

Step 1 Find and subtract the absolute values.

Step 2 Use the sign of the integer with the greater absolute value. Because the -7 has a greater absolute value, use the negative sign to get -2.

$+5 + (-7)$
$= 5 \ |-7| = 7 \ 7 > 5$

$7 - 5 = 2$
$+5 + (-7) = -2$

To summarize, to use rules to add integers with the same sign, follow the same sequence of steps each time. First, add the absolute values of the integers, and then use the sign of the integers.

To add integers with different signs, likewise follow the same sequence of steps each time. First, subtract the absolute values of the integers, and then use the sign of the integer with the greater absolute value.

THINK ABOUT MATH

Directions: Match each problem with its sum.

_____ 1. $+10 + (+2)$ **A.** -14

_____ 2. $-2 + (-12)$ **B.** -10

_____ 3. $+8 + (+8)$ **C.** -16

_____ 4. $-7 + (-9)$ **D.** 0

_____ 5. $-6 + (+6)$ **E.** $+12$

_____ 6. $-12 + (+2)$ **F.** $+16$

Vocabulary Review

Directions: Complete the sentences below using one of the following words:

addends   negative   positive   sign

1. Numbers to the left of 0 on a number line are _____ numbers.

2. Numbers to the right of 0 on a number line are _____ numbers.

3. _____ are the numbers to be added in an addition expression.

4. The _____ of a number tells whether it is positive or negative.

Skill Review

Directions: Copy the number line and illustrate how to find the sum for each expression below.

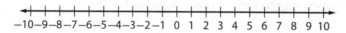

−10−9−8−7−6−5−4−3−2−1 0 1 2 3 4 5 6 7 8 9 10

1. +8 + (+1)   3. −10 + (+6)   5. −9 + (−1)

2. +7 + (−3)   4. +3 + (−5)   6. −3 + (+9)

Directions: Write the sequence of steps that must be followed to find the sum of the following expressions.

7. +8 + (+12)   8. +14 + (−9)   9. −15 + (+7)

Directions: Read each passage, then put the sentences in the order in which the action occurred. Write any clue words that helped you order the sentences.

10. (1) First, Samuel drew a number line. (2) Then he plotted the location of −3 on the number line. (3) After that, he counted to the right 5 units because he was adding a positive 5 to negative −3. (4) Samuel finally knew that the sum of −3 and +5 is +2.

11. (1) Julio knew the sum would have a positive sign. (2) He had earlier found the absolute value of +6 to be greater than the absolute value of −1. (3) Julio had wanted to find the sum of +6 + (−1). (4) After all his calculations, he found the sum to be +5. (5) He had subtracted 6 −1 and gotten 5.

Skill Practice

Directions: Choose the best answer to each question.

1. Which of the following explains why the sum of −9 and +16 is a positive number?

   A. +16 is to the right of 0 on a number line, and −9 is to the left of 0.

   B. A positive number plus a negative number is always a positive number.

   C. −9 has a greater absolute value than +16, and its sign is not positive.

   D. +16 has a greater absolute value than −9, and its sign is positive.

3. The solution to the problem Alba was working on is a negative integer. Which of the following problems could Alba be trying to solve?

   A. +8 + (−4)
   B. −8 + (+4)
   C. −6 + (+6)
   D. +7 + (−6)

2. The average temperature today was −5°F. Yesterday the average temperature was 6°F warmer than today. What was the average temperature yesterday?

   A. −11°F
   B. −1°F
   C. 1°F
   D. 11°F

4. A diver first dove to 35 feet below sea level to meet another diver. Then the diver came up 10 feet to untangle a line. At what depth did the diver stop to untangle a line?

   A. −45 feet
   B. −25 feet
   C. 25 feet
   D. 45 feet

Subtract Integers

KEY CONCEPT: Subtract two integers by adding the opposite of the integer that is being subtracted.

Use the number line below to find each sum.

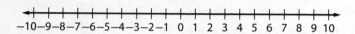

1. +1 + (−7)   2. −6 + (+4)   3. −10 + (+2)   4. +4 + (−10)

Find each sum.

5. +10 + (−15)   6. −7 + (+13)   7. −6 + (+8)   8. +5 + (−12)

Use a Number Line to Subtract Integers

You can use a number line to **solve**, or find the answer to, a subtraction problem involving integers. On the number line below, each **tic mark**, or division, indicates an integer.

Example 1 Subtract a Negative Integer

Find +4 − (−6).

Step 1 Locate the first number, +4, on the number line.

Step 2 Move 6 units to the right, since you are subtracting a negative number whose absolute value is 6. If you were adding −6, you would move 6 units to the left. Because you are performing the opposite of addition by doing subtraction, move 6 units to the right.

+4−(−6) = +10

Step 3 The number of the **point**, or specified place, where you end is the difference. In this case it is +10.

Example 2 Subtract a Positive Integer

Find −2 − (+6).

Step 1 Locate the first number, −2, on the number line.

Step 2 Move 6 units to the left, since you are subtracting a positive number whose absolute value is 6.

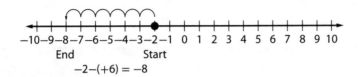

End   Start
−2−(+6) = −8

Step 3 The number of the point where you end is the difference. In this case, it is −8.

Use Addition to Subtract Integers

Subtracting integers is similar to adding integers. In fact, you can rewrite a subtraction problem as an addition problem and then add.

Example 3 Subtract a Positive Integer

Find +5 − (+7).

Step 1 Rewrite the problem so that you are adding the opposite of the integer being subtracted.

Step 2 Follow the rules for adding integers.

−7 is the opposite of +7
$+5 - (+7) \longrightarrow +5 + (-7)$
$|+5| = 5 \qquad |-7| = 7$
$7 - 5 = 2 \qquad 7 > 5$, so

$+5 + (-7) = -2$
  and
$+5 - (+7) = -2$

Example 4 Subtract a Negative Integer

Find +5 − (−9).

Step 1 Rewrite the problem so that you are adding the opposite integer.

Step 2 Follow the rules for adding integers.

+9 is the opposite of −9
$+5 - (-9) \longrightarrow +5 + (+9)$
$5 + 9 = 14$, so

$+5 + (+9) = +14$
  and
$+5 - (-9) = +14$

Core Skill
Perform Operations

You have already used number lines to compare fractions. You can also use number lines to show operations such as addition and subtraction.

When using number lines, you generally move left when subtracting one number from another— but not always. Recall that subtracting a number is the same thing as adding the opposite of the number. When you are subtracting a negative number, you are adding the opposite of the negative number—that is, a positive number.

In a notebook, draw a number line, and use it to show how to solve these two problems: −5 + (−1) and −5 − (+1). What does this problem show you about the relationship between subtracting a number and adding its opposite?

MATH LINK

Subtracting an integer is the same as adding its opposite.

Core Skill
Represent Real-World Problems

Consider the following problem involving operations of integers. Ginger ran 11 miles North to her friend's house for lunch. Afterward, she ran back South 6 miles to the park, where she teaches dog obedience classes. How many miles is Ginger from her house? In a notebook, create a number line and use it to determine the correct answer to the problem.

THINK ABOUT **MATH**

Directions: Match each problem with its difference.

_____ 1. +10 − (+2)   **A.** −12

_____ 2. −2 − (−12)   **B.** −10

_____ 3. +8 − (+8)   **C.** 0

_____ 4. −3 − (−9)   **D.** +6

_____ 5. −6 − (+6)   **E.** +8

_____ 6. −8 − (+2)   **F.** +10

Vocabulary Review

Directions: Complete each sentence with the correct word from the list below.

point(s)   solve   tic mark(s)

1. Number lines have evenly divided _____ to indicate certain numbers.

2. When using a number line to subtract, the _____ where you end is the difference.

3. To _____ an addition problem means to find its sum.

Skill Review

Directions: For each pair of expressions, copy the number line. Then use the number line to perform the operations and solve the expressions. If they are equal, add an equal sign to make an equation.

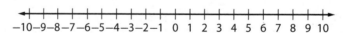

$$-10\,-9\,-8\,-7\,-6\,-5\,-4\,-3\,-2\,-1\ \ 0\ \ 1\ \ 2\ \ 3\ \ 4\ \ 5\ \ 6\ \ 7\ \ 8\ \ 9\ \ 10$$

1. +8 + (+2) and +8 − (−2)   **2.** +4 + (−5) and +4 − (+5)

Directions: Describe how to find each sum and difference without using a number line.

3. +12 + (+11) and +12 − (−11)   **4.** +7 + (−3) and +7 − (+3)

Directions: Explain the following.

5. Use what you have learned about the addition and subtraction of integers to explain how you know that +10 − (−3) equals a positive number without solving the problem.

Skill Practice

Directions: Choose the best answer to each question.

1. The temperature outside at 3:00 p.m was 3°F. By 9 p.m., the temperature had dropped 8°F. What was the temperature at 9 p.m.?

   A. +11°F   C. −5°F

   B. +5°F   D. −11°F

2. When solving the equation +7 − (−5) on a number line, which of the following steps would you take?

   A. Locate +7 on a number line, then move left 5 units.

   B. Locate +7 on a number line, then move right 5 units.

   C. Locate −5 on a number line, then move left 7 units.

   D. Locate −5 on a number line, then move right 7 units.

3. The basement of an office building is 40 feet below the surface street. The exterior wall of the foundation extends down an additional 8 feet. Where in relation to the surface street is the bottom of the exterior wall of the basement?

   A. +48 feet

   B. +32 feet

   C. −32 feet

   D. −48 feet

4. The balance in May's checking account was $245. She wrote a check for $302. If there is no fee charged to the account, what is the balance now?

   A. −$57

   B. −$55

   C. $57

   D. $547

5. The melting temperature for tin is about 231°C, and the melting temperature for oxygen is about −218°C. What is the difference in the melting temperatures for tin and oxygen?

   A. 453°C   C. −249°C

   B. 449°C   D. −449°C

6. What is the difference between −92 and 114?

   A. −206

   B. −22

   C. 22

   D. 206

7. Which of the following will have a negative answer?

   A. −23 + 24

   B. −23 − (−24)

   C. 24 + (−23)

   D. −23 − 24

8. Which of the following explains why −9 minus +16 is a negative number?

   A. +16 is to the right of 0 on a number line, and −9 is to the left of 0.

   B. −9 is the first number in the problem, and it is negative.

   C. A negative number minus a positive number is the same as adding two negative numbers.

   D. A negative number minus a positive number is a negative number whenever the second number is greater than the first number.

Multiply and Divide Integers

KEY CONCEPT: Use rules to find products and quotients of integers.

Find each sum or difference.

1. $-4 + (-4)$ **3.** $+9 + (+9)$ **5.** $-12 - (-6)$ **7.** $+14 - (+7)$

2. $-8 + (-8)$ **4.** $+4 + (+4)$ **6.** $-10 - (-5)$ **8.** $+20 - (+10)$

Multiply Integers

Recall that multiplication is repeated addition. A **repeated** action happens again and again. You can use your knowledge of adding integers with the same signs to help you when multiplying integers.

Example 1 Positive × Positive = Positive

Find $+4 \times (+3)$.

Step 1 Rewrite the problem as repeated addition.

Step 2 Add +3 four times. Recall that a positive number plus a positive number is a positive number.
$+4 \times (+3) = +12$

$4 \times (+3) =$
$(+3) + (+3) + (+3) + (+3) = +12$

Example 2 Positive × Negative = Negative

Find $+4 \times (-3)$.

Step 1 Rewrite the problem as repeated addition.

Step 2 Add −3 four times. Recall that a negative number plus a negative number is a negative number.
$+4 \times (-3) = -12$

$4 \times (-3) =$
$(-3) + (-3) + (-3) + (-3) = -12$

Recall that every integer has an opposite. Use this idea to help you see that a negative number times a positive number is a negative number, and that a negative number times a negative number is a positive number.

READ A TABLE

A **table** is a way to organize information in **rows** and **columns**. A table has cells of information arranged horizontally (rows) and vertically (columns) to make information easier to understand. A table has a title and labels. The **title** tells what the table is meant to show. The labels identify the information in the separate rows and columns.

Study the table below and then answer the questions.

ATIAN'S CELL PHONE MINUTES USED

	Wednesday	Thursday	Friday	Saturday
Daytime	23	46	10	0
Night & Weekend	113	38	24	176

The title tells that the table will show the cell phone minutes used by Atian over four days. You can find a specific piece of data. You can also find more than one piece of information and add, subtract, or compare the numbers. To find specific information, first identify which row and column the information is in. Then find the cell where the row and column intersect. To find the daytime minutes used on Thursday, find *Daytime*. Then follow the row across to the *Thursday* column.

1. How many total daytime minutes did Atian use for the 4 days?
   To find the total, add all the minutes from the *Daytime* row together: $23 + 46 + 10 + 0 = 79$ minutes, or 1 hour, 19 minutes.

2. How many more *Night & Weekend* minutes than *Daytime* minutes did he use on Wednesday?
   Find the number of minutes used on Wednesday: 113 night, 23 daytime. Because the question asks *how many more*, subtract the values: $113 - 23 = 90$ minutes, or 1 hour 30 minutes.

3. Did Atian use more *Daytime* minutes or *Night & Weekend* minutes over the four days?
   This question asks about total minute usage for each type. Add all of the *Daytime* minutes together: $23 + 46 + 10 + 0 = 79$. Then add all the *Night & Weekend* minutes together: $113 + 38 + 24 + 176 = 351$. Compare the two numbers. $351 > 79$, so Atian used more *Night & Weekend* minutes.

Tables are a handy way of organizing and displaying information and data. On the other hand, tables can sometimes present an overwhelming amount of information. Think back to the time when you learned how to read. Eventually, the black marks that stared up at you when you opened a book made sense. You unlocked their mystery. So too you will learn how to read tables.

Notice the arrangement of data in rows and columns. Also pay attention to a table's title and labels; they both provide clues that explain what specific information is being presented.

Read the table below. Then, in a notebook, write a paragraph describing the information presented in the table.

Park Attendance		
	Boys	Girls
June	77	80
July	110	157
August	191	142

What rows or columns might you add to the table above to provide more information?

It is easy to be distracted by all of the information in a table. Always spend a few moments to analyze, or study, the information there so that you understand what the table is telling you. Look at the title and labels. For example, a table titled *"Temperature (°F)"* might be reporting the "temperature in degrees Fahrenheit" recorded during a week.

Look at the table below and record your answers to the questions.

Temperatures (°F)

	High	Low
Monday	13°	2°
Tuesday	20°	3°
Wednesday	6°	−10°

1. Which temperature scale is being used?

2. What was the range of temperatures for Tuesday?

3. Which day had the greatest difference in high and low temperatures?

Example 3 Negative × Positive = Negative

Find $-4 \times (+3)$.

Recall that $+4$ is the opposite of -4. Think of $-4 \times (+3)$ as the *opposite of* $+4 \times (+3)$.
$+4 \times (+3) = +12$ (See Example 1.)
The *opposite of* $+4 \times (+3)$ = *opposite of* $+12$ (or -12), so $-4 \times (+3) = -12$.

Example 4 Negative × Negative = Positive

Find $-4 \times (-3)$.

Think of $-4 \times (-3)$ as the *opposite of* $+4 \times (-3)$.
$+4 \times (-3) = -12$, so the *opposite of* $+4 \times (-3)$
The *opposite of* $+4 \times (-3)$ = *opposite of* -12 (or $+12$), so $-4 \times (-3) = +12$.

When two integers have the same sign, their product will be positive, and when two integers have different signs, their product will be negative.

THINK ABOUT MATH

Directions: Match each problem with its product.

_____ 1. $+5 \times (-7)$	**A.** $+35$
_____ 2. $-5 \times (-7)$	**B.** -35
_____ 3. $+4 \times (+8)$	**C.** -32
_____ 4. $-4 \times (+8)$	**D.** $+32$

Divide Integers

Recall that division and multiplication are **inverse**, or opposite, operations. This means that you can use a related multiplication problem to help you solve a division problem.

Example 5 Positive ÷ Positive = Positive

Find $+10 \div (+5)$.

Think: What number can I multiply by $+5$ to get a product of $+10$?
Will that number be positive or negative?
Since $+5 \times +2 = +10$, then $+10 \div (+5) = +2$.

Example 6 Positive ÷ Negative = Negative

Find $+10 \div (-5)$.

Think: What number can I multiply by -5 to get a product of $+10$?
Since $-5 \times -2 = +10$, then $+10 \div (-5) = -2$.

Example 7 Negative ÷ Positive = Negative

Find $-10 \div (+5)$.

Think: What number can I multiply by $+5$ to get a product of -10?
Since $+5 \times -2 = -10$, then $-10 \div (+5) = -2$.

Example 8 Negative ÷ Negative = Positive

Find $-10 \div (-5)$.

Think: What number can I multiply by -5 to get a product of -10?
Since $-5 \times +2 = -10$, then $-10 \div (-5) = +2$.

When two integers have the same sign, their quotient will be positive, and when two integers have different signs, their quotient will be negative.

MATH LINK

The table below shows the rules for dividing integers.

Divide Integers		
Dividend	Divisor	Quotient
+	+	+
+	−	−
−	+	−
−	−	+

THINK ABOUT **MATH**

Directions: Match each problem with its product.

_____ 1. $+24 \div (+8)$ **A.** $+5$

_____ 2. $-24 \div (+8)$ **B.** -5

_____ 3. $+45 \div (-9)$ **C.** -3

_____ 4. $-45 \div (-9)$ **D.** $+3$

Vocabulary Review

Directions: Complete the sentences below using one of the following words.

columns   inverse   repeated   rows   table   title

1. The _____ of a table is usually found at the top of the table.

2. The _____ of multiplying by -4 is dividing by -4.

3. -6 is _____ three times when writing -6×3 as an addition problem.

4. _____ are the vertical divisions in a table.

5. _____ are the horizontal divisions in a table.

6. A _____ is a tool for organizing information.

Directions: The table below shows each of the three integers assigned to various students. Use the table to answer the following questions.

INTEGERS ASSIGNED TO EACH STUDENT

Student	First Integer	Second Integer	Third Integer
Mandisa	−1	+8	−5
Tuan	+6	−2	+12
Reyna	−14	−7	−9
Wil	+2	+9	+18

1. What is the product of Reyna's second and third integers?

2. What is the product of Mandisa's second integer and Wil's first integer?

3. What is the quotient of Wil's third integer divided by his first integer?

4. What is the quotient of Tuan's third integer divided by his first integer?

5. What is the quotient of Tuan's third integer divided by his second integer?

Directions: Use the table below to answer the following questions.

Population of Smithville (in hundreds)

Year	1990	2000	2010
Population	42	55	68

6. What does the second column tell you?

7. What does the second row tell you?

8. How much greater was the population of Smithville in 2000 than in 1990?

9. If the population of Smithville increases by the same amount every 10 years, what would you expect the population of Smithville to be in 2020?

Skill Practice

Directions: Choose the best answer to each question.

1. Which of the following explains why the product of −3 and +2 is a negative number?

   A. +2 is to the right of 0 on a number line, and −3 is to the left of 0.
   B. A negative number times a number with the greater absolute value is negative.
   C. The sum of +2 + (+2) is a positive number.
   D. The sum of −3 + (−3) is a negative number.

2. When Marcie plugs in her new freezer, the freezer's temperature is 20°C. The temperature of the freezer descends by 2° every hour. Which of the expressions represents how many hours it takes for the freezer's temperature go down 36°?

   A. 20 + (−36) ÷ (−2)
   B. −36 ÷ (−2)
   C. 36 ÷ (−2)
   D. (20 − (−36)) ÷ (−2)

3. What is −12 × (−7)?

   A. 84
   B. 72
   C. −72
   D. −84

4. A diver dives 10 feet every 20 seconds. What is the location of the diver in feet with relation to the surface after 120 seconds?

   A. −60
   B. −16
   C. 6
   D. 60

The Coordinate Grid

KEY CONCEPT: Coordinate grids are a method of locating points in the plane by means of directions and numbers.

Answer the following questions.

1. Where are the positive numbers located on a horizontal number line?

2. Where are the negative numbers located on a horizontal number line?

The Coordinate Plane

A **coordinate plane** is formed by two number lines perpendicular to each other. The horizontal number line is called the **x-axis**, and the vertical number line is called the **y-axis**. The point where the two lines meet, (0, 0), is called the **origin**.

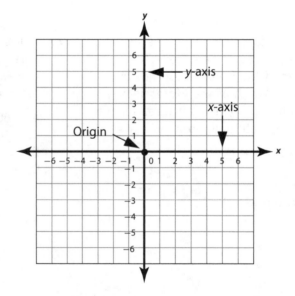

A point in the coordinate plane is named by two coordinates (numbers): an **x-coordinate** and a **y-coordinate.** Together the coordinates form an **ordered pair**. The pair is *ordered* because the *order* of the two coordinates matters. The x-coordinate is always first, and the y-coordinate is always second. They are written as (*x*, *y*). The ordered pair tells exactly where the point lies in the coordinate plane.

In the coordinate plane below, point *P* is the origin: (0, 0). The *x*-coordinate tells how far right or left a point is from the origin. The *y*-coordinate tells how far up or down the point is from the origin. Point *A* is at (2, 5): 2 units right and 5 units up from the origin. Point *B* is at (–3, 2): 3 units left and 2 units up from the origin.

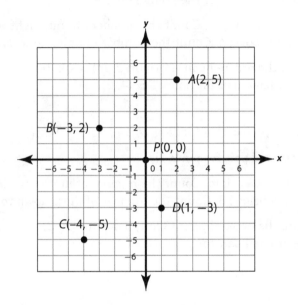

Use the coordinate plane below for the examples that follow.

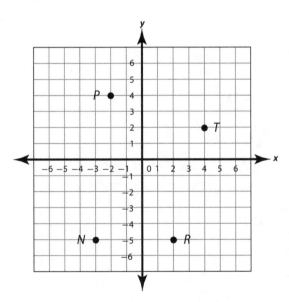

Core Skill
Interpret Data Displays

A grid is one of the many ways you can display data. The type of grid that appears throughout this lesson is called a coordinate grid. Knowing how to "read" the information in a coordinate grid is an important skill.

The scales in a coordinate grid are the same ones found on integer number lines. The two axes in a coordinate grid have names: the *x*-axis and the *y*-axis. You can use these two axes to help you locate points anywhere within the grid. Each point in the grid is identified by an ordered pair of numbers: first, the *x*-value, then the *y*-value.

Work with a partner to practice plotting points on a coordinate plane. Each partner take turns drawing a grid and plotting two or three coordinates. The other partner names the ordered pairs that lie on the coordinate plane.

During a day of running errands, Javier stopped at three places. After leaving home, he drove 3 miles west to Zeeb's Auto Shop, 4 miles south to Nelson's Book Nook, and 6 miles north and 5 miles east to the D&L Market. Using the origin as his home, determine how many miles north, south, east, west, or any combination of those Javier is from home.

Using your notebook and a coordinate grid, draw the path Javier took to help find his current location.

Example 1 Plot a Point on a Coordinate Grid

What is the letter of the point at (2, −5)?

Step 1 Start at the origin. Count 2 units right and 5 units down.

Step 2 Identify the point for R. The point at (2, −5) is R.

Example 2 Identify the Ordered Pair for a Point

What is the ordered pair for the point T?

Step 1 Start at the origin. Count the number of units right until T is directly above: 4. Count the number of units up to T: 2.

Step 2 Write the ordered pair: (4 right, 2 up) = (4, 2). The ordered pair for point T is (4, 2).

Example 3 Identify the Ordered Pair for a Point

What is the ordered pair for point N?

Step 1 Start at the origin. Count the number of units left until N is directly below: 3. Count the number of units down to N: 5.

Step 2 Write the ordered pair: (3 left, 5 down) = (−3, −5). The ordered pair for point N is (−3, −5).

Example 4 Use a Coordinate Plane

Use the coordinate plane below to help you answer the question.

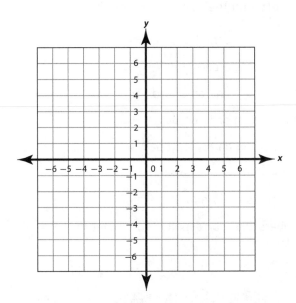

The coordinates of point *W* are (−2, 5). What is the location of point *M*?
Mark your answer on the graph above.

 THINK ABOUT **MATH**

Directions: Write the letter of the point at each ordered pair on the coordinate grid.

1. (3, −2)

2. (8, 5)

3. (−9, −4)

4. (−4, 2)

Directions: Write the ordered pair for each point on the coordinate grid at the right.

5. *K*(,)

6. *E*(,)

7. *F*(,)

8. *P*(,)

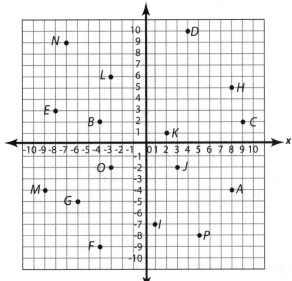

Directions: Match the words in the column on the left with their description on the right.

1. _____ coordinate plane
2. _____ ordered pair
3. _____ origin
4. _____ perpendicular
5. _____ x-axis
6. _____ x-coordinate
7. _____ y-axis
8. _____ y-coordinate

A. a grid made up of horizontal and vertical lines

B. the vertical number line on a grid

C. the horizontal number line on a grid

D. the first number in an ordered pair

E. the second number in an ordered pair

F. the point where the x- and y-axes intersect

G. two lines that cross to form a 90° angle

H. a pair of numbers that corresponds to a point in the coordinate plane

Skill Review

Directions: Use the coordinate grid below to answer each question.

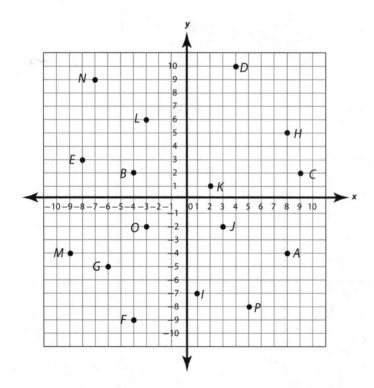

1. For which point would you start at the origin, move right 8 units, and then move down 4 units?

2. Explain how you would find the ordered pair that locates point C.

3. Which points have a positive x-coordinate and a positive y-coordinate?
   A. C, D, H, K   **B.** F, G, M, O   **C.** B, E, L, N   **D.** A, J, L, P

Skill Practice

Directions: Choose the best answer to each question.

1. Which two points, when graphed on the coordinate grid and connected, form a line segment that is a vertical line?

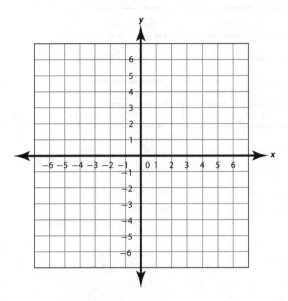

   A. (2, 6), (8, 6)
   B. (2, −4), (0, 6)
   C. (−4, −4), (−6, −6)
   D. (2, 4), (2, −6)

2. Which best describes how to plot the point at (3, −7) on a coordinate grid?

   A. Start at 3. Move 7 units left.
   B. Start at the origin. Move 3 units to the left and 7 units up.
   C. Start at the origin. Move 3 units to the right and 7 units down.
   D. Start at −7. Move 3 units down.

Directions: Mark each answer on the corresponding coordinate grid.

3. The coordinates of point A are (−6, 5). What is the location of point A?

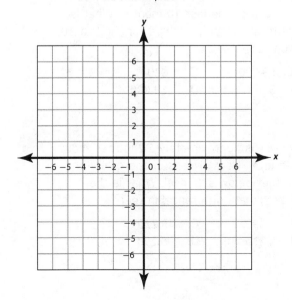

4. The coordinates of point B are (4, −3). What is the location of point B?

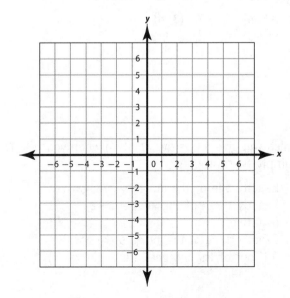

Directions: Choose the best answer to each question.

Questions 1 and 2 refer to the following coordinate grid.

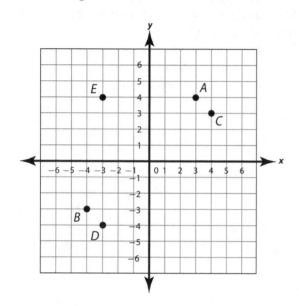

1. Which point is located at (−3, −4)?

   A. A   **C.** C

   B. B   **D.** D

2. What are the coordinates of point E?

   A. (−3, 4)   **C.** (3, 4)

   B. (4, −3)   **D.** (−4, 3)

3. Which will give a positive solution when you multiply four integers?

   A. Exactly one of the integers is negative.

   B. Exactly three of the integers are negative.

   C. Exactly three of the integers are positive.

   D. All four of the integers are negative.

4. Kofi tracked the gains and losses of his inventory for four months.

Month	Change
June	+102
July	−24
August	+89
September	−225

What was the net change for the four months he tracked his inventory?

   A. −440   **C.** 58

   B. −58   **D.** 440

5. Which describes how to locate the point (−8, 9) in the coordinate plane?

   A. From the origin, go 8 units to the right. Then from there, go 9 units up.

   B. From the origin, go 8 units to the right. Then from there, go 9 units down.

   C. From the origin, go 8 units to the left. Then from there, go 9 units down.

   D. From the origin, go 8 units to the left. Then from there, go 9 units up.

6. Sovann's investment portfolio has lost $300 for each of the last 10 months. What is his total change in dollars for the last 10 months?

   A. $3,000   **C.** −$30

   B. $30   **D.** −$3,000

7. Denver is approximately 5,280 feet above sea level. New Orleans is approximately 10 feet below sea level. What is the approximate difference, in feet?

   A. 5,290   **C.** −5,270

   B. 5,270   **D.** −5,290

Review

8. A lobster fishing boat drops its traps to a depth of about 20 meters. A crab fishing boat drops its traps to a depth of about 120 meters. Which expression can be used to find about how many times deeper the crab trap goes than the lobster trap?

    A. $-20 \div (-120)$

    B. $20 \div (-120)$

    C. $-120 \div (-20)$

    D. $-120 \div 20$

9. Faiza had a starting balance of $178 in her checking account. She deposited $250, withdrew $60 at the ATM, got charged a $2 fee for the withdrawal, and wrote a check for $187. Which of the following shows how to find her ending balance?

    A. $178 + (+250) + (-60) + (-2) + (-187)$
       $= 179$

    B. $178 + (+250) - (-60) - (-2) - (-187)$
       $= 677$

    C. $-178 + (+250) + (-60) + (-2) + (-187)$
       $= -177$

    D. $-178 + (+250) - (-60) - (-2) - (-187)$
       $= 321$

10. Which list is in order from least to greatest?

    A. 0, −2, 14, 31, −35

    B. −35, −2, 0, 14, 31

    C. 14, −2, −35, 0, −31

    D. 31, 14, 0, −2, −35

11. Absolute value is used to determine distance from zero. It can also be used to determine distance from other numbers. For example $|7 - 3| = 4$ shows that 7 is a distance of 3 units from 4. What pair of numbers are equal distance from 2?

    A. −5 and 9

    B. −4 and 4

    C. 7 and 5

    D. 0 and −2

Check Your Understanding

On the following chart, circle the number of any item you answered incorrectly. Near each lesson title, you will see the pages you can review to learn the content covered in the question. Pay particular attention to reviewing lessons in which you missed half or more of the questions.

Chapter 4: Integers	Procedural	Conceptual	Application/ Modeling/ Problem Solving
Introduction to Integers and Absolute Value pp. 104–107	10		11
Add Integers pp. 108–113	9		4
Subtract Integers pp. 114–117			7
Multiply and Divide Integers pp. 118–123		3	6, 8
The Coordinate Grid pp. 124–129	1, 2	5	

Expressions and Equations

Algebra is a language you can use to solve real-life problems. A word expression can be translated into an algebraic expression. A used-car dealer has a sale in which he will sell a car for *$1,500 less than the sticker price of any car on the car lot*. The phrase *$1,500 less than the sticker price of any car on the car lot* can be translated to the algebraic expression, $n - 1,500$, in which n represents *the sticker price of any car on the lot*, and $-1,500$ represents *$1,500 less than*. The expression for the sale price of any car can then be given value by replacing n with the sticker price of the car.

Real-life situations involving numbers and relationships can be translated into equations and inequalities. A pattern can be generalized by using expressions and equations. Patterns can be numerical (with numbers) or visual (a pattern that repeats images). What patterns do you see in your daily life?

The Key Concepts you will study include:

Lesson 5.1: Expressions
Mathematical and real-world situations can be represented by expressions that can be evaluated and simplified.

Lesson 5.2: Solve One-Step Equations
Use equations to represent situations and use inverse operations to solve one-step equations.

Lesson 5.3: Solve Two-Step Equations
Use two inverse operations to solve two-step equations.

Lesson 5.4: Solve One- and Two-Step Inequalities
Use inverse operations to solve one- and two-step inequalities.

Lesson 5.5: Identify Patterns
Identify, represent, and generalize patterns using expressions and equations.

Goal Setting

Before starting this chapter, set goals for your learning. Think about the ways that strengthening your understanding of expressions, equations, and patterns will benefit you.

Look for patterns each day. Identify the pattern as one that uses numbers or pictures (numeric or visual), and then determine what algebraic expression you could write to describe the pattern. One is done as an example.

Pattern	Number	Visual	Expression
Two ducks and one bunny in each square of a child's quilt		√	The number of ducks in any number of squares s is $2s$.

Expressions

Lesson Objectives

You will be able to

- Translate between verbal and symbolic representations of expressions

- Simplify expressions

- Evaluate expressions

Skills

- **Core Skill:** Evaluate Expressions

- **Core Practice:** Make Sense of Problems

Vocabulary

algebraic expression
coefficient
constant term
mathematical expression
symbolic expression
variable
verbal expression

KEY CONCEPT: Mathematical and real-world situations can be represented by expressions that can be simplified and evaluated.

Find each sum or difference.

1. $1.24 + 8.3$ **2.** $-12 - (-3)$ **3.** $-18 + 3$

Find each product or quotient.

4. 1.2×3.5 **5.** $-56 \div (-4)$ **6.** $-3 \times (-6)$

Verbal and Symbolic Representations of Expressions

A numeric or **mathematical expression** is any combination of symbols, numbers, and operations. Mathematical expressions can be used to represent many different things. Also, multiple expressions can be used to describe the same object. For example, 2×5 and $6 + 4$ both represent the number 10. However, the expression 2×5 may represent the total number of players on the court at the same time during a basketball game, whereas the expression $6 + 4$ may represent the total number of fruit bought at a store (6 apples and 4 oranges). An expression can be thought of as a whole or it can thought of as a collection of parts.

Example 1 Mathematical Expressions

-3 $5 \times (-8)$ $78 - 23(2 + 1)$

A numeric expression may be simplified by using the order of operations.

Example 2 Simplify a Mathematical Expression

Simplify $21 - 5(3)$.

Use the order of operations and multiply 5 times 3. $5 \times 3 = 15$
Then subtract that result from 21. $21 - 15 = 6$

A variable or **algebraic expression** is a combination of numbers, one or more variables, and operations. A **variable** is a symbol that stands for an unknown number or value and is written in italics, such as x. A number that multiplies the variable is called the **coefficient**. A number that is added or subtracted in a variable expression is called a **constant term**.

Algebraic expressions can also be used to represent real-world scenarios. For example, the expression $2l+2w$ can represent the perimeter of a rectangle because both lengths need to be added to both widths. Also, using the Distributive Property, the expression can be rewritten as $2(l+w)$, saying that the sum of the length and width need to be doubled to find the perimeter.

Example 3 Algebraic or Variable Expressions

Identify the variable, coefficient, and constant term for each algebraic or variable expression.

expression	variable	coefficient	constant term
$2 + x$	x	1	2
m	m	1	0
$3n - 8$	n	3	−8

A **symbolic expression** is an expression that uses variables, numbers, and symbols for operations.

MATH LINK

Multiplication in expressions can be shown in several ways. Each of the following represents multiplication.

$8n$ $8 \cdot n$ $8(n)$

IDENTIFY KEY WORDS

Some sentences can be confusing or complicated. Focusing on the most important words in the sentence, or key words, can help the reader to better understand the meaning of that sentence. To identify key words, look for the subject (*who* or *what*) and the action (what the subject did). Consider the following sentence: *Marion, by adding, subtracting, multiplying, and simplifying several numbers, found the value of the expression.*

Who? Marion
Did what? found the value of the expression

Marion found the value of the expression. Everything else is just details.

Read the passage below and identify the key words in each sentence.

> (1) Variables are a part of math expressions that stand for an unknown value. (2) They can be a letter or other symbol as long as the same variable is used consistently in a problem. (3) Variables can be positive or negative numbers, decimals, fractions, or other type of real numbers. (4) There are many rules for learning to work with variables.

1. variables, unknown value

2. letter, other symbol

3. real number

4. rules

Core Skill
Evaluate Expressions

Often, as you have learned, solving a math problem requires you to follow a series of steps in a logical order. The step-by-step approach to solving problems in this book provides a **framework,** or structure, for explaining how you arrived at a particular solution. Use the same step-by-step approach to explain a solution. For example, to explain the solution for Example 4, you might say, "First, I assigned the variable n to represent the unknown number. Then I identified the correct operation in the verbal phrase, *5 less than*. This phrase told me that I should subtract..."

After you complete question 1 in Skill Practice on p. 139, write in your notebook the steps you took to solve the problem.

Example 4 Write a Variable Expression That Represents a Verbal Phrase

Write the variable expression for *5 less than twice a number*.

Step 1 Assign a variable to the unknown number.
Let n represent the unknown number.

Step 2 Identify the operation or operations in the verbal phrase.
The phrase *5 less than* means that 5 is subtracted from a number, so the operation is subtraction.
The phrase *twice a number* means 2 times a number, so the other operation is multiplication.

Step 3 Write the variable expression: $2n - 5$.

Example 5 Write an Expression That Represents a Real-World Situation

Tomas has a cell phone plan that charges a flat rate of $35.00 per month plus $0.25 for each text message. Write an algebraic expression that shows the monthly fee for Tomas's cell phone plan.

Step 1 Assign a variable to the unknown value.
The unknown value is the number of text messages.
Let t represent the number of text messages.

Step 2 Identify the operation or operations in the situation.
The word *plus* indicates addition. The phrase *$0.25 for each text message* indicates multiplication.

Step 3 Write the algebraic expression.
$35.00 is the monthly charge, and $0.25t$ is the charge for text messages.
$35.00 + 0.25t$

A **verbal expression** is written in words and numbers. The verbal representation might be mathematical or it might refer to a real-world situation.

Example 6 Translate an Expression Into a Verbal Phrase

$3m + 4$

Step 1 Identify the variable.
The variable is m.

Step 2 Identify the operation or operations.
$3m$ means 3 times $m + 4$ means add 4.

Step 3 Write a verbal phrase.
One way to write a verbal phrase for $3m + 4$ is *three times a number m plus four.*
Another way is to write a real-world situation for $3m + 4$:
Martin paid a one time fee of $4 to join the coffee club at work and paid $3 each of the following weeks.

THINK ABOUT MATH

Directions: Write each verbal phrase as a variable expression.

1. a number increased by 12 _____

2. $250 less than twice Rahman's salary _____

Directions: Write each variable expression as a verbal phrase.

3. $4t \div 2$ _____   4. $c - 9$ _____

Evaluate Expressions

When you **evaluate**, or find the value of, an expression, you substitute a given value for the variable and then perform the operation.

Example 7 Evaluate an Expression

Evaluate $x + 7$ when $x = -2$.

Step 1 Substitute the given value for the variable.
$x + 7 = -2 + 7$

Step 2 Perform the operation.
$-2 + 7 = 5$
The value of $x + 7$ when $x = -2$ is 5.

When there is more than one operation, use the order of operations:

1. Do operations within parentheses.

2. Do multiplication and division from left to right.

3. Do addition and subtraction from left to right.

Because real-world world problems are set within situations like the ones we experience in everyday life, they provide a way to add, subtract, multiply, and divide real objects. You are dealing with actual quantities. Suppose you wanted to write such a problem of your own. Somebody reading your problem would have to add in order to solve it. What would you say? Fortunately, there are key words that tell readers what operation they should use to solve this kind of problem. See the chart below, which lists key words for the four major operations.

Operation	Key word
Addition	add, sum, total, altogether, increased, in all
Subtraction	subtract, difference, more than, less than, farther than
Multiplication	multiply, total, product, times, twice
Division	divide, quotient, each, average, split

Milena walked 3 miles on Saturday and 2 miles on Sunday. How much farther did she walk on Saturday? The key words how much farther and walk indicate that you should subtract the amounts that Milena walked.

In a notebook, make a list of key words that appear in the problems in this lesson. Write whether they successfully achieved their goal—namely, to lead a reader to add, subtract, multiply, or divide in order to solve the problem.

Example 8 Evaluate an Expression Using the Order of Operations

Evaluate $a + 4b$ when $a = -10$ and $b = 1.5$.

Step 1 Substitute the given values for the variables.
$a + 4b = -10 + 4 \times 1.5$

Step 2 Use the order of operations.
First perform the multiplication: $-10 + 4 \times 1.5 = -10 + 6$
Then perform the addition: $-10 + 6 = -4$
The value of $a + 4b$ when $a = -10$ and $b = 1.5$ is -4.

THINK ABOUT **MATH**

Directions: Evaluate each expression. Use the order of operations if necessary.

1. $13.6 - a$ when $a = 2.9$ _____

2. $-3m + n$ when $m = 4$ and $n = 6$ _____

3. $c + d$ when $c = -5$ and $d = 2$ _____

4. $-5x \div 2y$ when $x = 8$ and $y = -5$ _____

Vocabulary Review

Directions: Match each word to the letter that describes the word.

1. _____ algebraic expression

2. _____ coefficient

3. _____ constant term

4. _____ mathematical expression

5. _____ symbolic expression

6. _____ variable

7. _____ verbal expression

A. a number that is added or subtracted in a variable expression

B. a symbol that represents an unknown number or value

C. an expression written using a combination of words and numbers

D. an expression that uses variables, numbers, and symbols for operations

E. the numerical factor of a variable

F. another name for a variable expression

G. another name for a numeric expression

Directions: Identify the key words in each passage.

1. (1) To write an algebraic expression from a verbal expression, identify the key words. (2) The unknown quantity in the expression is what the variable represents. (3) Key words also indicate what numbers and operations should be used.

2. (1) When evaluating the expression $g + 2h$ when $g = 4$ and $h = -3$, first substitute the given values for the variables. (2) To avoid errors, make sure you substituted the correct values for each variable, especially if there is more than one variable. (3) The second step is to perform the operation or use the order of operations, if there is more than one operation. In this case, multiply first and then add.

Directions: Answer the following question using what you have learned about evaluating expressions.

3. What ideas would you apply to evaluate the expression $2x + y$ when $x = -3$ and $y = 4$? Evaluate the expression.

Skill Practice

Directions: Choose the best answer to each question.

1. What is the value of $s + 5t$ when $s = 3$ and $t = -2$?

   A. −16
   B. −7
   C. 9
   D. 13

2. Which phrase best describes the expression $3x - 8$?

   A. the number x less eight times three
   B. eight minus three times a number x
   C. three times a number x decreased by eight
   D. three times the number x and negative 8

3. Jamila bought a ladder and rented a rototiller to do some work in her backyard. The ladder cost $48.75, and the cost to rent the rototiller is $18 per day. Which expression can Jamila use to find the cost of doing the work in her yard?

   A. $48.75 + 18$
   B. $(48.75 + 18)d$
   C. $48.75 - 18d$
   D. $48.75 + 18d$

4. Which expression represents −17 less than the product of −12 and some number?

   A. $-17 - (-12)x$
   B. $-12 - 17x$
   C. $-12x - (-17)$
   D. $-12x + (-17)$

Solve One-Step Equations

Lesson Objectives

You will be able to
- Understand and write equations
- Solve one-step equations

Skills

- **Core Practice:** Make Sense of Problems
- **Core Skill:** Represent Real-World Arithmetic Problems

Vocabulary

equal sign
equation
equivalent equation
inverse operations
solution

MATH LINK

The Addition, Subtraction, Multiplication, and Division Properties of Equality state that when you add, subtract, multiply, or divide each side of an equation by the same value, you produce an equivalent equation. An **equivalent equation** is one that has the same solution as the original equation.

KEY CONCEPT: Use equations to represent situations, and use inverse operations to solve one-step equations.

Write an expression for each situation. Use n as the variable.

1. the sum of a number and four _____

2. three times a number decreased by one _____

3. a number split into eight equal parts _____

Evaluate each expression when x = −1 and y = 3.

4. $x - y$ _____   5. $2y + 1$ _____   6. $3(x + y)$ _____

Understand and Write Equations

An **equation** is a mathematical statement that two expressions are equal. In algebra, an equation has at least one variable, which represents an unknown value. It is usually a letter. An **equal sign** (=) is placed between the two expressions to show a mathematical statement of equivalence.

Example 1 Write an Equation for a Situation

Bao works 4 hours and earns $72. Write an equation that can be used to find how much he earns per hour.

Step 1 Identify the variable.
Let d represent the dollar amount Bao earns per hour.

Step 2 Identify the operation.
Multiply the number of hours by dollars per hour to get total earnings. $4d$ represents the operation.

Step 3 Write the equation using the two quantities.
$4d$ is equal to $72. The equation is $4d = 72$.

A **solution** of an equation is the value of the variable that makes the equation a true statement.

Example 2 Check for Solutions

Is 13 a solution of the equation $x + 12 = 25$?

Step 1 Substitute the number, or solution, for the variable.
$13 + 12 = 25$

Step 2 Determine whether the solution results in a true statement.
$25 = 25$, so 13 is a solution of $x + 12 = 25$.

THINK ABOUT MATH

Directions: Write an equation for each situation.

1. the sum of a number and two is three _____

2. Anya's age decreased by five is twelve _____

Directions: Explain whether the value given for the variable is a solution of the equation.

3. $7c = 42$; $c = 8$ _____

4. $y + 7 = 3$; $y = -4$ _____

Solve Equations

Use **inverse operations** to solve an equation. Inverse operations are operations that are the opposite of each other and undo each other's results.

- The inverse of addition is subtraction.

- The inverse of subtraction is addition.

- The inverse of multiplication is division.

- The inverse of division is multiplication.

Example 3 Solve an Addition Equation

Solve the equation $y + 4 = 21$.

Step 1 Identify the operation used in the equation.
There is an addition sign, so the operation is addition.

Step 2 Identify the inverse operation.
The inverse of addition is subtraction.

Step 3 Use the inverse operation to solve the equation.
In this case, subtract 4 from each side of the equal sign.

$$\begin{array}{r} y + 4 = 21 \\ \underline{-4 \quad -4} \\ y = 17 \end{array}$$

Step 4 Check the solution.
$17 + 4 = 21$ is a true statement, so the solution is 17.
$y = 17$.

Core Practice
Make Sense
of Problems

Math questions can sometimes be tricky, because they may contain information that you do not need to know in order to solve a problem. For example, consider this problem: "Mariska's school ordered 100 box lunches for a student field trip. Twenty-five were vegetarian lunches, and 75 of the box lunches contained meat sandwiches. The cost of each lunch was $4.50. How much did the school spend to buy the lunches?" Before considering the information you were given, you should focus on the actual question, which appears in the last sentence: "How much did the school spend to buy the lunches?" Then look back at the question to see what information you need to know in order to solve the problem. Do you need to know the number of vegetarian or meat lunches? No, you can ignore those details, because you do not need them to solve the problem.

Before you answer question 4 in Skill Practice on page 143, make a two-column chart in your notebook: "Information I Need" and "Information I Can Ignore." Then fill out the two columns in your list. Finally, solve the problem.

Learning how to perform operations in the classroom or when doing your homework prepares you for solving the kinds of real-world arithmetic problems you will encounter in everyday life. You can use addition, for example, to calculate the cost of two pairs of jeans when the price tag in the clothing store says that each pair costs $23.95. First, translate the problem into an equation: $23.95 + $23.95 = x. On the other hand, since the price of each pair was the same, you might realize that you can use multiplication: $23.95 × 2 = x. Both equations will produce the same answer: $47.90.

Consider the following problem. The local library charges $0.10 a copy to make photocopies. You have designed a flyer announcing an upcoming Carnival Night that will be held at the school, and you need to make 150 copies. How much will you spend to make the copies? In a notebook, first write an equation for the problem.

Example 4 Solve a Multiplication Equation

Solve the equation $3a = 24$.

Step 1 Identify the operation used in the equation.
$3a$ means $3 \times a$, so the operation is multiplication.

Step 2 Identify the inverse operation.
The inverse of multiplication is division.

Step 3 Use the inverse operation to solve the equation. In this case, divide each side of the equal sign by 3.
$3a = 24 \qquad \frac{3a}{3} = \frac{24}{3} \qquad 1a = 8 \qquad a = 8$

Step 4 Check the solution.
$3 \times 8 = 24$ is a true statement, so the solution is 8. $a = 8$

Example 5 Solve a Subtraction Equation

Solve the equation $x - 8 = 18$.

Step 1 Identify the operation used in the equation.
There is a subtraction sign, so the operation is subtraction.

Step 2 Identify the inverse operation.
The inverse of subtraction is addition.

Step 3 Use the inverse operation to solve the equation. In this case, add 8 to each side of the equal sign.
$$x - 8 = 18$$
$$\underline{+8 \quad +8}$$
$$x = 26$$

Step 4 Check the solution.
$26 - 8 = 18$ is a true statement, so the solution is 26. $x = 26$.

Example 6 Solve a Division Equation

Solve the equation $\frac{z}{6} = 5$.

Step 1 Identify the operation used in the equation.
$\frac{z}{6}$ means $z \div 6$, so the operation is division.

Step 2 Identify the inverse operation.
The inverse of division is multiplication.

Step 3 Use the inverse operation to solve the equation. In this case, multiply each side of the equal sign by 6.
$$\frac{z}{6} \times 6 = 5 \times 6$$
$$z = 30$$

Step 4 Check the solution.
$\frac{30}{6} = 5$ is a true statement, so the solution is 30. $z = 30$.

THINK ABOUT MATH

Directions: Solve each equation.

1. $5y = 35$ _____

2. $n - 1 = 10$ _____

3. $\frac{z}{3} = 2$ _____

4. $y + 20 = 21$ _____

Directions: Match each word to one of the statements below.

_____ 1. equal sign   _____ 3. equivalent equation   _____ 5. solution

_____ 2. equation   _____ 4. inverse operations

A. has the same solution as another equation
B. the value of a variable that makes an equation a true statement
C. a statement that two expressions are equal
D. used between the two expressions in an equation
E. operations that are opposites and undo each other

Skill Review

Directions: Use what you learned about determining important information to answer the following questions.

1. What part of a problem do you focus on to determine what information you need to know in order to solve the problem?

_____

2. What tool allows you to undo the result of another operation?

_____

Skill Practice

Directions: Choose the best answer to each question.

1. Maemi ordered a lunch for each attendee at her meeting. Each lunch cost $16. The total cost was $80. Which of the following equations can be used to find how many lunches Maemi ordered?

   A. $16t = 80$ C. $80 - t = 16$
   B. $16 + t = 80$ D. $\frac{t}{16} = 80$

2. Which of the following is a solution of $2 + b = 14$?

   A. $b = 28$
   B. $b = 16$
   C. $b = 12$
   D. $b = 7$

3. Nizhoni wants to check the solution of $\frac{n}{6} = 6$. What should she do to check the solution?

   A. Multiply the solution by 6.
   B. Add 6 to the solution.
   C. Subtract 6 from the solution.
   D. Substitute the solution for n in the equation.

4. Tamika is a computer solutions associate who works in a call center. She usually works a 7-hour day and handles 84 calls in an average shift. She uses the equation $7x = 84$ to find the number of calls she answers in an hour. How many calls can Tamika answer in an hour?

   A. 12   C. 77
   B. 18   D. 588

Solve Two-Step Equations

KEY CONCEPT: Use two inverse operations to solve two-step equations.

Match each equation to its solution.

———— **1.** $x + 3 = 12$ ———— **3.** $4x = 24$

———— **2.** $x - 3 = -4$ ———— **4.** $\frac{x}{3} = 4$

A. $x = 6$ **B.** $x = 9$ **C.** $x = 12$ **D.** $x = -1$

Translate Verbal Sentences into Two-Step Equations

A **two-step equation** is an equation that contains two different operations, such as multiplication and subtraction in the equation $3t - 7 = 14$. To translate a verbal sentence into an equation, represent the unknown with a variable, identify the two operations, and then write the equation.

Example 1 Translate a Sentence into an Equation

Four times a number plus four is twelve.

Step 1 Choose a variable to represent the unknown value. Let n represent the unknown number.

Step 2 Identify the expressions on each side of the equal sign. The word *is* stands for the equal sign, so *four times a number plus four* is on one side, and *12* is on the other side.

Step 3 Identify the two operations. The phrase *four times a number* indicates multiplication, or $4n$. The phrase *plus four* indicates addition, or $+ 4$.

Step 4 Write the equation. The equation is $4n + 4 = 12$.

Example 2 Translate a Real-World Situation into an Equation

Mai-Ling is saving $2,000 for a down payment on a car. She has saved $500 so far and plans to save the rest in 6 months. Write an equation that shows how much she needs to save per month to have a $2,000 down payment.

Step 1 Choose a variable for the unknown value. Mai-Ling wants to know how much to save each month for the down payment. This is the unknown. Let d represent the dollar amount she needs to save each month.

Step 2 Identify the expressions on each side of the equal sign. $2,000 is the total Mai-Ling needs to save, so it goes on one side of the equal sign. On the other side of the equal sign is the unknown dollars she needs to save each month for the next 6 months *and* the $500 she has already saved.

Step 3 Identify the two operations on one side of the equation. 6 months $\times$ d indicates multiplication, or $6d$. The phrase *and $500* indicates addition, or $+ 500$

Step 4 Write the equation. $6d + 500$ represents one side of the equation. 2,000 represents the other side. It shows that the money Mai-Ling saves in 6 months plus the money she already has saved is equal to $2,000. The equation is $6d + 500 = 2,000$.

MATH LINK

Remember that the variable in an equation can be any letter. In general, the letter n or x is used to represent an unknown number, but any letter can be used. In real-world situations, the variable is often the first letter of the unknown, such as c for *cost*. Even though this is often the case, it is not always the case. What is important to remember is that the variable always stands for the unknown value or number.

Solve Two-Step Equations

The goal in solving any equation is to **isolate** the variable. To isolate the variable means to get the variable by itself on one side of the equation. In a two-step equation, perform two inverse operations to isolate the variable.

Solving two-step equations involves a sequence of steps that must be performed in a certain order. When solving two-step equations, therefore, perform the operations in the reverse order to the order of operations.

Example 3 Solve an Equation That Involves Addition and Multiplication

Solve the equation $3x + 6 = 24$.

Step 1 Identify each operation and its inverse in the equation. $3x$ means $3 \times x$, so the first operation is multiplication. Its inverse is division. There is a plus sign, so the second operation is addition. Its inverse is subtraction.

Step 2 Identify the order in which the inverse operations should be performed. First, use subtraction to undo addition. Then use division to undo multiplication.

Step 3 Perform the first inverse operation on each side of the equation. Subtract 6 from each side of the equation.

$$\begin{array}{r} 3x + 6 = 24 \\ \underline{-\ 6\ \ = -6} \\ 3x\ \ = 18 \end{array}$$

Step 4 Perform the second inverse operation on each side of the equation. Divide each side of the equation by 3.
$$3x = 18 \qquad \frac{3x}{3} = \frac{18}{3} \qquad x = 6$$

Step 5 Check the solution by substituting the solution for the variable in the original equation.
$3(6) + 6 = 24$ $18 + 6 = 24$ The statement is true, so $x = 6$.

Example 4 Solve an Equation That Involves Subtraction and Multiplication

Solve the equation $-2x - 4 = -8$.

Step 1 Identify each operation and its inverse in the equation. The first operation is multiplication. Its inverse is division. The second operation is subtraction. Its inverse is addition.

Step 2 Identify the order in which the inverse operations should be performed. First, use addition to undo subtraction. Then use division to undo multiplication.

Step 3 Perform the first inverse operation on each side of the equation. Add 4 to each side of the equation.

$$\begin{array}{r} -2x - 4 = -8 \\ \underline{+4\ \ = +4} \\ -2x\ \ = -4 \end{array}$$

Step 4 Perform the second inverse operation on each side of the equation. Divide each side of the equation by −2.
$$-2x = -4 \qquad \frac{-2x}{-2} = \frac{-4}{-2} \qquad x = 2$$

Step 5 Check the solution by substituting the solution for the variable in the original equation.
$-2(2) - 4 = -8$ $-4 + (-4) = -8$
The statement is true, so $x = 2$.

Notice that in Example 4, you added 4 to both sides of the equal sign to solve the problem. However, this is the same as subtracting its negative. That way, in both examples, subtraction took place first. This is because addition and subtraction are essentially the same operation, since subtracting a negative number is the same as adding a positive number.

Example 5 Solve an Equation That Involves Addition and Division

Solve the equation $\frac{x}{-8} + 2 = 5$.

Step 1 Identify each operation and its inverse in the equation. The first operation is division. Its inverse is multiplication. The second operation is addition. Its inverse is subtraction.

Step 2 Identify the order in which the inverse operations should be performed. First, use subtraction to undo addition. Then use multiplication to undo division.

Step 3 Perform the first inverse operation on each side of the equation. Subtract 2 from each side of the equation.

$$\frac{x}{-8} + 2 = 5 \qquad \frac{x}{-8} + 2 - 2 = 5 - 2 \qquad \frac{x}{-8} = 3$$

Step 4 Perform the second inverse operation on each side of the $\neq$ equation. Multiply each side of the equation by -8.

$$\frac{x}{-8} = 3 \qquad \frac{x}{-8} \times (-8) = 3(-8) \qquad x = -24$$

Step 5 Check the solution by substituting the variable into the original equation.

$$\frac{-24}{-8} + 2 = 5 \quad 3 + 2 = 5 \quad \text{The statement is true, so } x = -24.$$

Example 6 Solve an Equation That Involves Subtraction and Division

Solve the equation $\frac{x}{4} - 7 = 3$.

Step 1 Identify each operation and its inverse in the equation. $\frac{x}{4}$ means $x \div 4$, so the first operation is division. Its inverse is multiplication. There is a minus sign, so the second operation is subtraction. Its inverse is addition.

Step 2 Identify the order in which the inverse operations should be performed. First, use addition to undo subtraction. Then use multiplication to undo division.

Step 3 Perform the first inverse operation on each side of the equation. Add 7 to each side of the equation.

$$\frac{x}{4} - 7 = 3 \qquad \frac{x}{4} - 7 + 7 = 3 + 7 \qquad \frac{x}{4} = 10$$

Step 4 Perform the second inverse operation on each side of the equation. Multiply each side of the equation by 4.

$$\frac{x}{4} = 10 \qquad \frac{x}{4} \times 4 = 10 \times 4 \qquad x = 40$$

Step 5 Check the solution by substituting the solution for the variable in the original equation.

$$\frac{40}{4} - 7 = 3 \qquad 10 - 7 = 3 \quad \text{The statement is true, so } x = 40.$$

Just as addition and subtraction are very similar (as discussed on the previous page), multiplication and division operations are just as similar. Notice that in both division examples, the division could have been written as a product of the variable and a fraction. For instance, $x \div 4$ could have be rewritten as $\frac{1}{4} \times x$. Therefore you could have divided by $\frac{1}{4}$, which is the same as multiplying by 4. This is because multiplication and division are just as similar as addition and subtraction.

Core Skill
Evaluate Expressions

Arithmetic and algebraic expressions are similar in some ways. Both contain numbers, symbols, and operations. Of course, there are some important differences. Arithmetic expressions such as $5 \times (-8)$ or $78 - 23(2 + 1)$ are pretty easy to solve. You merely add, subtract, multiply, or divide the numbers to get a solution. But algebraic expressions involve unknowns that are represented by a mix of both numbers and variables. What must you do to solve the algebraic equation $5x + 4 = 24$? In a notebook write the equation and the solution.

THINK ABOUT MATH

$\sqrt{x}$ **Directions:** Solve each equation.

1. $5x + 1 = 41$ _____

2. $\frac{x}{6} - 12 = -10$ _____

3. $\frac{x}{-8} + 6 = 14$ _____

4. $-2x + 3 = 1$ _____

Vocabulary Review

Directions: Fill in the blanks with one of the words or phrases below.

affect   isolate   two-step equation

1. A(n) _____ requires that you do two operations to solve it.

2. When you _____ a variable, you get the variable by itself on one side of the equation.

3. If you undo operations in the wrong order, it will _____ your answer.

Skill Review

Directions: Answer the questions below.

1. Explain how the skill of understanding sequence helped you solve two-step equations.

_____

_____

_____

2. Describe the sequence you would use to solve the equation $6x + 3 = -9$.

_____

_____

_____

3. Describe a sequence that you could use to solve any two-step equation.

_____

_____

_____

Skill Practice

Directions: Choose the best answer to each question.

1. Which equation best describes this sentence: *A number divided by eight plus three is fifty-one?*

   A. $n + \frac{3}{8} = 51$

   B. $\frac{8}{n} + 3 = 51$

   C. $\frac{n}{8} + 3 = 51$

   D. $8 + \frac{3}{n} = 51$

2. Dan drives 49 miles to work, which is 7 miles less than twice the number of miles that Eduardo drives. Which equation shows the number of miles m that Eduardo drives to work?

   A. $2m - 7 = 49$

   B. $2(m - 7) = 49$

   C. $2m + 7 = 49$

   D. $2(m + 7) = 49$

Directions: Solve the problems.

3. What is the solution for the equation $\frac{n}{7} + 12 = 58$?

   _____

4. Ibrahim has saved $2,000. He is planning on buying a new computer and accessories that cost $3,650. Ibrahim earns $150 in commission for each elliptical trainer that he sells. He writes this equation to figure out how many trainers he will need to sell this month in order to be able to buy his computer.

   $150t + 2,000 = 3,650$

   How many elliptical trainers does Ibrahim need to sell?

   _____

Solve One- and Two-Step Inequalities

KEY CONCEPT: Use inverse operations to solve one- and two-step inequalities.

Compare each pair of numbers, using $<$, $>$, or $=$.

1. 0.10 _____ 0.1   3. 186 _____ 183   5. -54 _____ -50

2. 14 _____ 23   4. -2 _____ -5 6. 8.57 _____ 8.5

Translate Verbal Statements into Inequalities

An **inequality** is a statement in which an inequality symbol is placed between two expressions, such as $t + 6 \geq 2$. The inequality symbols are $<$ (less than), $\leq$ (less than or equal to), $>$ (greater than), and $\geq$ (greater than or equal to).

Inequalities and equations are similar in some ways. For example, both contain expressions that combine numbers and variables. In other words, there is a connection, or relationship, between inequalities and equations. There are, however, differences. The inequality symbols tell you the expressions that are being compared may not be equal.

Because there is this connection between inequalities and equations, translating a verbal statement into an inequality is similar to translating a statement into an equation. First, choose a variable for the unknown. Next, identify the operations in the statement, and choose the inequality symbol that represents the statement. Then write the inequality.

Example 1 Translate a Verbal Statement into a One-Step Inequality

A number increased by 5 is no more than 9.

Step 1 Choose a variable to represent the unknown number. Let n represent the unknown number.

Step 2 Identify the operation and write the expression. The phrase *increased by* 5 indicates addition. The expression is $n + 5$.

Step 3 Identify the inequality symbol in the verbal statement. The phrase *no more than* means less than or equal to, so the symbol is $\leq$.

Step 4 Write the inequality. $n + 5 \leq 9$

Example 2 Translate a Verbal Statement into a Two-Step Inequality

Antonio will pay no less than $115 total for two tickets to the football game plus $35 for snacks.

Step 1 Choose a variable to represent the unknown.
The unknown is the cost of a ticket.
Let t represent the cost of one ticket.

Step 2 Identify the operation and write the expression.
The phrase *two tickets* indicates multiplication, since $2 \times t$ will be the cost of two tickets. The expression *plus $35* indicates addition.
The expression is $2t + 35$.

Step 3 Identify the inequality symbol in the verbal statement.
The phrase *no less than* means greater than or equal to, so the symbol is $\geq$.

Step 4 Write the inequality.
$2t + 35 \geq 115$

THINK ABOUT MATH

Directions: Translate each verbal statement into a one-step or two-step inequality.

1. A number minus 8 is more than 12.

2. The cost of lunch plus a $3 tip is at least $10.

3. Four more than twice a number is less than 25.

4. A representative at Ace Plumbing told a customer that three hours of labor and a service fee of $25 is no less than $310.

Solve One-Step Inequalities

Solve an inequality in the same way that you solve equations. Use inverse operations to isolate the variable.

The **solution of an inequality** is the set of all of the numbers that make the inequality true. These sets contain an **infinite**, or endless, number of numbers.

MATH LINK

The solution $r > 12$ means that all numbers greater than 12 are a solution for the inequality. That means there are infinitely many correct solutions to this inequality.

Core Skill
Solve Inequalities

You have some experience already with solving one- and two-step equations with one variable. Much of what you have learned about solving such equations applies to the process of solving inequalities. For example, you want to isolate the variable. You must also follow the order of operations. There is one important difference between solving equations and solving inequalities. When multiplying or dividing both sides of an inequality by a negative number, you must reverse the inequality symbol.

In a notebook solve the equations $14y > -2$ and then solve $-14y > 2$.

Example 3 Use Addition to Solve an Inequality

Solve the inequality $r - 2 > 10$.

Step 1 Identify the operation in the inequality and its inverse. The operation is subtraction, and its inverse is addition.

Step 2 Use the inverse operation to solve the inequality. In this case, add 2 to each side of the inequality.

$$\begin{array}{r} r - 2 > 10 \\ \underline{+2 \quad +2} \\ r > 12 \end{array}$$

Step 3 Check the solution. Since r is greater than 12, substitute a number greater than 12, such as 12.5, for the variable in the original inequality.

$12.5 - 2 > 10 \qquad 10.5 > 10 \qquad$ This is true, so $r > 12$.

Example 4 Use Subtraction to Solve an Inequality

Solve the inequality $h + 6 \leq 3$.

Step 1 Identify the operation in the inequality and its inverse. The operation is addition, and its inverse is subtraction.

Step 2 Use the inverse operation to solve the inequality.

$$\begin{array}{r} h + 6 \leq 3 \\ \underline{-6 \quad -6} \\ h \leq -3 \end{array}$$

Step 3 Check the solution. Since h is less than or equal to -3, substitute -3 for the variable in the original inequality.

$-3 + 6 \leq 3 \qquad 3 \leq 3 \qquad$ This is true, so $h \leq -3$.

As you might have noticed in the two previous examples, both involved subtracting the number that was being added (or subtracted). In Example 3, you subtracted a -2 from both sides to solve the inequality (subtracting a negative is the same as adding). Similarly, in Example 4, you subtracted 6 from both side to solve the inequality. There are also similarities between solving inequalities involving multiplication and division.

One major difference in solving equations and solving inequalities is that if you divide or multiply by a negative number to solve an inequality, you must **reverse**, or change the direction, of the inequality symbol. If a symbol is $\leq$, it becomes $\geq$ when you reverse its direction. If you multiply or divide by a positive number, the direction of the symbol is not reversed.

Example 5 Use Division to Solve an Inequality

Solve the inequality $-4b \geq 20$.

Step 1 Identify the operation in the inequality and its inverse. The operation is multiplication, and its inverse is division.

Step 2 Use the inverse operation to solve the inequality. In this case, the inverse operation is division by a negative integer. You will need to reverse the direction of the inequality symbol from $\geq$ to $\leq$.

$$\frac{-4b}{-4} \geq \frac{20}{-4} \qquad\qquad b \leq -5$$

Step 3 Check the solution. Since b is less than or equal to –5, substitute both –5 and a number less than –5, such as –6, for the variable in the original inequality.

$$-4(-5) \geq 20 \qquad 20 \geq 20$$
$$-4(-6) \geq 20 \qquad 24 \geq 20 \qquad \text{This is true, so } b \leq -5.$$

THINK ABOUT MATH

Directions: Solve each inequality.

1. $-3t > 15$

2. $x - 14 \leq -12$

3. $\frac{a}{6} \leq -1$

4. $c + 2 < 9$

Example 6 Use Multiplication to Solve an Inequality

Solve the inequality $\frac{y}{3} < -12$.

Step 1 Identify the operation in the inequality and its inverse. The operation is division, and its inverse is multiplication.

Step 2 Use the inverse operation to solve the inequality. The inverse operation is multiplication by a positive number, so you will *not* need to reverse the direction of the inequality symbol.

$$\frac{y}{3} < -12 \qquad \frac{y}{3} \times 3 < -12 \times 3 \qquad y < -36$$

Step 3 Check the solution. Since y is less than –36, substitute a number less than –36 in the original inequality, such as –39.

$$\frac{-39}{3} < -12 \qquad -13 < -12 \qquad \text{This is true, so } y < -36.$$

In Examples 3 and 4, you noticed that in both inequalities, subtraction was used to solve the inequality. The same is true of Examples 5 and 6. In both examples, you can divide by the number being multiplied to the variable. Example 5 has the variable being multiplied by –4 and in Example 6, the variable is multiplied by $\frac{1}{3}$. To solve those examples, you divided by –4 and 3 (the reciprocal of $\frac{1}{3}$), respectively.

Solve Two-Step Inequalities

You use two inverse operations to solve a two-step inequality. In general, perform addition or subtraction before multiplication or division. As with one-step inequalities, reverse the direction of the inequality symbol if you multiply or divide by a negative number.

Core Practice
Evaluate Reasoning

Precision is all-important when solving two-step inequalities. First, you must do the two steps in the correct order: addition and subtraction before multiplication and division. You must also be mindful of those cases in which your calculations lead to a reversal of the direction of the inequality symbol.

Enesta solves the following problem: $-5x - 25 \geq 50$. First, she divides each side by –5, but she doesn't reverse the direction of the inequality symbol when dividing by a negative number. As a result, she gets $x + 5 \geq -10$. Then she subtracts 5 from both sides to get $x \geq -15$. She checks her solution by replacing x with –14. $-5(-14) - 25 \geq 50$; $70 - 25 \geq 50$; $45 \not\geq 50$.

What did she do wrong? In her first step, she divided both sides of the inequality by –5 correctly, but she should have remembered to reverse the inequality sign. She divided first, then added, which was a valid path to finding the solution. Her error occurred during the division step.

Do Enesta's problem again, but this time, add first and then divide. Check your solution.

Example 7 Use Addition and Multiplication to Solve a Two-Step Inequality

Solve the inequality $\frac{k}{-8} - 1 < 2$.

Step 1 Identify each operation in the inequality and its inverse.
The first operation is division, and its inverse is multiplication.
The second operation is subtraction, and its inverse is addition.

Step 2 Identify the order in which the inverse operations should be performed.
Use addition to undo subtraction, and then use multiplication to undo division.

Step 3 Perform the first inverse operation on each side of the inequality.
Add 1 to each side of the inequality.
$$\frac{k}{-8} - 1 < 2 \qquad \frac{k}{-8} - 1 + 1 < 2 + 1 \qquad \frac{k}{-8} < 3$$

Step 4 Perform the second inverse operation on each side of the inequality.
Multiply each side by –8. Since you are multiplying by a negative integer, you need to reverse the direction of the inequality symbol from < to >.
$$\frac{k}{-8} < 3 \qquad \frac{k}{-8} \times (-8) > 3 \times (-8) \qquad k > -24$$

Step 5 Check the solution. Substitute a number greater than –24 into the original inequality. Choose a number that can be easily divided by –8, such as –16.
$$\frac{-16}{-8} - 1 < 2 \qquad 2 - 1 < 2 \qquad 1 < 2 \qquad \text{This is true, so } k > -24.$$

Example 8 Use Subtraction and Division to Solve a Two-Step Inequality

Solve the inequality $6p + 21 \geq 39$.

Step 1 Identify each operation in the inequality and its inverse.
The first operation is multiplication, and its inverse is division.
The second operation is addition, and its inverse is subtraction.

Step 2 Identify the order in which the inverse operations should be performed.
Use subtraction to undo addition and then division to undo multiplication.

Step 3 Perform the first inverse operation on each side of the inequality.
Subtract 21 from each side of the inequality.
$$6p + 21 \geq 39 \qquad 6p + 21 - 21 \geq 39 - 21 \qquad 6p \geq 18$$

Step 4 Perform the second inverse operation on each side of the inequality.
Divide each side by 6. Since you are dividing by a positive number, you do *not* need to reverse the direction of the inequality symbol.
$$6p \geq 18 \qquad \frac{6p}{6} \geq \frac{18}{6} \qquad p \geq 3$$

Step 5 Check the solution. Since p is greater than or equal to 3, choose a number greater than 3, such as 4, and substitute it into the original inequality.
$$6(4) + 21 \geq 39 \qquad 24 + 21 \geq 39 \qquad 45 \geq 39$$
This is true, so $p \geq 3$.

THINK ABOUT **MATH**

Directions: Solve each inequality.

1. $3x + 10 < 7$ _____

2. $\frac{y}{5} - 7 \leq 2$ _____

3. $\frac{m}{-2} + 6 \geq 20$ _____

4. $-7b - 11 > 10$ _____

Vocabulary Review

Directions: Match each word to a statement.

1. _____ inequality
2. _____ infinite
3. _____ reverse

A. extends indefinitely

B. type of mathematical sentence that contains $<$, $>$, $\leq$, or $\geq$ between the two expressions

C. what happens to an inequality symbol when multiplying or dividing by a negative number to solve the inequality

Skill Review

Directions: Answer the following questions.

1. Translate $x - 4 > 6$ into a verbal statement.

_____

2. Describe how the connections you made between solving equations and solving inequalities helped you learn how to solve inequalities.

_____

Skill Practice

Directions: Choose the best answer to each question.

1. Which statement best describes the inequality $x + 4 \leq 10$?

   A. A number increased by 4 is at least 10.
   B. A number plus 4 is at most 10.
   C. A number and 4 is less than 10.
   D. A number plus 4 is no less than 10.

2. Nina used the inequality $4s + 35 \geq 180$ to find the least amount that she will have to pay to learn how to use a database. What is the least she will have to pay per session s?

   A. $80.00
   B. $53.75
   C. $36.25
   D. $10.00

3. Which inequality is equivalent to the inequality $-5x > 5$?

   A. $x < -25$
   B. $x > -25$
   C. $x > -1$
   D. $x < -1$

4. A print shop manager estimated that it would cost no more than $65 to print 100 flyers, plus $15 for paper. Which inequality can be used to find the greatest cost per flyer?

   A. $100f + 15 \leq 65$
   B. $100f + 15 \geq 65$
   C. $100f - 15 \leq 65$
   D. $100f - 15 < 65$

Identify Patterns

KEY CONCEPT: Identify, represent, and generalize patterns using expressions and equations.

Translate each verbal phrase into an expression or equation.

1. Twice a number is ten.   **2.** three times a number minus five

Evaluate each expression when $x = 4$.

3. $12x$ **4.** $7x + 2$

Write Expressions to Represent Patterns

A **numerical pattern** is a set of numbers related by a rule. The **rule** is the operation or operations used to form the pattern. Once you know the rule, you can **generalize** it. That means that you can write an expression or equation that describes the pattern. The rule must apply to all of the numbers in the pattern. Sometimes, the pattern you discover is a **common difference**, which is when the difference between consecutive terms in a sequence of numbers is the same, or common.

A **sequence** is a set of numbers in a specific order. A **term** in the sequence is a number in the sequence. For example, 3 is the first term in the sequence 3, 6, 9, 12, 15, ….

Example 1 Write a Rule with One Operation

Write an expression to represent the pattern in the sequence 3, 6, 9, 12, 15, 18, ….

> **Step 1** Make a table that shows the position and number of each term in the sequence. Label the position of the term *n*.

Position of Term **n**	1	2	3	4	5	6
Number in Sequence	3	6	9	12	15	18

> **Step 2** Find the common difference between consecutive numbers in each row.
> In the first row, add 1 to the previous number to get the next number, so the common difference is 1.
> In the second row, add 3 to the previous number to get the next number, so the common difference is 3.

> **Step 3** Use the common difference to find and write a rule for the pattern.
> The common difference in the second row is 3 times the difference in the first row, so the rule could be 3 times *n* or $3n$.

> **Step 4** Try the rule on at least three numbers in the table. Substitute 1, 4, and 6 for *n* to see if you get 3, 12, and 18 in the sequence.
> $3n = 3(1)$, or 3 $3n = 3(4)$, or 12 $3n = 3(6)$ or 18
> These are true, so the rule is $3n$.

MAKE A TABLE

Tables usually have rows and columns that contain specific types of information. A passage may contain a lot of information, and it may not always be organized. Tables allow the reader to track information and review it quickly in an organized way.

Often, the most difficult part of making a table is deciding which information is to be included and how best to arrange it. The first step is to decide how many categories of information are being presented and how it would be best to organize them.

Read the following passage, and create a table of the information.

> Catalina found three apartments that she liked. The first had 1 bedroom and 1 bathroom. The rent was $600 per month, and heat and hot water were included. Parking was $50 per month in a covered garage. The second had 2 bedrooms and 1 bathroom. The rent was $750 per month. Heat and hot water were included, but parking was $75 per month. The last apartment had 1 bedroom, 1 bathroom, a deck, and a dishwasher. The rent was $700 per month, and parking was included. Heat and hot water were not included in the rent.

The passage describes three apartments, so the table should have three rows. Now, reread the passage and look for details about each apartment, such as the rent, number of bedrooms and bathrooms, and extra fees (for example, heat and parking). Two of the apartments also have extra information, so the table should also have a place for "other."

	Number of beds/ bathroom	Rent ($)	Parking ($)	Heat/hot water included	Other
Apartment 1	1/1	600	50	Yes	Covered parking
Apartment 2	2/1	750	75	Yes	None
Apartment 3	1/1	700	0	No	Deck, dishwasher

A passage in a math problem may contain a lot of information that is not organized. By organizing the information presented in a real-world problem, tables reveal patterns or relationships between numbers that may not have been obvious. Solving problems becomes easier when you've used a table to arrange or structure the important data.

Katiah recorded the high and low temperatures every day for a week. The high temperatures were Sunday 74°, Monday 76°, Tuesday 78°, Wednesday 72°, Thursday 71°, Friday 81°, and Saturday 80°. The low temperatures were Sunday 59°, Monday 61°, Tuesday 63°, Wednesday 57°, Thursday 56°, Friday 66°, and Saturday 65°. What is the difference between the high and low for each day? Is there a common difference?

To answer the question, complete the table below.

Temperatures (°F)

Day	High Temperature	Low Temperature
Sun.	74	59
		62
Tues.	78	
Wed.		
Fri.		
Sat.		

At some point, step
back and take the time
to think about what
you have learned. When
solving problems, you
don't want to simply
repeat the series of
steps from memory
that you encountered in
each one of the lesson
examples you have
studied. You want to
be able to understand
why each step is
important as well as
the connections among
the various steps. If
you truly understand
the math, you can solve
problems that may differ
slightly from the ones
you have seen before.
Critical reflection
provides a way to
achieve this level of
understanding.

Consider the following
sequence that begins
with 1, 4, 9, 16, 25, …
Unlike other patterns
you have seen, this
pattern does not have
a common difference
(3, 5, 7, 9, …). Can you
still find a pattern in
this sequence? What
rule is occurring? In a
notebook, write out
what is happening to
each term to help find
a pattern.

THINK ABOUT MATH

Directions: Make a table for each sequence. Then write
a rule to represent the sequence.

1. 4, 8, 12, 16, 20, 24, …   2. 5, 8, 11, 14, 17, 20, …

A two-variable equation describes the pattern between two quantities and
has two variables, such as x and y. The variable that you apply the rule
to is the **input variable**. The result is the **output variable**. Write the
output variable equal to the rule, such as $y = 2x$.

Example 2 Write a Rule with Two Operations

Write an equation to represent the pattern shown in the table.

x	1	2	3	4	5	6
y	9	14	19	24	29	34

Step 1 Find the common differences for the values of x and y.
The common difference for the x-values is 1.
The common difference for y-values is 5.

Step 2 Use the common difference to find and write a rule for
the pattern. The common difference for the y-values is 5,
so the rule could be $y = 5x$.

Step 3 Try the rule on at least three numbers in the table.
If $x = 1$, then $y = 9$, and $5(1) = 5$.
$5 \neq 9$. The rule is not correct.

Step 4 Try adding or subtracting a number from the rule to get
the value for y. If this works, try the new rule on two
more numbers. The difference between 9 and 5 is 4,
so add 4 to the rule to give a new rule of $y = 5x + 4$.
If $x = 1$, then $y = 9$, and $5(1) + 4 = 9$.
If $x = 4$, then $y = 24$, and $5(4) + 4 = 24$.
If $x = 6$, then $y = 34$, and $5(6) + 4 = 34$.
These are true, so the rule is $y = 5x + 4$.

Example 3 Find the Next Numbers in a Pattern

Find the next three numbers in the sequence 1, 3, 5, 7, 9, 11, ...

Step 1 Make a table that shows the position and the number of each term in the sequence. Label the position of the term n.

Position of Term, n	1	2	3	4	5	6	7	8	9
Number in Sequence	1	3	5	7	9	11			

Step 2 Write a rule for the sequence. Look at the first three numbers in each row. If you multiply 1 by 2 and subtract 1, you will get 1. Multiply 2 by 2 and subtract 1 to get 3. Multiply 3 by 2 and subtract 1 to get 5. So the rule is $2n - 1$.

Step 3 To find the next three numbers in the sequence use the rule $2n - 1$:

$$2 \times 7 - 1 = 13 \qquad 2 \times 8 - 1 = 15 \qquad 2 \times 9 - 1 = 17$$

Position of Term, n	1	2	3	4	5	6	7	8	9
Number in Sequence	1	3	5	7	9	11	13	15	17

Example 4 Find Any Number in a Pattern

What is the 20th term in the pattern 3, 7, 11, 15, 19, 23, ...?

Step 1 Make a table.

Position of Term, n	1	2	3	4	5	6
Number in Sequence	3	7	11	15	19	23

Step 2 Write a rule. There is a common difference of 1 for the first row and a common difference of 4 for the second row. If you multiply each term in the first row by 4 and then subtract 1, you will get the numbers in the pattern. The rule is $4n - 1$.

Step 3 Apply the rule to the term you need to find. You need to find the 20th term. Substitute $n = 20$ in the rule, $4n - 1$.
$4(20) - 1 = 79$; the 20th term of the pattern is 79.

Vocabulary Review

Directions: Fill in the blanks with the correct word or phrase below.

common difference numerical pattern generalize input variable term output variable

1. When you _____ a pattern, you write an expression or equation to describe the rule of the pattern.

2. A(n) _____ is one of the numbers in a sequence.

3. When analyzing a pattern, the _____ is the result of applying the rule to the _____.

4. The difference between consecutive terms in a sequence is called the _____.

5. A(n) _____ is a set of numbers related by a rule.

Skill Review

Directions: Use what you learned about finding patterns to answer the following questions.

1. Explain the process you would use to find a pattern for the sequence 1, 4, 7, 10, 13, …. Then find the pattern and write the rule.

   _____

   _____

   _____

   _____

2. Explain why, after finding a pattern, it is useful to write the rule for the pattern using an expression or an equation.

   _____

   _____

   _____

   _____

Skill Practice

Directions: Choose the best answer to each question.

1. Dae-Jung is thinking about switching mail carriers for his business. The table below shows the costs for shipping and handling of different weights. Which equation shows the pattern between pounds p and cost c?

Number of pounds, **p**	1	2	3	4	5
Total Cost, **c**		8	10	12	14	16

A. $c = 8p$
B. $c = p + 7$
C. $c = 2p + 7$
D. $c = 2p + 6$

2. Elias says that the expression $4n$ describes the pattern of the sequence shown in the table.

Position of Term, **n**	1	2	3	4	5
Number in Sequence	5	9	13	17	21

Which of the following best describes Elias's statement?

A. He is correct.
B. He only used the common difference instead of the common difference plus 1, or $4n + 1$.
C. He should have added 4 to n for the rule $n + 4$.
D. The rule should be $5n$ because 1×5 is 5.

3. What is the next number in the pattern?

39, 77, 115, 153, 191

A. 209
B. 219
C. 229
D. 239

4. Zoey wrote the equation $n = 12m$ to describe the pattern in one of the tables below. Which table did she use?

A.

m	1	2	3	4	5
n	12	24	36	48	60

B.

m	1	2	3	4	5
n	11	22	33	44	55

C.

m	1	2	3	4	5
n	12	20	28	36	44

D.

m	1	2	3	4	5
n	12	22	32	42	52

Directions: Choose the best answer to each question.

1. What is the solution for x in the equation?

   $-8x + 11 = 35$

   A. $x = -24$
   B. $x = -3$
   C. $x = 3$
   D. $x = 5$

2. What is the value of $5x - 3y$ when $x = 7$ and $y = -8$?

   A. -61
   B. 11
   C. 59
   D. 61

3. Atian is 7 years younger than 5 times his son's age. If Atian is 43, how old is his son?

   A. 9 years old
   B. 10 years old
   C. 11 years old
   D. 12 years old

4. Chuma rents a jackhammer for a rental fee of $45 and $13 per day. Which equation will give Chuma the total cost t of renting the jackhammer for d days?

   A. $t = 13d + 45$
   B. $t = 45d - 13$
   C. $d = 45t + 13$
   D. $t = 45d + 13$

5. How can $\frac{y}{-12} = -36$ be solved?

   A. Divide both sides of the equation by -12.
   B. Multiply both sides of the equation by 12.
   C. Divide both sides of the equation by -36.
   D. Multiply both sides of the equation by -12.

Directions: Use the following table for Questions 6 and 7.

x	1	2	3	4	5
y	7	12	17	22	27

6. Which equation describes the rule for the table?

   A. $y = 7x + 5$
   B. $y = 4x + 3$
   C. $y = x + 6$
   D. $y = 5x + 2$

7. If the pattern in the table above is followed, what is the tenth term of the sequence?

   A. 16   C. 52
   B. 43   D. 75

8. What are the next three terms for the sequence, 4, 7, 10, 13, …?

   A. 14, 15, 16
   B. 15, 16, 17
   C. 16, 19, 22
   D. 19, 22, 25

9. What is the value of $-5n - 23$ when $n = -11$?

   A. -88
   B. -32
   C. 32
   D. 88

10. Diego will charge no more than $265 for 2 sets of brakes and an $85 labor fee. What is a possible value of one set of brakes?

   A. $200
   B. $190
   C. $100
   D. $90

11. $x \geq -8$ is the solution to which inequality?

   A. $-4x \leq 32$
   B. $-4x \leq -32$
   C. $-4x \geq -32$
   D. $4x \leq -32$

12. The retail price on a watch is twice the wholesale price plus $15. The price of the watch, after markup, is $465. What is the wholesale cost of the watch?

   A. $215   C. $450
   B. $225   D. $915

13. The prices of tickets for the theater are shown in the chart.

Number of tickets	1	2	3	4
Cost ($)	45	90	135	180

What sentence is the rule for the number of tickets (t) and the cost of the tickets (c)?

   A. $c = 2t$
   B. $2c = t$
   C. $45c = t$
   D. $c = 45t$

14. What are the two steps to find the solution for $\frac{a}{6} + 17 \leq 53$?

   A. Subtract 17, then multiply by 6 on both sides of the inequality.
   B. Add 17, then divide by 6 on both sides of the inequality.
   C. Multiply by 6, then subtract 17 on both sides of the inequality.
   D. Divide by 6, then add 17 on both sides of the inequality.

Check Your Understanding

On the following chart, circle the number of any item you answered incorrectly. Near each lesson title, you will see the pages you can review to learn the content covered in the question. Pay particular attention to reviewing lessons in which you missed half or more of the questions.

Chapter 5: Expressions and Equations	Procedural	Conceptual	Application/ Modeling/ Problem Solving
Expressions pp. 134–139	2, 9		
Solve One-Step Equations pp. 140–143		5	
Solve Two-Step Equations pp. 144–149	1		3, 4, 12
Solve One- and Two-Step Inequalities pp. 150–155	11	14	10
Identify Patterns pp. 156–161	7, 8	6, 13	

Linear Equations and Functions

Linear equations occur in multiple places in everyday life. The amount of water that comes out of a faucet fills a sink in a linear fashion. The cost of multiple items are represented linearly.

Linear equations can be used to model information as well as solve these models. You will learn how to solve systems of linear equations and learn what the solution means graphically, numerically, and algebraically.

Another tool used for modeling is scatter plots. Scatter plots can be used to represent information linearly, and they can also help predict more information that isn't given.

The Key Concepts you will study include:

Lesson 6.1: Linear Equations
Graph linear equations and determine if a relationship between two quantities is linear.

Lesson 6.2: Graphing Linear Equations
Graph linear equations written in multiple forms.

Lesson 6.3: Pairs of Linear Equations
Recognize that the solution to a pair of linear equations is the intersection point of those lines as well as solve the pair algebraically.

Lesson 6.4: Scatter Plots
Understand that a scatter plot represents data graphically, can be approximated by a function, and can be used to predict other data depending on the strength of the approximating function.

Lesson 6.5: Functions
Recognize functions graphically and algebraically.

Goal Setting

Before starting this chapter, set goals for your learning.

- What do you hope to learn about linear equations, solving linear equations, and scatter plots?

  _____

  _____

  _____

- Where in life have you encountered linear functions?

  _____

  _____

  _____

- How can you use the knowledge you have previously learned to help understand linear functions?

  _____

  _____

  _____

Linear Equations

KEY CONCEPT: A variable is something you are trying to measure. There are two kinds of variables, independent and dependent. An **independent variable** has a value that remains the same. That is, it is not affected by a **dependent variable**. A dependent variable is a value that depends on other factors.

An independent variable, such as how much time you spend preparing for a math test, is something you control. It affects the dependent variable, which is your test score. Or in another example, the kind of exercise you choose to do is an independent variable. It affects your heart rate, which is the dependent variable.

*Two variables have a **linear relationship** if their corresponding points lie on the same line in the coordinate plane. This means that as the independent variable increases (or decreases), the dependent variable increases (or decreases) proportionally.*

A coordinate plane is a two-dimensional surface on which you can plot points. Each point is located by a pair of x- and y-coordinates.

Plot the following coordinate pairs on the coordinate plane below.

(1, 2), (3, 3), (6, 0), (0, 7), (5, 8), (9, 10)

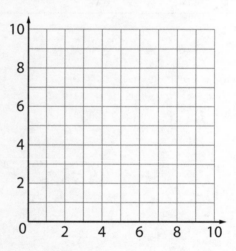

Linear Relationships

Linear relationships are a very important concept in mathematics and science. They also appear in many applications that you come across in everyday life—sometimes without even noticing them!

Imagine that you have saved enough money to buy a new cell phone. As part of your new cell phone plan, you are allowed to send up to 200 text messages per month for a flat fee of $5.00. For each text message over the 200-message limit, you pay an additional 20 cents, or $0.20.

In January, you sent 210 text messages, which means that you sent 10 extra messages. You can calculate the text-message charge.

Total text-message charge = $5.00 + ($0.20 × 10) = $5.00 + $2.00 = $7.00

flat fee   charge for extra messages

So, for 10 extra messages, your monthly text-message charge is $7.00.

In February, you sent 215 text messages, which means that you sent 15 extra messages. You can calculate the total text-message charge.

Total text-message charge = $5.00 + ($0.20 × 15) = $5.00 + $3.00 = $8.00

The following table displays the total text-messaging costs for two months.

Number of Extra Text Messages	Total Text-Message Charge
10	7
15	8

The same information can be expressed as two coordinate pairs, where the first number is the number of extra text messages, and the second number is the text-message charge.

(10, 7)   (15, 8)

You can plot these points on the coordinate plane, where the horizontal axis represents the number of extra text messages and the vertical axis represents the text-message charge in dollars.

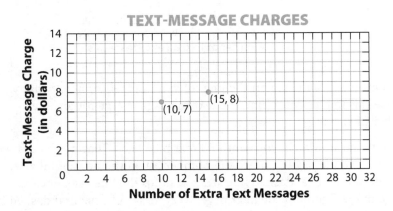

TEXT-MESSAGE CHARGES

Mathematics is used to solve a variety of everyday problems. Say, for example, that a credit counselor suggests that her clients spend no more than 25 percent of their monthly income on rent. She can use mathematics to model costs. This allows her to help her clients limit their apartment searches to affordable apartments. How much would a person need to earn each month to afford a monthly rent of $600.00?

Monthly Salary (in dollars)	Maximum Rent (25% of monthly salary in dollars)
1,000	250.00
1,250	312.50
1,500	375.00
1,750	437.50
2,000	500.00
2,250	562.50
2,500	625.00

Remember from geometry that if you have two points on a plane, you can draw a line that goes through both of them.

You can use this line to determine the text-message charge for any number of extra text messages *without* having to do the calculations.

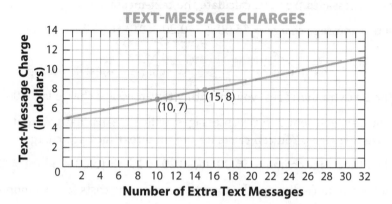

How do you know that the line you drew will give you the correct text-message charge for any number of extra text messages? You can check it against another piece of information that you already know. Remember that your cell-phone plan says that you pay $5.00 if you send 200 text messages or less, which means that there are zero extra text messages.

$$(0, 5)$$

You can express this information as a coordinate pair and plot it on the coordinate plane.

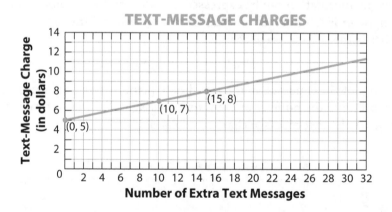

The point (0, 5) is indeed on the line, so you can feel confident that you can use this line to determine the text-message charge for any given number of extra text messages.

Let's say that in March, you sent 230 text messages. That means that you sent 30 extra messages. Locate the number 30 on the horizontal axis. Then, draw a vertical dashed line straight up to the line, as shown in orange in the figure on the next page. Next, draw a horizontal dashed line back to the vertical axis, as shown in the figure at the top of page 169.

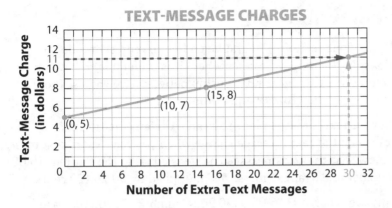

TEXT-MESSAGE CHARGES

The graph tells you that if you sent 30 extra text messages in March, your text-message charge is $11.00. This corresponds to the point (30, 11) on the coordinate plane.

Complete a Data Table

Use the plot to complete the data table.

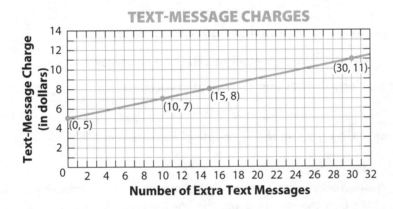

TEXT-MESSAGE CHARGES

The points, or coordinate pairs, are on the same line. This means that there is a linear relationship between the number of extra text messages and the text-message charge.

In mathematics, special terms are used to describe two values that have a linear relationship. The number of extra text messages is the independent variable. The text-message charge is the dependent variable. The text-message charge is the dependent variable because its value *depends* on the number of extra text messages. In other words, you need to know the number of extra text messages before you can determine the total text-message charge.

You can also identify whether there is a linear relationship between the number of extra text messages and the text-message charge. Look at the data table that you completed. A pattern in the data exists. Each time the number of extra text messages increases by 5, the text-message charge increases by one dollar. This indicates that there is a linear relationship between these two variables.

Number of Extra Text Messages	Text-Message Charge (in dollars)	Coordinate Pair
0	5	(0, 5)
5		
10	7	(10, 7)
15	8	(15, 8)
20		
25		
30	11	(30, 11)

Linear Equations

21st Century Skills
Critical Thinking and
Problem Solving

Applications of linear equations are widely encountered in business, social sciences, economics, science, and engineering fields. In business, for example, linear equations model total costs after sales tax is added to purchases. In engineering, linear equations show the relationship between speed and time in calculations of distance. In science, linear equations show relationships between animal behavior and the environment, such as how often crickets chirp in different outdoor temperatures. Linear equations model relationships between variables in social data, too, such as the frequency of cell-phone use and times of day. People use linear equations to predict events and understand and solve difficult problems more easily.

Choose an electronic device that you use often throughout the day. Record the times you use the device and how long each use occurs over a period of several days. Then graph the results. Describe any patterns that emerge from the data.

In the previous section, you graphed a line to represent a linear relationship. There may be cases where it is preferable to have an equation that represents the linear relationship between two values. These equations are called **linear equations**.

Recall your text-message plan. You can write an equation that will help you calculate the text-message charge when you have extra text messages. You know that the text-message charge is equal to the flat fee ($5.00) plus the number of extra text messages multiplied by $0.20 (20 cents.)

text-message charge = $5.00 + $0.20 × number of extra messages

flat fee   charge for extra messages

Before continuing, remove the dollar sign ($) and assign letter symbols to the values, or variables, that you are going to work with. This will make it easier to write the equation.

Use the letter y to identify the dependent variable.

$$y = \text{text-message charge}$$

Use the letter x to identify the independent variable.

$$x = \text{number of extra text messages}$$

After making substitutions, the equation becomes:

$$y = 5 + 0.2x$$

Rearrange the terms on the right side of the equation.

$$y = 0.2x + 5$$

This equation represents the line that you plotted in the previous section. Look at the how the equation and the line are related.

The y-axis corresponds to the values of text-message charge. Each charge is represented by the letter y. The number 5 is called the **y-intercept** of the line. It represents the point where the line crosses the y-axis (vertical axis).

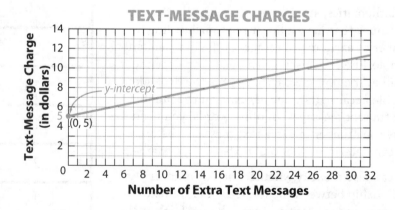

The number 0.2 is called the **slope** of the line. The slope of a line represents how steep it is. A horizontal line has a slope of zero. A line that slants upward to the right has a positive slope. A line that slants downward to the right has a negative slope.

You can calculate the slope of a line by comparing any two points on the line. For example, use the points (15, 8) and (30, 11), as shown on the graph below.

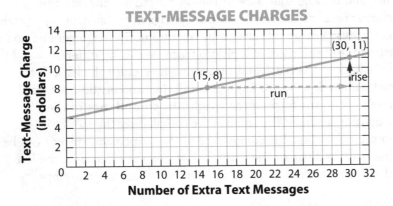

TEXT-MESSAGE CHARGES

The slope of a line is defined as the ratio between the rise and the run, or

$$slope = \frac{rise}{run}$$

The **rise** is the vertical distance between two coordinates. It is calculated by finding the difference between the vertical, or y-coordinates.

$$rise = 11 - 8 = 3$$

In this case, the rise is positive.

The **run** is the horizontal distance between two coordinates. It is calculated by finding the difference between the horizontal, or x-coordinates.

$$run = 30 - 15 = 15$$

So, the slope of this line is:

$$slope = \frac{3}{15} = 0.2$$

In general, the equation of a line, or linear equation, is written as

$$y = mx + b,$$

where m is the slope and b is the y-intercept. So, for this example

$$m = 0.2$$
$$b = 5$$

THINK ABOUT **MATH**

Directions: Use the graph to determine two coordinate points. Use the coordinate points to calculate the slope.

What is m, or the slope of the line?

_____

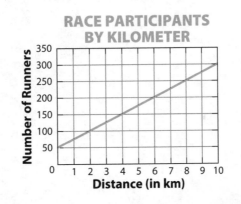

RACE PARTICIPANTS BY KILOMETER

The data for calculating the compliance of the suspension springs can be plotted as the slope of the graph. Follow these steps to calculate the slope.

1. Pick two points on the line. It will be easier to perform the calculations if you pick points whose values are easy to work with, such as (300, 1.2) and (500, 2).

2. The slope is the rise divided by the run. First, determine the rise by calculating the vertical distance between the points, which is equal to the difference between the y-coordinates of each point. So, the rise $= 2 - 1.2 = 0.8$.

3. Next, determine the run by calculating the horizontal distance between the points, which is equal to the difference between the x-coordinates of the each point. So, the run $= 500 - 300 = 200$.

4. Now, calculate the slope $= \frac{rise}{run} = \frac{0.8}{200} = 0.004$. The suspension springs have a low compliance, which means it takes a lot of weight to compress them.

Have you ever put something heavy in the back of a car or pickup truck and noticed that back of the car moved downward? This happens because the weight causes springs in the suspension system to compress. Imagine that you are an automotive engineer and you're testing the suspension of a pickup truck by placing different weights in the back of the truck and measuring how much the back of the truck moves downward. Engineers call the ratio of the downward movement to the force of the weight the **compliance** of the suspension springs. The smaller the compliance, the harder it is to compress a spring. The compliance of the suspension springs can be shown as the slope of the graph below. See the sidebar on the left titled Solve Linear Equations for steps in calculating the slope.

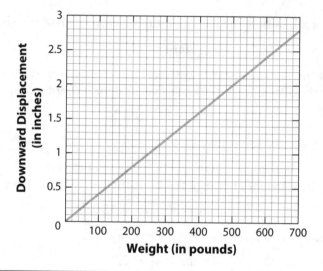

Vocabulary Review

Directions: Match each term to its example.

1. _____ slope

2. _____ *y*-intercept

3. _____ rise

4. _____ run

5. _____ linear relationship

6. _____ dependent variable

7. _____ independent variable

A. the horizontal distance between two points on a line

B. the value where a line crosses the vertical axis

C. a variable that remains the same, or unaffected by other variables

D. the vertical distance between two points on a line

E. the steepness of a line

F. a variable whose value depends on other factors

G. a relationship between two variables on a graph that can be shown by drawing a straight line between them

Skill Review

Directions: Indicate whether the lines on the plots below are linear or nonlinear. Explain your answers.

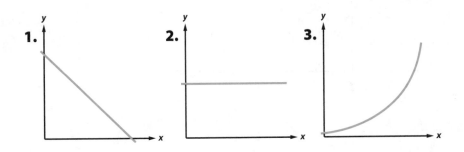

1. _____

_____

2. _____

_____

3. _____

_____

Earlier in this lesson, we used the graph of a linear equation to determine the value of the dependent variable if we were given the value of the independent variable. We can also determine the value of these variables from the linear equation itself.

In the example where we examined the relationship between the weight applied to the back of a pickup truck and the downward movement of the suspension springs, we can write the linear equation as

$$y = 0.004x$$

where *y* is the downward movement of the suspension in inches and *x* is the weight in pounds. If we want to calculate the downward movement, *y*, for a given weight, *x*, we plug in the value of *x* and solve for *y*. For example, if we put an object that weighs 200 pounds in the back of the pickup, the downward movement will be

$$y = 0.004 \times 200 = 0.8 \text{ inches}$$

Directions: Read the problem below. Then answer items 5–8 that follow.

> You are planning a trip to Canada, where the temperature is measured in degrees Celsius instead of degrees Fahrenheit. You find out that the linear equation that determines the temperature in degrees Fahrenheit from the temperature in degrees Celsius is
>
> $$F = 1.8C + 32,$$
>
> where F is the temperature in degrees Fahrenheit and C is the temperature in degrees Celsius. Before you go, you create a linear graph that you can use to convert degrees Celsius to degrees Fahrenheit quickly. This will help you know how to dress appropriately for outside temperatures.

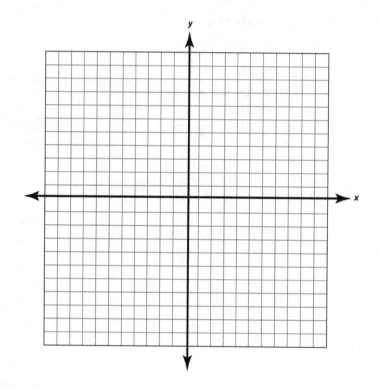

5. What is the independent variable?

_____

6. What is the dependent variable?

_____

7. What will the slope of the line be?

_____

8. What will the y-intercept be?

_____

Skill Practice

Directions: Use the flow chart below to write your answers.

1. Show the steps for determining the slope of a line that is drawn on a coordinate plane. Number each stage.

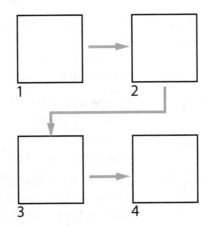

Directions: Use the graph below to choose the best answer for each question.

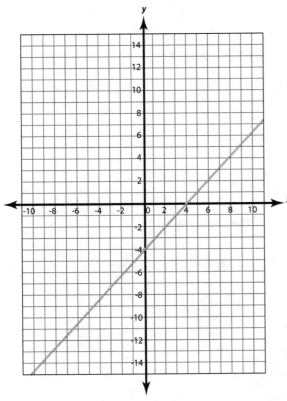

2. What is the slope of the line?

   A. 4   C. 1

   B. −4   D. −1

3. What is the y-intercept?

   A. 4   C. 0

   B. −4   D. 1

Directions: Use the graph below to choose the best answer.

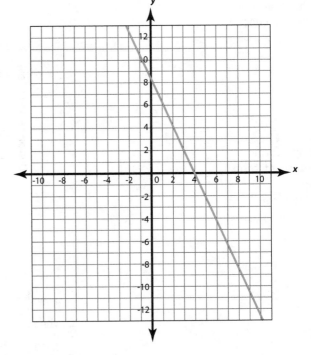

4. What is the equation for this line?

   A. $y = 2x + 8$

   B. $y = 2x - 8$

   C. $y = -2x - 8$

   D. $y = -2x + 8$

Graphing Linear Equations

Lesson Objectives

You will be able to

- Use the point-slope form to graph the equation of a line

- Use the slope-intercept form to graph the equation of a line

- Use the two-point form to graph the equation of a line

Skills

- **Core Skill:** Perform Operations

- **Core Skill:** Interpret Graphs and Functions

Vocabulary

intersect
point-slope form
slope-intercept form
subscript
two-point form

KEY CONCEPT: There are two ways to graph a linear equation. (1) If two coordinate pairs that lie on the line are known, then the graph of the line can be constructed, or (2) if one coordinate pair that lies on the line and the slope of the line are known, then the graph of the line can be constructed.

If an independent variable increases (or decreases), and the dependent variable increases (or decreases) proportionally, the two variables are linearly related. If the increase and decrease are disproportional, there is no linear relationship.

The data from the following tables were graphed. Examine the line of the equation in the first graph. The straight line shows a proportional relationship between the variables of Time and Distance.

Now examine the second graph. You can see that it is impossible to connect the points on a straight line, meaning there is no proportional, or linear, relationship between altitude and temperature in this particular example.

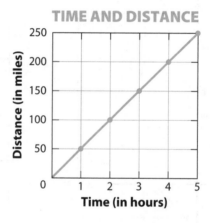

TIME AND DISTANCE

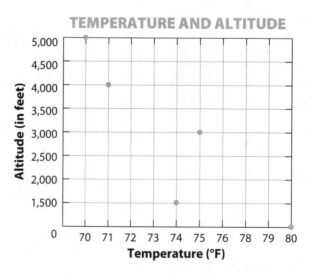

TEMPERATURE AND ALTITUDE

Graphing Linear Equations

To plot the graph of a linear equation, you need to know the line's slope and one point on the line. The slope, which is also called the **gradient**, is how steep a line on a plane is.

Say, for example, that the slope of a linear relationship between two variables is 5. Say also that the point (3, 4) lies on the line. How can you graph the line that represents the linear equation?

First, plot the point (3, 4) on a coordinate plane. Recall that a coordinate plane is a two-dimensional surface on which you plot points located by their x- and y-coordinates.

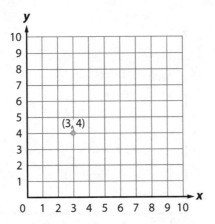

The slope is said to represent the "rise over run." Recall from the previous lesson that the rise is the change in y, which goes up or down. The run is the change in x, which goes left or right. The slope is the rise divided by run.

$$slope = \frac{rise}{run} = 5 = \frac{5}{1}$$
$$rise = 5$$
$$run = 1$$

In the figure to the right, start from the point (3, 4) and go to the point (4, 4). Notice that the run, or horizontal distance from point (3, 4) to point (4, 4), equals 1.

Then, go to the point (4, 9). Notice that the rise, or vertical distance from the point (4, 4), equals 5.

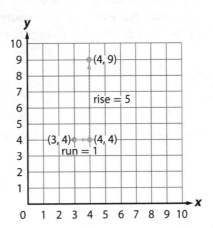

You decided to save money so that you can buy a new smartphone. You start with 20 dollars in your bank account, and each month, you add an additional 30 dollars. Suppose you wanted to graph the line that represents the total amount in your bank account each month.

You know one point on the line. It is (0, 20), where the first number is the number of months—you start at zero months, and the second number is the total amount in your bank account—you start at 20 dollars.

You can also determine the slope of the line. You save 30 dollars each month, which means that the total amount in your bank account *increases* by 30 dollars for every month that goes by. Think of these values as the rise and run.

$$\text{slope} = \frac{\text{rise}}{\text{run}} = \frac{30}{1} \text{ month}$$

Use the point-slope formula to graph this line: $y - y_1 = m(x - x_1)$

Use the variable x to represent the number of months, and the variable y to represent the total amount in your bank account. Let the horizontal axis be the number of months and the vertical axis be the total amount in your bank account.

Use the graph of the line to predict how much money you will have in seven months.

THINK ABOUT MATH

Directions: Two sets of data have a linear relationship. A graph of this relationship includes the points (5, 35) and (10, 45). What is the slope of the line that joins these points?

_____

Point-Slope Form

Recall that you must have two points to define a line. Now that you have plotted (3, 4) and (4, 9), you can draw the line that connects them. This line is the graph of the linear equation.

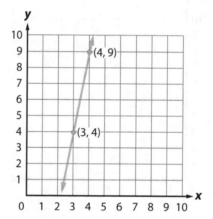

How can you determine the equation of this line? You can use the **point-slope form** of the equation of a line:

$$y - y_1 = m(x - x_1)$$

In the point-slope form of the equation of a line, x_1 is the x-coordinate of the known point of the line, whose coordinates you were initially given. You were also given y_1, the y-coordinate of that point, and m is the slope of the line. So, for this example, you have:

$$x_1 = 3$$
$$y_1 = 4$$
$$m = 5$$

Putting these numbers into the point-slope form, you get:

$$y - 4 = 5(x - 3)$$
$$y - 4 = 5x - 15$$
$$y = 5x - 11$$

Slope-Intercept Form

There is a special case of point-slope form for the equation of a line. If the x-coordinate of the point provided is zero, then that point is the y-intercept of the line. Recall that the y-intercept is where the line **intersects**, or crosses, the y-axis. In this case, the form of the equation is called the **slope-intercept form**, and it is represented by the equation

$$y = mx + b,$$

where m is the slope of the line and b is the y-intercept. The coordinate pair representing the y-intercept is $(0, b)$.

This equation is very useful when you are given the equation of a line and want to plot it. For example, say you are given the following equation:

$$y = 2x - 7$$

You can use the same procedure for graphing the line as you learned for the point-slope form. The slope, $m = 2$, and the y-intercept, $b = -7$, which means that the graph of the line crosses the y-axis at the point $(0, -7)$, as shown below.

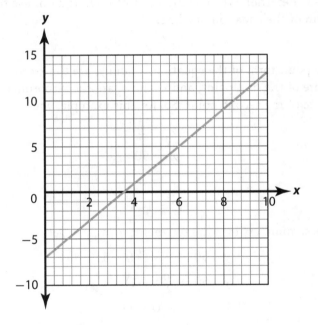

21st-Century Skills
Critical Thinking and Problem Solving

Graphs of linear equations are very useful in predicting trends that follow a linear pattern. For example, an economist determines that the increase in the cost of living in a particular city has followed a linear pattern for the past thirty years. She may use this trend to reasonably predict the cost of living in that city over the next several years. City managers can use this data to plan the city's budgets in preparation for rising costs.

Think about a regular cost you have each month. Perhaps it is purchases you make at an online music or bookstore. Or perhaps you pay a cell phone bill each month. Record the payments you have made over the last six months. Graph the data to determine if your costs are increasing, decreasing, or staying about the same each month. How can the results help you plan for future months?

Two-Point Form

You have already learned that if you know two points on a line, you can plot the graph of the linear equation. Let's say that you are given two points (1, 50) and (3, 100), and you draw the line that goes through both of them, as shown below.

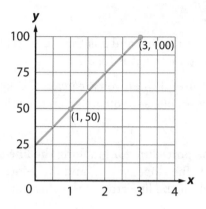

How can you determine the equation of this line? You can use the **two-point form** of the equation of a line:

$$y - y_1 = \frac{y_2 - y_1}{x_2 - x_1}(x - x_1)$$

In the two-point form of the equation, x_1 and y_1 are the x-coordinate and y-coordinate of the first point, and x_2 and y_2 are the x-coordinate and y-coordinate of the second point. So, for this example, you have:

$$x_1 = 1$$
$$y_1 = 50$$
$$x_2 = 3$$
$$y_2 = 100$$

Putting these numbers into the two-point form, you get

$$y - 50 = \frac{100 - 50}{3 - 1}(x - 1)$$
$$y - 50 = \frac{50}{2}(x - 1)$$
$$y - 50 = 25(x - 1)$$
$$y - 50 = 25x - 25$$
$$y = 25x + 25$$

If you compare this two-point form of the equation to the slope-intercept form, you see that the slope is 25 and the y-intercept is 25. Check the graph of the line in the figure above to see if the slope and y-intercept are correct.

Vocabulary Review

Directions: Use one of the words below to complete each sentence.

intersects   point-slope form   slope-intercept form
subscript   two-point form

1. When you know one point on a line and the line's slope, you can use the
_____ to find the equation of the line.

2. The y-intercept of a line is the point where the line's graph
_____ the y-axis.

3. When you use the form $y = mx + b$ to find the equation of a line,
you use the _____.

4. When you have two points on a coordinate plane, you can use the
_____ to find the equation of a line.

5. A letter, figure, or number is a _____ if it is set below
the line.

Skill Review

Directions: Match each form for the equation of a line with when to
use it.

1. _____ slope-intercept form

2. _____ point-slope form

3. _____ two-point form

A. Use this to determine the equation of a line when you know its
y-intercept and slope.

B. Use this to determine the equation of a line when you know two points
on a line.

C. Use this to determine the equation of a line when you know the slope
of a line and a point on the line.

Recall that in point-slope
form, you can use a point
(x_1, y_1) and a given slope
m to find the equation of
a straight line using the
formula $y - y_1 = m(x - x_1)$.
It is important not to
be distracted by the
subscript, the number
written below and to the
right of each coordinate
in the coordinate pair.
Remember that the
subscript only indicates
the coordinate pair you
are given that you will
use to find the equation.

For example, say you
want to find the equation
of the straight line that
has a slope (m) of 3 and
passes through the point
(2, 4). You would use
the point-slope form
$y - y_1 = m(x - x_1)$ to find
the equation:

$y - y_1 = m(x - x_1)$
$y - 4 = 3(x - 2)$
$y - 4 = 3x - 4$
$y = 3x - 2$

4. You're on your first trip to New York City and you take a taxi from the airport to your hotel. At the taxi stand, you see a sign that provides information about taxi fares into the city.

> **NYC TAXI SERVICE FARES**
>
> $2.00 Airport charge
> $5.00 Starting fare
>
> PLUS
>
> $2.00 per mile

Plot the line that represents the total fare, where the total fare values are on the vertical axis and the number of miles is on the horizontal axis. Use the variable y to represent the total fare and use the variable x to represent the number of miles. Plot the graph for $x = 0$ to 10 miles. What is the equation for this line? Express the equation using slope-intercept form.

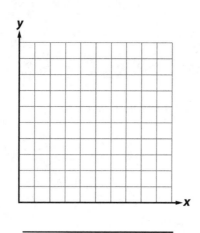

_____

Skill Practice

Directions: Answer the following.

1. A classmate tells you that it is possible to graph a line given only its slope. Explain whether this statement is correct.

_____

_____

_____

Directions: Use the graph below to answer the questions.

You recently took advantage of a zero-interest credit card offer to purchase a new mountain bike that costs $700. Each month you pay either $150 or the balance, whichever amount is lower. The graph of the line is plotted, where the vertical axis is the balance owed and the horizontal axis is the number of months.

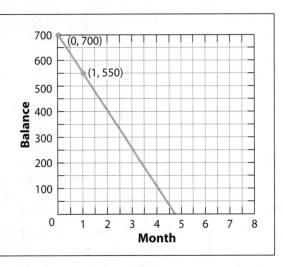

2. How many months it will take to pay off the credit card? Refer to the graph to explain your answer.

_____

_____

_____

3 Calculate the slope of the graph. Explain what the slope represents. Use the terms *rise* and *run* in your explanation.

_____

_____

_____

4 What is the *y*-intercept of the graph? What does it represent?

_____

Pairs of Linear Equations

KEY CONCEPT: A pair of linear equations forms a system of two simultaneous linear equations. The solution to a system of two linear equations in two variables corresponds to a point of intersection of their graphs, because points of intersection satisfy both equations simultaneously.

In previous lessons, you graphed lines in separate graphs to represent specific linear relationships and represented the linear relationship between two variables as a linear equation.

Two variables have a linear relationship if their corresponding points lie on the same line in the coordinate plane, the two-dimensional surface on which you can plot points by a pair of x- and y-coordinates. Every pair of (x, y) coordinate points on the line is a solution to the linear equation that represents the line. These values make the relationship between the left and right sides of the equation sign true. These points satisfy the equation.

Simultaneous Linear Equations

Two equations that are satisfied by the same set of variables form a system of two simultaneous linear equations. The word **simultaneous** means "occurring at the same time." A **system of simultaneous linear equations** is a collection of linear equations. To solve a system, look for the values of the variables that make all of the equations true simultaneously. If you graph each of the linear equations in the set, the point where they intersect is a single common solution, meaning that the point lies on each of the lines. Simultaneous linear equations can be mathematical objects or models of the real world.

A Real-World Model

You may find yourself making a decision that requires you to compare two financial alternatives. Often, these alternatives can be represented by a linear relationship. For example, imagine you are purchasing a text-messaging plan. You have two options:

Option 1: Pay a flat fee of $5.00 for the first 200 text messages, and $0.20 for each message thereafter.

Option 2: Pay a flat fee of $20.00 for the first 200 text messages, and $0.05 cents for each message thereafter.

Which option is better for you? How can you tell? You can use the descriptions of each option to write the equation of a line.

Option 1: $y = 0.20x + 5$
Option 2: $y = 0.05x + 20$

To tell which option is better, you calculate costs for a given number of messages. You select the number 210 to represent the number of text messages sent in a given month, meaning you exceeded the text-message allowance by 10 messages.

Option 1: $y = 0.20x + 5$
Total text-message charge $= \$5.00 + (\$0.20 \times 10) = \$5.00 + \$2.00 = \$7.00$

Option 2: $y = 0.05x + 20$
Total text-message charge $= \$20.00 + (\$0.05 \times 10) = \$20.00 + \$0.50$
$= \$20.50$

Option 1 appears to be a better plan if you send 210 messages. However, you decide to test the plans using a different number of text messages. This time, you decide to send a total of 400 text messages. That's 200 more messages than your plan includes without an extra cost. You can write the number 200 in the place of x in the equation for each option and solve each equation to determine the cost.

Option 1: $y = 0.20x + 5$
Total text-message charge $= \$5.00 + (\$0.20 \times 200) = \$5.00 + \$40.00$
$= \$45.00$

Option 2: $y = 0.05x + 20$
Total text-message charge $= \$20.00 + (\$0.05 \times 200) = \$10.00 + \$20.00$
$= \$30.00$

If you send lots of text messages per month, Option 1 will cost \$15.00 more than Option 2. It seems that the number of text messages you anticipate sending will determine which option is more economical for you.

You can look at these alternative options as a system of related equations.

Now suppose that you decide that a graph of the alternative cost plans will help you compare them more closely. You can use lines on the graph to determine the text-message charge for any number of text messages *without* doing calculations. You know that a point on a line satisfies its linear equation. So, a common point of intersection will tell you how many sent messages will result in equal costs.

The following table displays the information for the total text-messaging costs for the two plans you are considering.

Number of Messages Above 200	Option 1 Cost ($)	Option 2 Cost ($)
0	5.00	20.00
100	25.00	25.00
150	35.00	27.50
200	45.00	30.00

The **substitution method** provides a direct way to solve a pair of linear equations without graphing. In the substitution method, you solve one of the equations for one of the variables and then substitute that solution into the other equation. For example, given $3a + 4b = 5$ and $a - b = 3$, you can solve for a in terms of b in the second equation. Then, you can substitute the value of a into the first equation to solve for b. Finally, you can substitute the value found for b into an equation to solve for a.

1. $3a + 4b = 5$

2. $a - b = 3$
   $a = 3 + b$

3. $3(3 + b) + 4b = 5$
   $9 + 3b + 4b = 5$
   $7b + 9 = 5$
   $7b = -4$
   $b = -\frac{4}{7}$

4. $a - (-\frac{4}{7}) = 3$
   $a + \frac{4}{7} = 3$
   $a = 3 - \frac{4}{7}$
   $a = 2\frac{3}{7}$

Use the substitution method to find the solution to this linear pair:

1. $3a + 4b = 5$

2. $a + b = 3$

To draw their lines, plot two points for each plan on the coordinate plane. Plot the points where the horizontal axis represents the number of text messages exceeding each plan's 200-message plan, and the vertical axis represents the cost in dollars. Note the point of intersection of the lines.

You can use the lines to determine the text-message charge for any number of text messages greater than 200 *without* having to do the calculations. The lines intersect at the point (100, 25). This point is the solution of both lines, and it is called the **equilibrium point**. It shows where both plans cost the same amount, or $25.00. If you follow the lines upward from this point, the difference in costs becomes more obvious.

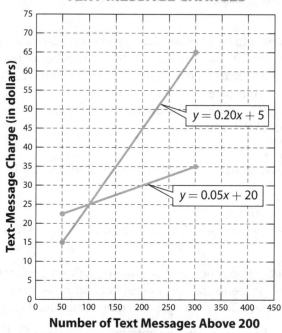

TEXT-MESSAGE CHARGES

$y = 0.20x + 5$

$y = 0.05x + 20$

Number of Text Messages Above 200

THINK ABOUT MATH

Directions: Say that you planned to share your phone and the phone cost with a family member. Each of you is likely to send 250 messages per month. If you pay half of the phone bill, what would you pay under each plan?

Option 1: _____

Option 2: _____

Combining Methods to Solve Pairs of Linear Equations

Earlier in the lesson, you graphed lines to represent a system of simultaneous linear relationships. There may be cases where it is preferable to solve the system without graphing.

The **addition method** is based on the principle that adding the same value to each side of an equation does not change the equality of that relationship. The addition method is also known as the elimination method because it uses a process to **eliminate**, or cancel out, variables to solve an equation. Consider this pair of equations:

$2x + 3y = -12$
$2x - 3y = 4$

If you add the equations, the terms with y cancel out, or add to 0:

$2x + 3y + 2x - 3y = -12 + 4$
$$4x = -8$$
$$x = -2$$

To find y, use the substitution method you applied earlier in the lesson. Substitute $x = -2$ in the first equation:

$2(-2) + 3y = -12$
$-4 + 3y = -12$
$3y = -8$
$y = -\frac{8}{3}$

Now substitute $(-2, -\frac{8}{3})$ in the second equation to check your answer:

$2(-2) - 3(-\frac{8}{3}) = 4$
$-4 + 8 = 4$
$4 = 4$

Core Skill
Solve Simple Equations by Inspection

Not all pairs of equations have a single solution. Consider the equations $3p + 2q = 4$ and $3p + 2q = 5$. There is no single solution because $3p + 2q$ cannot simultaneously equal both 4 and 5. If you graphed these equations, you would draw parallel lines. Their slopes would be the same, but their y-intercepts would differ.

Now consider the equations $3p + 2q = 4$ and $6p + 4q = 8$. Notice that you can divide both equations by the common factor 2. So, these equations are equivalent. Consequently, there are an infinite number of solutions. If you graphed the equations, you would see they are identical lines, meaning their slopes and intercepts are the same.

Examine the following pair of equations: $x + y = 1$ and $3x + 3y = 3$. How would you describe the lines that would appear on a graph of these equations?

MATH LINK

To determine if using the addition method is an appropriate way to eliminate a variable, think first about whether you can manipulate one or both of the equations to eliminate one of the variables. Consider the following example:

$3a + 2b = 4$
$4a + 2b = 5$

To eliminate a variable, you can multiply one equation by -1.

$-(3a + 2b) = -4$
$-3a - 2b = -4$
$4a + 2b - 3a - 2b = 5 - 4$
$4a - 3a = 1$
$a = 1$

Vocabulary Review

Directions: Match each term to its example.

1. _____ addition method

2. _____ eliminate

3. _____ equilibrium point

4. _____ simultaneous

5. _____ substitution method

6. _____ system of simultaneous linear equations

A. cancel out

B. keeps the equality relationship between both sides of an equation

C. occurring at the same time

D. a set of equations that are satisfied by the same set of variables

E. the coordinate pair that represents a common solution to two simultaneous equations in two variables

F. a technique for solving simultaneous equations by first solving for one variable in terms of the other

Skill Review

Directions: Indicate whether the systems of two linear equations have 0, 1, or an infinite number of solutions. Explain your answers.

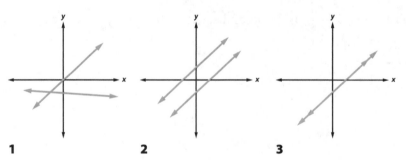

1   2   3

1. _____

2. _____

3. _____

Directions: Read the problem. Then answer the questions that follow.

> A portion of $100,000 ($x$) was invested with a return of 3 percent after one year. The remainder of the investment (y) was invested at a return of 1 percent. The total return on the investment was $1,800.

4. What is the equation that shows the investment yield?

_____

5. What is the equation that shows the way the $100,000 was split?

_____

6. How much money was invested at a 3 percent rate of return?

_____

7. How much money was invested at a 1 percent rate of return?

_____

Skill Practice

Directions: Solve each pair of linear equations.

1. $x + y = 20$
$x + y = 40$

   A. $x = 120; y = -100$
   B. $x = 30; y = -10$
   C. infinite solutions
   D. no solution

3. $5a - 3b = 12$
$3a - 5b = 14$

   A. $a = \frac{17}{8}; b = \frac{29}{8}$
   B. $a = \frac{29}{8}; b = \frac{17}{8}$
   C. $a = \frac{-9}{8}; b = \frac{17}{8}$
   D. $a = \frac{9}{8}; b = -\frac{17}{8}$

2. $5a - 4b = 12$
$3a + 4b = 20$

   A. $a = 2; b = 4$
   B. $a = 4; b = 2$
   C. $a = 4; b = -2$
   D. $a = -2; b = 4$

Directions: Read the problem. Find the solution.

4. Part of a $20,000 investment was invested at 6 percent return rate. The remainder of the investment was invested at 4 percent return rate. The total return was $1,000. Write a pair of equations that can be used to answer how much was invested at each rate.

_____

_____

Scatter Plots

Lesson Objectives

You will be able to

- Describe the information that a trend line provides about two correlated variables

- Describe various aspects of the correlation between two variables

Skills

- **Core Skill:** Represent Real-World Problems

- **Core Skill:** Interpret Data Displays

Vocabulary

cluster
correlation
outlier
scatter plot
trend line

KEY CONCEPT: We can use the concept of correlation to describe the relationship between two variables that generally follow a linear pattern but cannot be described by a linear equation. Plotting data on a scatter plot and constructing a trend line can determine the strength and direction of the correlation between such variables.

Up to now, we have used equations to define linear relationships between two variables. For example, say you begin saving money to buy a new set of noise-canceling headphones. You start with $50 and then save $20 per month. You can plot the data as shown below.

Month	Total Savings (in dollars)
0	50
1	70
2	90
3	110
4	130
5	150
6	170

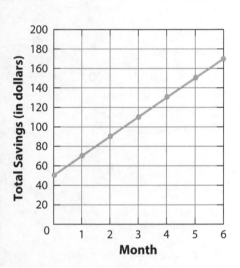

You can write the equation for this line:

$$y = 20x + 50$$

where y is the total savings and x is the month. If you look carefully at the graph, you will notice that all of the points lie exactly on a line that connects them.

Scatter Plots

Sometimes you will encounter two sets of variables that result in a graph whose data points do not lie *exactly* on a line but may be close to a line. Let's look at an example.

The owner of a snow cone stand records the average number of snow cones that she sells, based on the forecasted high temperature of the day. The data are shown on page 191 in a table and a plot of each coordinate pair.

The plot makes it easy to see that the data points do not all lie on the same line. Instead, they appear to be scattered. Therefore, graphs such as these are called **scatter plots**.

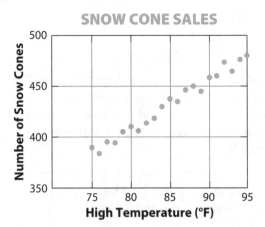

SNOW CONE SALES

In this scatter plot, all of the data points are so close you can draw a line right through them. Such a line is called a **trend line**, and it graphically shows the relationship between daily high temperatures and the number of snow cones sold.

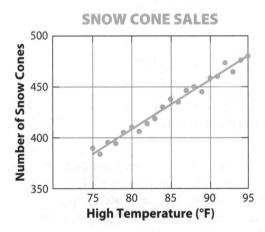

SNOW CONE SALES

Daily High Temperature (°F)	Average Number of Snow Cones Sold
75	389
76	384
77	395
78	394
79	405
80	410
81	406
82	413
83	418
84	430
85	437
86	435
87	446
88	450
89	445
90	459
91	460
92	473
93	465
94	476
95	481

THINK ABOUT MATH

Directions: Draw an approximate trend line for the following scatter plot that shows the results of a study on the daily amount of exercise and the amount of sleep that a person gets.

EXERCISE AND SLEEP

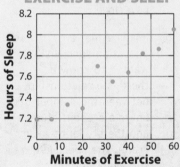

If you look at all of the data points on the scatter plot, you see that some are above the trend line, some are below the trend line, and some lie almost exactly on the trend line.

Now, imagine the values of the vertical distance between all of the points and the trend line. The trend line is the line for which the overall distance between the points and the trend line is minimized.

The data for the daily high temperature and the number of snow cones that are sold follow a close linear pattern, or **correlation**. Therefore, you can say that there is a **linear correlation** between the variables of daily high temperature and the number of snow cones sold.

Scatter plots are one means of recording data, but the data have value only if you analyze them and look for relationships between variables. The positions of points on a scatter plot tell a story. It is a story of relationship. Drawing a trend line through the points is a quick and effective means of determining whether a relationship exists and how close it is.

Imagine that an education specialist wanted to know if there is a relationship between the number of hours a student studies and the student's exam grade. The specialist surveyed a group of math students to collect data and recorded the data in scatter plot.

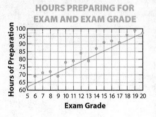

Examine how the points fall along or near the trend line. These points and the trend line tell a story about the relationship between two variables. How would you describe that relationship?

Linear Correlations

When relationships exist between two variables, those relationships can be described as positive or negative. Let's consider positive correlations first.

Positive and Negative Correlations

Recall the scatter plot that showed the relationship between the daily high temperature and the number of snow cones sold. As temperatures increased, so did sales. Since both variables increased together, the slope of the trend line is positive—therefore, you can say there is a **positive correlation** between the two variables.

It is possible for two variables to be negatively correlated, also. Consider the relationship between the variables of the number of hours a student watches television and the student's overall grade point average.

The following scatter plot shows the data collected from a student survey. Look closely at the points and the trend line.

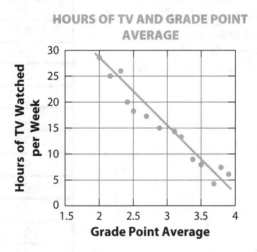

A relationship between the two variables is plain to see. The points lie close to or on the line. However, the trend line for this set of data has a negative slope, meaning that as the number of hours of watching television increases, the grade point average decreases. The scatter plot shows a **negative correlation** between the variables.

Nonlinear Correlations

You can investigate data that approximately follow a linear pattern, revealing a linear correlation between the variables. It is also possible to investigate and display data that approximately follow nonlinear patterns. Here are two examples.

Quadratic Model: Sizes of Rocks Moved by River Currents

A hydrologist works in a national park. She wants to know how the speed of a river's current determines the sizes of rocks that the river's current can carry downstream. She collects and graphs the data.

Speed of the River Current (m/sec)	Diameter of Rocks Carried Downstream (mm)
0.5	10
1.0	19
1.5	40
2.0	72
2.5	96
3.0	140
3.5	175

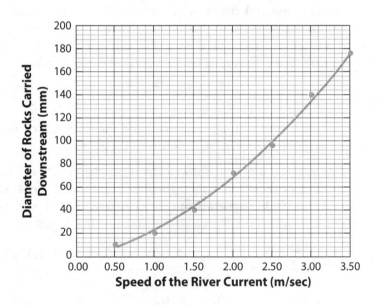

Core Skills
Interpret Data Displays

Imagine as an example that a web-based advertising company used two different styles of advertisements to promote the sales of their online products. The company wanted to know which advertisement style resulted in more sales. So, they collected data for each and plotted the data in a scatter plot.

You can draw an approximate trend line for each scatter plot. Explain what the shape of each trend line tells you about the relationship between the number of websites that uses each advertising style and sales.

Just as with data that closely follow a linear pattern, we can draw a trend line. Here, the trend line is a **quadratic curve**, which means that the size of the rocks that the river can carry downstream is proportional to the speed of the river raised to the second power. *Quadratic* is an algebraic term that refers to the square of an unknown quality. Thus, the size of the rock will increase much more quickly than the speed of the river.

Exponential or Power Model: Bacteria Response to an Antibiotic Treatment

A medical laboratory assistant tests how bacteria respond to a new type of antibiotic. He wants to know how quickly the antibiotic kills bacteria once it is added to a bacterial culture. He records and graphs the results.

Time (hours)	Number of Bacteria in the Culture
0	10,000
2	7,075
4	4,018
6	2,818
8	2,333
10	1,071
12	908

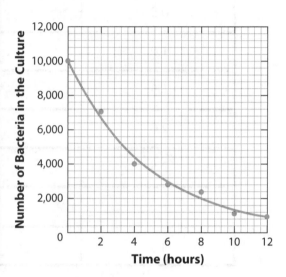

Here, the trend line is a "decaying" **exponential curve**, which means that the trend line decreases more rapidly at first and decreases less slowly as time increases. It is called an exponential curve because, when written as a formula, the variable is not in the base of the function ($x2$) but in the exponent ($2x$). The exponential curve can also be called a power curve.

Correlation Strength

In addition to the direction of a correlation (positive or negative) and the form of a correlation (linear or nonlinear), you can also describe a correlation in terms of its strength.

Earlier in this lesson, you looked at a scatter plot that showed the relationship between the daily high temperature and the number of snow cones sold. The data points were very close to the trend line. Therefore, we say that a **strong correlation** exists between the two variables. Let's look at a different example to distinguish between a strong and a weak correlation.

A physical education specialist surveys students to find the relationship between students' height and their weight.

The data points are not as close to the trend line as they were for the snow cone example above. Therefore, we say that a **weak correlation** exists between the two variables.

You may also encounter data that does not seem to follow any pattern. For example, you decide to do a survey to see if there is any relationship between students' grade point average and their height.

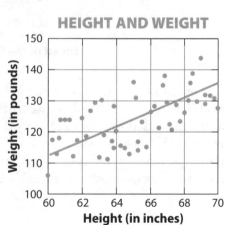

HEIGHT AND WEIGHT

The data in the scatter plot show **no correlation**. Therefore, there is no trend line. There is no relationship between the variables of grade point average and height.

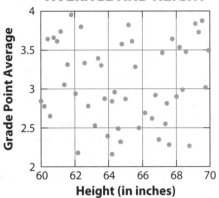

GRADE POINT AVERAGE AND HEIGHT

Outliers and Clusters

Sometimes the points in a scatter plot either do not appear where you expect them to be, or they do not appear uniformly along the trend line.

A business manager sells active wear for young adults. He studies purchasing records to determine if there is a relationship between the height and weight of his top customers. He recorded the data in a scatter plot like the one below.

The data are positively correlated, but there are several interesting features of this scatter plot. Notice the point that is clearly far away from the trend line. Such a point is called an **outlier**. Also, notice that the data points fall within two distinct groups, or **clusters**.

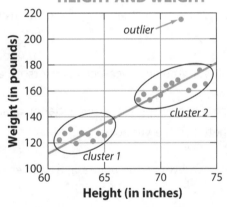

HEIGHT AND WEIGHT

MATH LINK

Sometimes outliers and clusters appear on a scatter plot as the result of an error in recording the data. However, this may not always be true, so it is important to investigate outliers and clusters carefully to be sure that they are the result of errors.

In the manager's scatter plot, the outlier is not an error. It represents a young man who is a freshman linebacker for a high school football team. The young man's weight is greater than a typical person of his height.

Also, the clusters in the scatter plot are not errors. After double-checking, the manager was surprised to learn that none of his customers has a height between 65.5 and 68 inches.

THINK ABOUT **MATH**

Directions: Look again at the scatter plot of the weight and height of the young adult customers. Although there are clusters and an outlier in the scatter plot, these hardly affect the trend line. Identify the pattern the data points follow and explain the pattern's meaning. In other words, explain how height and weight change together.

_____

_____

_____

_____

Directions: Choose a term to complete each sentence.

cluster   correlation   linear correlation   negative correlation   nonlinear correlation
outlier   positive correlation   scatter plot   trend line

1. If the value of one variable increases while the value of the second variable decreases, there is a _____ between the variables.

2. If the value of one variable increases as the value of the other variable increases, there is a _____ between the variables.

3. A _____ is a visual display of the relationship between two variables.

4. If two variables follow a clearly recognizable pattern, then there is a _____ between the two variables.

5. If points in a scatter plot increase or decrease proportionally, then there is a _____ between the variables that they represent.

6. A _____ is located further away from the trend line than the other points in a scatter plot.

7. A _____ is a grouping together of points on a scatter plot.

8. If the trend line on a scatter plot is exponential or quadratic, then there is a _____ between the variables.

9. The line or curve around which the points in a scatter plot appear is called a _____.

Directions: Read the text. Then complete the activity.

1. How many outliers appear in the following scatter plot? Explain how you determined which points were outliers.

_____

_____

_____

_____

_____

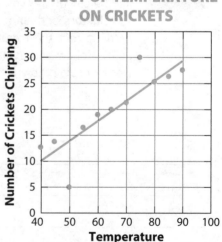

EFFECT OF TEMPERATURE ON CRICKETS

2. The following scatter plot shows the length and the width, measured in centimeters, of clamshells found along a beach.

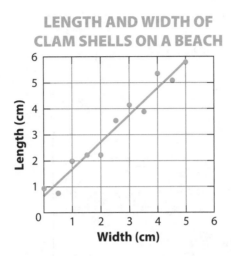

**LENGTH AND WIDTH OF
CLAM SHELLS ON A BEACH**

Describe the correlation between the variables of length and width.

_____

_____

3. The data in the following table show the number of online posts to a music website and the number of recordings customers download daily. Create a scatter plot of these data and draw the trend line that you feel best fits the data.

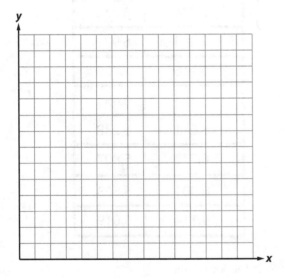

Social Media Posts	Music Downloads
10,000	900
20,000	1,300
25,000	1,800
30,000	1,500
35,000	1,600
40,000	2,100
50,000	2,200
55,000	2,700
60,000	2,500
65,000	2,600
70,000	2,800
75,000	2,900
80,000	2,800
90,000	3,000
95,000	3,100

Directions: Read the text. Then complete the activity.

1. Match the descriptions of correlations with their corresponding scatter plots.

   _____ **A.** strong negative correlation

   _____ **B.** no correlation

   _____ **C.** weak positive correlation

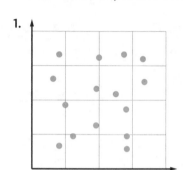
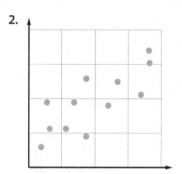
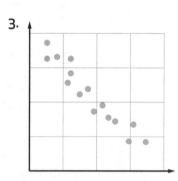

2. A health official collected data on the number of flu shots clinics gave in a month and the number of patients clinics treated for flu in the same month. The official put the data in the table on the right. Use the data to build a scatter plot. Draw a trend line through the data points.

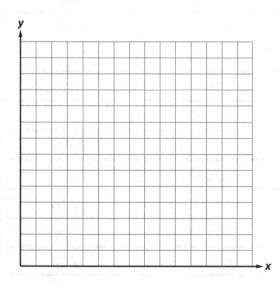

Flu Shots Given	People with Flu
7,846	6,832
7,994	6,971
8,216	6,726
8,364	6,453
8,631	6,238
8,772	5,807
8,901	5,563
9,114	5,019

Use the trend line to identify the kind of relationship between the variables.

_____

_____

_____

3. A Reading Specialist is doing research on the age and reading levels of children and adolescents. He collects the data shown in the scatter plot below. Identify any clusters in the data by circling them. Also, identify any outliers. Summarize the meaning of each outlier. Provide a possible explanation of what the outliers might indicate.

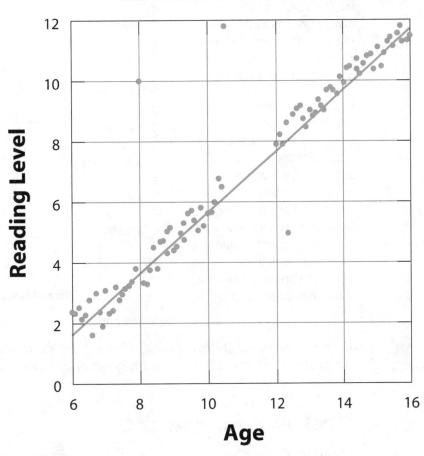

READING LEVEL AND AGE

_____

_____

_____

_____

_____

_____

_____

_____

Functions

Lesson Objectives

You will be able to
- Identify a function
- Determine whether an equation represents a function

Skills

- **Core Skill:** Build Lines of Reasoning
- **Core Skill:** Interpret Graphs and Functions

Vocabulary

function
input
linear function
nonlinear function
output
vertical line test

KEY CONCEPT: You can look at a function as a set of instructions that tells you what to do with the input, or values you put in. The result of the instructions is called the output. Functions are equations that provide only one output for each input.

In a previous lesson, you learned about linear equations and some of the ways that they are used in different applications. For example, imagine you are a builder. You can use a linear equation to determine the cost of construction materials (in dollars) for any given size house (in square feet). The linear equation is

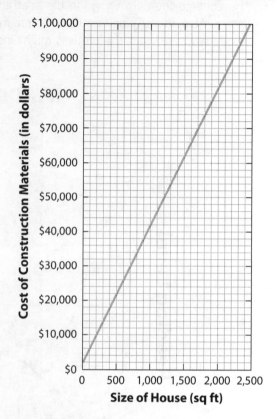

$$y = 40x + 2,000$$

where x is the size of the house (in square feet) and y is the cost of the construction materials. Examine the graph of this linear equation.

What Is a Function?

A **function** is a mathematical equation that has two variables, an **input** variable that goes into the equation and an **output** variable that results from the input. A rule to remember is that for each input, a function has only one output.

You can think of a function as a set of instructions that tells you how to take the input and use it to calculate the output. Mathematicians often describe a function as a "black box," like the one in the drawing below. Think of it as a computer that takes in the input value, follows the instructions on what to do with the input value, and produces an output value. In the drawing, the input is labeled *x* and the output is labeled *y*.

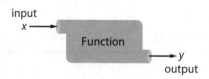

Is It a Function?

Remember that the definition of a function states that for each input, there is only one output. This definition will help you determine whether or not an equation represents a function.

At the beginning of this lesson, we looked at a linear equation that related the size of a house to the cost of the construction materials required to build it:

$$y = 40x + 2000$$

where x is the size of the house (in square feet) and y is the cost of the construction materials. If you select a value for the size of the house, you can calculate the cost of the materials. Say the value of $x = 1,500$ square feet. Now you can calculate the value of y.

$$y = (40 \times 1500) + 2000 = 60,000 + 2,000 = 62,000$$

So, the cost of the construction materials to build a 1,500 square foot house is $62,000. The equation provides only one possible answer for each size of house. This is also evident from the graph of the linear equation. If you select any value of x, there is only one possible value of y. So, for each input, or house size, there is only one output, or cost of materials. Therefore, the linear equation $y = 40x + 2,000$ is a function.

Now, consider the following equation

$$y = \sqrt{x}$$

If $x = 25$, then the value of y is 5, since $5 \times 5 = 25$. But the value of y could also be –5, since $-5 \times -5 = 25$. So, for this equation, there are two possible answers. In other words, for any input x, there are two possible values for the output, y. This violates the definition of a function, which states that there is only one output for each input. Therefore, this equation is **not** a function.

Look at the graph of this equation, which shows the two values of y when $x = 25$. Note that because these points have the same x coordinate, they lie on the same vertical line.

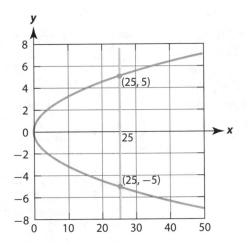

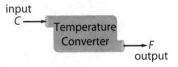

You can do a **vertical line test** to determine if an equation is a function. Examine the line of the graph in the figure on the previous page. Then, observe the vertical line $x = 25$. Notice that it intersects the graph of the equation at two points. This indicates that there are two output values, 5 and -5, for one input value, 25. Because the vertical line crosses the graph at more than one point, the equation is not a function. It violates the definition of a function that there can be only one output for each input.

Function Categories

Functions can be classified into two broad categories—linear functions and nonlinear functions. You have been working with linear functions throughout this chapter. A **linear function** has the form

$$y = mx + b$$

where m is the slope of the line and b is the y-intercept.

If a function does not have this form, then it is a **nonlinear function**. Let's look at two examples to illustrate the difference between linear and nonlinear functions.

Perimeter of a Square

The perimeter of a square is the sum of the lengths of all its sides. The equation that represents the perimeter, P, is

$$P = s + s + s + s$$
$$P = 4s$$

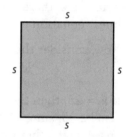

where s is the length of each side. Because it has the form $y = mx + b$, this equation represents a linear function.

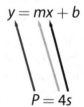

The slope is 4 and the y-intercept is 0. Look at the graph of this function. Note that all of the points lie on a straight line, thus confirming that the formula for finding perimeter, $P = 4s$, is a linear function.

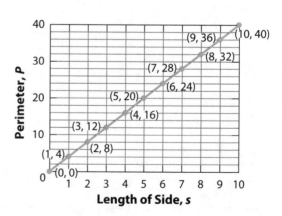

Area of a Square

The area of a square is the product of the length of two of its sides. The equation that represents the area, A, is

$$A = s \times s$$
$$A = s^2$$

You can graph this function, as you see. Note that the points do not lie on a straight line. Thus the equation $A = s^2$ is a nonlinear function. More specifically, it is a quadratic function.

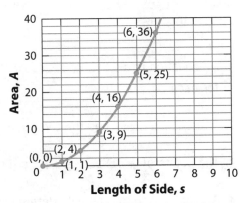

Core Skill
Interpret Graphs and Functions

When you are asked to graph a function or interpret the graph of a function, it is always helpful to look at its equation, if it is available. You can look at the form of the function to determine whether it is linear or nonlinear. Remember, if the function has the form

$$y = mx + b$$

then the function is linear—if it does not have that form, then the function is nonlinear.

THINK ABOUT MATH

Directions: Label each graph as a Function or Not a Function.

A. _____

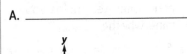

C. _____

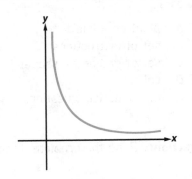

B. _____

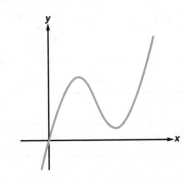

D. _____

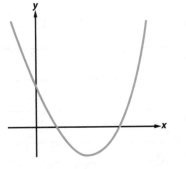

MATH LINK

The circumference of a circle, C, is given by the equation $C = 2\pi r$, and the area of a circle, A, is given by the equation $A = \pi r^2$. Explain how you can determine whether each of these functions is linear or nonlinear, without performing any calculations or plotting their graphs.

Hint: You can compare each of these equations with the form of a linear function. Remember that the equation of a linear function is $y = mx + b$, where y is the dependent variable and x is the independent variable. If you can rewrite one or both of the equations in the form $y = mx + b$, it is a linear function. If you cannot rewrite one or either of the equations in the form $y = mx + b$, it is not a linear function.

Vocabulary Review

Directions: Use the following terms to complete each sentence. Note that some terms may be used in more than one sentence.

function   input   linear function   nonlinear function   output
vertical line test

1. An equation is a _____ if there is only one _____ for each _____.

2. A _____ has the form $y = mx + b$.

3. The points of a _____ are not all on a straight line.

4. A _____ can help determine whether the graph of an equation is a _____ or not.

Skill Review

Directions: Circle the best answer to each question.

1. You can think of a function as a
   A. point on a line
   B. set of instructions
   C. set of geometric shapes
   D. collection of mathematical questions

2. The vertical line test helps you determine whether
   A. a function is linear or nonlinear
   B. a graph represents a function
   C. a function is a "black box"
   D. an equation is graphed correctly

Directions: Read the problem. Then follow the directions.

3. Imagine you are programming a computer. Write the step-by-step "instructions" for the following nonlinear function. Include the words *input* and *output* in your instructions.

$$y = x^2 + 2x + 1$$

_____

_____

_____

_____

_____

_____

_____

Skill Practice

Directions: Read the problem. Then follow the directions.

1. Label each graph of an equation as a Linear Function or a Nonlinear Function.

A. _____   B. _____   C. _____

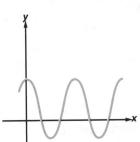   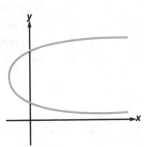   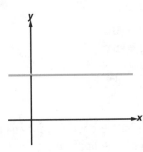

2. The potential energy (PE) of an object is related to its vertical position. In other words, it is related to the object's height above the ground. The potential energy (in kilojoules) of an object is given by the equation

$$PE = \frac{mgh}{1000}$$

where m is the mass of the object, g is a constant related to gravity, or 9.8, and h is the height above the ground.

Use the data in the table to calculate the PE of someone with a mass of 90 kg traveling upward in an elevator 10 meters at a time.

Mass (in kg)	g	Height (in m)	PE (in kj)
90	9.8	0	
90	9.8	10	
90	9.8	20	
90	9.8	30	
90	9.8	40	
90	9.8	50	

3. Draw a graph to plot the PE values you calculated in the previous problem.

4. Does the graph that you drew represent a function? If so, is it linear or nonlinear? Explain each of your answers.

_____

_____

Directions: Choose the best answer to each question.

1. Which of the following equations is NOT a linear equation?

   A. $y = 0$
   B. $y = 5x + 2$
   C. $y = x + 5$
   D. $y = \frac{1}{x} + 4$

2. The monthly cost of your cell phone bill, C, is given by the linear equation $C = 0.1 D + 5$, where D is the number of calls made in a month. If this line is plotted with C on the vertical axis and D on the horizontal axis, which of the following is the correct slope, m, and y-intercept, b?

   A. $m = 0.15, b = 5$
   B. $m = 1, b = 5$
   C. $m = 5, b = 0.1$
   D. $m = 0.1, b = 5$

4. A line with a positive slope
   A. slants upward to the left.
   B. is vertical.
   C. is horizontal.
   D. slants upward to the right.

5. What is the slope of the line that passes through the points (5, 9) and (–1, –3)?

   A. $\frac{1}{2}$
   B. -2
   C. 2
   D. $\frac{1}{2}$

6. Use the substitution method to solve for the value of each variable. Then insert the values you find to check your work.

   $2x - 2y = -2$
   $4x + y = 16$

3. Which of the following data sets follows a linear pattern?

A.

Time (a.m.)	6	7	8	9	10	11
Temperature	60	64	68	70	74	80

B.

Time (a.m.)	6	7	8	9	10	11
Temperature	60	64	68	72	76	80

C.

Time (a.m.)	6	7	8	9	10	11
Temperature	60	60	61	61	62	62

D.

Time (a.m.)	6	7	8	9	10	11
Temperature	60	62	64	65	69	72

Review

7. What is the slope of the line shown in the graph?

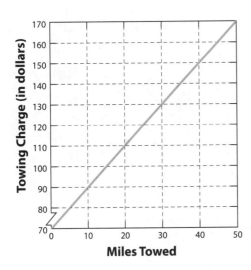

Miles Towed

A. 9

B. 2

C. 70

D. −2

8. Which graph represents a function?

A.

B.

C.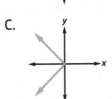

D.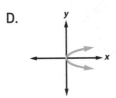

9. Which of the following sets of data represent a nonlinear function?

A.

Week	1	2	3	4	5	6
Account Balance	10	20	30	40	50	60

B.

Week	1	2	3	4	5	6
Account Balance	20	40	60	80	100	120

C.

Week	1	2	3	4	5	6
Account Balance	1	4	9	16	25	36

D.

Week	1	2	3	4	5	6
Account Balance	10	10	10	10	10	10

10. Which of the following equations represents a linear function?

   A. the area of a square, $A = s^2$
   B. the area of circle, $A = \pi r^2$
   C. the circumference of a circle, $C = 2\pi r$
   D. the volume of a cube, $V = s^3$

11. The data shown in the scatter plot are

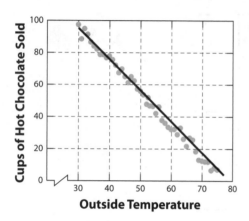

   A. linearly and positively correlated.
   B. linearly and negatively correlated.
   C. not correlated.
   D. nonlinearly correlated.

12. Which trend line best fits the correlated data shown in the scatter plot?

   A. A
   B. B
   C. C
   D. D

Check Your Understanding

On the following chart, circle the number of any item that you answered incorrectly. Near each lesson title, you will see the pages you can review to learn the content covered in the question. Pay particular attention to reviewing those lessons in which you missed half or more of the questions.

Chapter 6: Linear Equations and Functions	Procedural	Conceptual	Application/ Modeling/ Problem Solving
Linear Equations pp. 166–175	1, 3	4	2
Graphing Linear Equations pp. 176–183	5		7
Pairs of Linear Equations pp. 184–189			6
Scatter Plots pp. 190–199		11, 12	
Functions pp. 200–205	10	8	9

UNIT 3

More Number Sense and Operations

Ratios, Proportions, and Percents

Many real-life problems involve ratios, proportions, or percents. You may be given a ratio of water to orange juice concentrate to make a pitcher of orange juice, or you may need to use a proportion to find the dimensions of a room from a blueprint. Percents are very common in daily life. Advertisements indicate percent discounts. Taxes are based on percents of income, property value, or amount of sale.

In this chapter you will see and use the important relationships among fractions, decimals, and percents.

The Key Concepts you will study include:

Lesson 7.1: Ratios and Rates
Understand how to write ratios to understand the meaning of a unit rate.

Lesson 7.2: Unit Rates and Proportional Relationships
Understand how to convert to unit rates and determine if two ratios are in proportion.

Lesson 7.3: Solve Proportions
Understand how to use proportions to solve problems.

Lesson 7.4: Introduction to Percents
Percents, like decimals and fractions, represent part of a whole.

Lesson 7.5: Solve Percent Problems
Decimals, fractions, and proportions can be used to solve percent problems.

Lesson 7.6: Use Percents in the Real World
Simple interest can be calculated using a formula and percents.

Goal Setting

Before starting this chapter, set goals for your learning. Think about the places and times you can use ratios, proportions, or percents at home and at work.

Percents abound in real-world situations. The best way to master percents is to take all of the situations in which you see them on a daily basis and work them out. Use a notebook to record all of the times you use percents in the course of a week.

- Where did you find percents used most often?

  _____

  _____

  _____

- Did you find any percents that were worked out incorrectly? For example, was a discount actually 25% when the sign said 15% off?

  _____

  _____

  _____

- Would you prefer that a sign say $10 dollars or 10% off? In what situation is it better to have a money amount off? In what situation is it better to have a percent off?

  _____

  _____

  _____

Ratios and Rates

Lesson Objectives

You will be able to
• Understand and write ratios
• Understand and find unit rates and unit prices

Skills

• **Core Skill:** Understand Ratio Concepts
• **Core Skill:** Use Ratio and Rate Reasoning

Vocabulary

paraphrase
rate
ratio
unit price
unit rate

KEY CONCEPT: Understand how to write ratios to understand the meaning of a unit rate.

Write each fraction in lowest terms.

1. $\frac{15}{25}$ 2. $\frac{14}{28}$ 3. $\frac{8}{12}$ 4. $\frac{9}{15}$

5. Draw a picture that shows that 3 out of 4 equal parts are shaded.

Understand Ratios

A **ratio** is a comparison of two numbers. Use a ratio to compare the numbers 2 and 7. There are three different ways to write the ratio:

Use the word *to*.	Use a colon (:).	Write a fraction.
2 to 7	2:7	$\frac{2}{7}$

Think of a ratio as a fraction. Just as the fractions $\frac{2}{7}$ and $\frac{7}{2}$ are not the same, the ratios "2 to 7" and "7 to 2" are not the same. So the order of the two numbers in a ratio is important. A ratio is usually reduced to lowest terms.

Example 1 Write a Ratio

In Roberto's family, there are 2 boys and 4 girls. What is the ratio of boys to girls in Roberto's family?

Step 1 Write the ratio in one of these three ways. Follow the order in the phrase *ratio of boys to girls* for the order of numbers in the ratio. *Boys* is first, so the number representing the number of boys goes to the left of the word *to* or colon or in the numerator of the fraction. The number representing the number of *girls* will be on the right of the word *to* or colon or in the denominator of the fraction.

Use the word *to*.	Use a colon.	Write a fraction.
2 boys to 4 girls	2 boys:4 girls	$\frac{2 \text{ boys}}{4 \text{ girls}}$
2 to 4	2:4	$\frac{2}{4}$

Step 2 Reduce to lowest terms.

Use the word *to*.	Use a colon.	Write a fraction.
1 boy to 2 girls	1 boy:2 girls	$\frac{1 \text{ boy}}{2 \text{ girls}}$
1 to 2	1:2	$\frac{1}{2}$

Step 3 Make a statement about the ratio.
The ratio of boys to girls in Roberto's family is 1 to 2 or 1:2 or $\frac{1}{2}$.

RESTATE OR PARAPHRASE INFORMATION

When you read, you may come across a long or complicated sentence or paragraph. To make sure you have understood what you just read, go back and read it again slowly. **Paraphrase**, or use your own words, to restate the information.

Look at each key word or important idea and imagine that you are explaining it to someone unfamiliar with the information. Then use clear, simple language to restate the ideas in an understandable way.

Read the following paragraph. Then read the two paraphrases below, and choose the best one.

> Ratios and rates are two ways of comparing quantities. A ratio is used when two items have different measures of the same thing. For example, a scale drawing might use a scale of 1 inch:5 feet. Because feet and inches are both units of length, this is a ratio. A rate compares two items that are of a different nature. Miles per hour is a rate, because miles are a unit of length and hours are a unit of time. There are many rates that people use every day: hourly wages ($15 per hour) and fuel efficiency (20 miles per gallon) are both examples of rates. There are three ways to express ratios or rates: 3:5, $\frac{3}{5}$, or 3 to 5.

1. You can compare with ratios and rates. You use ratios if you are comparing two things that are of the same type, as in a scale drawing. You use rates if you are comparing two things that are different, such as miles per hour, miles per gallon, or dollars per hour. You can write a ratio or rate as 3:5, $\frac{3}{5}$, or 3 to 5.

2. Both ratios and rates compare things. Rates are common in everyday life. Ratios and rates can be written in different ways.

Paraphrase 1 restates the information in the original paragraph in different words. Paraphrase 2 leaves out several details, such as the difference between a ratio and a rate and the different ways to write them. Paraphrase 1 is the better choice.

Core Skill
Understand Ratio Concepts

Often, you can demonstrate your understanding of a problem and then solve it by restating the information in a way that breaks it down into simpler terms.

Some ratio problems are a good place to practice this skill. Look at Example 1 on the previous page. You are told the number of boys and girls in Roberto's family and asked to find the ratio of boys to girls. What if, instead, you were asked to find the ratio of boys to all of the children? You can restate the problem as two simpler problems: "First, I need to find the total number of children by adding the number of boys and girls. Then I can find the ratio of boys to total children."

Select a partner with whom you can practice calculating ratios. Each of you write several situations that would require finding a ratio. Swap the situations that each of you wrote. Then restate each problem before solving.

Math offers ways to exercise your creative skills. For instance, in Example 2, you are asked to find the ratio of boys to girls in a class of 30 that has 14 boys. However, what if the ratio of boys to total students was $\frac{7}{15}$?

This fraction of boys tells you two things: 1) the fraction of boys, which was given, and 2) the fraction of girls. Since $\frac{7}{15}$ is boys, the rest is girls. The question asks for the ratio of boys to girls, not girls to total students, which is what $\frac{8}{15}$ represents. But 7 parts of the whole are boys, and 8 parts of the whole are girls. So the ratio is 7:8. Without even knowing the number of students you can determine the appropriate ratio.

In a notebook, write down the ratio of boys to total students if the ratio of boys to girls is 3:7.

MATH
L I N K

A ratio is a comparison of two different quantities. A ratio can be written in fraction form, but a ratio cannot be written as a decimal.

Some ratio problems require calculation before you write the ratio.

Example 2 Solve a Multistep Ratio Problem

In a class of 30 students, there are 14 boys. What is the ratio of boys to girls in the class?

Step 1 Calculate the number of girls in the class because this number is not given.
30 students − 14 boys = 16 girls

Step 2 Write the ratio of boys to girls.
$\frac{14}{16}$

Step 3 Reduce the ratio to lowest terms.
$\frac{14}{16} = \frac{7}{8}$

Step 4 Make a statement about the ratio.
The ratio of boys to girls in the class is 7 to 8.

To write a ratio comparing two numbers, do the following:

- Use the word *to*, use a colon, or write a fraction.
- Remember that the order of the numbers in a ratio is important.
- Always reduce the ratio to lowest terms.

THINK ABOUT MATH

$\sqrt{x}$

Directions: Write each ratio in two other ways.

1. 6 to 7   **2.** 1:50   **3.** $\frac{10}{19}$

Directions: Write a ratio for each situation described.

4. To make orange juice, combine 1 can of concentrate with 3 cans of water. Write the ratio of concentrate to water.

5. A ski resort advertised that it had 23 snow days during January. Write a ratio of snow days to days without snow (there are 31 days in January).

Understand Unit Rates

A **rate** shows a relationship between two quantities measured in different units. Rates are a commonly used type of ratio. A **unit rate** is the rate for one unit of a quantity. To find a unit rate, divide the two numbers in the ratio.

Example 3 Find a Unit Rate

Huy and Kim used 16 gallons of gasoline to drive 500 miles. Find the unit rate of gas mileage (miles per gallon).

Step 1 Write the ratio of miles to gallons: $\frac{500 \text{ miles}}{16 \text{ gallon}}$.

Note that this is a rate, but not a unit rate.

Step 2 Divide so that you have a unit rate.
500 miles ÷ 16 gallons = 31.25 miles to 1 gallon = 31.25 miles per gallon

A **unit price** is the price per unit of an item. Divide to find a unit price.

Example 4 Find a Unit Price

Soap is priced at $1.19 for 4 bars. Find the unit price.

Step 1 Write the ratio of price to number of bars: $\frac{\$1.19}{4 \text{ bars}}$.

Step 2 Divide so that you have a unit price:
$1.19 ÷ 4 bars = $0.2975 for 1 bar.
The unit price is about $0.30 per bar.

To find a unit rate or a unit price, do the following:

• Write a ratio of the two quantities.

• Divide the numerator by the denominator to find the unit rate or unit price.

THINK ABOUT **MATH**

Directions: Write the unit rate for each situation described.

1. Julietta and Won drove 135 miles and used 6 gallons of gasoline. Find the gas mileage rate of their car (miles per gallon).

2. This week, Ravi worked 20 hours and earned $130. What is his earning rate (dollars per hour)?

Directions: Write each unit price. Round your answer to the nearest cent.

3. 5 pounds of potatoes for $1.49   4. 12 eggs for $1.09

Directions: Complete each sentence with the correct word.

rate   ratio   unit price   unit rate

1. A _____ shows a relationship between two quantities measured in different units.

2. To find a _____, divide the price by the number of units.

3. To find the _____ of gas mileage, you find the number of miles driven per gallon.

4. A _____ can be written with the word *to*, with a colon (:), or as a fraction.

Skill Review

Directions: Match each description of a ratio on the left with its unit rate on the right.

1. _____ A store is selling 3 pairs of pants for $72.

2. _____ Camille walked 2.5 miles in a half-hour.

3. _____ Francis made $21 in 1.5 hours.

4. _____ The school club received $36 in dues for every 4 students that joined.

5. _____ Sasha baked 147 cupcakes in 7 hours.

A. 9:1

B. 24:1

C. 21:1

D. 5:1

E. 14:1

Skill Practice

Directions: Choose the best answer to each question.

1. What is the unit price of shampoo if a 15-ounce bottle costs $2.79?

   A. about $0.19 per ounce
   B. about $0.29 per ounce
   C. about $1.90 per ounce
   D. about $5.38 per ounce

3. In a local election, Marshall received 582 votes, and Emilio received 366 votes. Which shows a ratio for the number of votes for Emilio to the number of votes for Marshall?

   A. 366 to 948
   B. 97:61
   C. $\frac{61}{97}$
   D. 588 to 948

2. Which statement has the same meaning as "the ratio of males to females is 1 to 2"?

   A. Half are male, and half are female.
   B. There are twice as many males as females.
   C. Two out of three are female.
   D. One out of three is female.

4. Briana ran 6 miles in 50 minutes. How many miles per minute did she run?

   A. 0.083
   B. 0.12
   C. 1.2
   D. 8.3

Unit Rates and Proportional Relationships

KEY CONCEPT: A unit rate is a special example of a ratio. When it is expressed in fractional form, the denominator equals one. When expressed verbally, a ratio is an example of a unit rate if the second value being compared is one.

Recall that a ratio is comparison of two values. For example, in a college mathematics class, 20 students may be math majors and 8 students may be majors of a different kind. You can say that the ratio of math majors to nonmath majors is "20 to 8." You can also represent this ratio as a fraction, $\frac{20}{8}$. When you rewrite the fraction in simplest form, the ratio becomes $\frac{5}{2}$. This ratio means that for every five students who are math majors, there are two students who are majoring in a different field.

$$20 \text{ to } 8 = \frac{20}{8} = \frac{5}{2}$$

What Is a Unit Rate?

Recall that a unit rate is a ratio that is used to compare two different types of quantities. You encounter unit rates every day. You may see a road sign indicating that the speed limit is 30 miles per hour $\left(\frac{30 \text{ miles}}{\text{hour}}\right)$. You may receive a flyer from your local grocery store advertising tomatoes on sale for 69 cents per pound $\left(\frac{69¢}{\text{lb}}\right)$. You may read a financial news report that reports the stock price of a software company is 15 dollars per share.

Not all ratios you hear or read about represent unit rates. What are some examples of ratios that are *not* unit rates? Perhaps you hear a commercial that declares that 8 out of every 10 dentists surveyed recommends a particular brand of toothpaste. Maybe a digital music website announces a special deal encouraging you to buy four songs for $3.00. Or a newspaper cites a statistic that 89 out of every 100 people nationwide passed a newly developed test for getting a driver's license.

Let's express all of these examples in fractional form, as shown in the figure below. Notice that unit rates are always ratios with a denominator of 1. In other words, ratios that are unit rates have 1 unit, such as 1 hour, 1 pound, or 1 share, in their denominators.

Unit Rates	Not Unit Rates
$\dfrac{30 \text{ miles}}{1 \text{ hour}}$	$\dfrac{8 \text{ dentists recommend}}{10 \text{ dentists surveyed}}$
$=1 \leftarrow \dfrac{69 \text{ cents}}{1 \text{ pound}}$	$\neq 1 \leftarrow \dfrac{3 \text{ dollars}}{4 \text{ songs}}$
$\dfrac{15 \text{ dollars}}{1 \text{ share}}$	$\dfrac{89 \text{ passed test}}{100 \text{ took test}}$

Converting Ratios to Unit Rates

Recall the special digital music deal you read about. For $3.00, you can buy four songs. What if you wanted to convert this ratio into a unit rate? In other words, if you took advantage of this deal, how much would you pay for each song?

To convert this ratio into a unit rate, the denominator must be 1. So, divide both the numerator and denominator in $\frac{3}{4}$ by the denominator, 4.

$$\frac{3 \text{ dollars}}{4 \text{ songs}} = \frac{\frac{3}{4} \text{ dollars}}{\frac{4}{4} \text{ songs}} = \frac{0.75 \text{ dollars}}{1 \text{ song}} = \frac{0.75 \text{ dollars}}{\text{song}}$$

The unit rate is $0.75, or 75 cents per song. Notice that when you divide the numerator and denominator by the denominator, you use only the number, not the units (songs) associated with the denominator.

THINK ABOUT MATH

Directions: Let's look at a more complicated ratio, where there is a fraction in the denominator. Your friend tells you that he can run three miles in a half hour. You want to calculate his running speed in miles per hour. The ratio is:

$$\frac{3 \text{ miles}}{\frac{1}{2} \text{ hour}}$$

Convert this ratio into a unit rate.

_____

_____

If you know a unit rate, you can use it to generate an infinite number of value pairs. Consider high-definition television (HDTV) screens, for example. These televisions have different shapes from earlier television models. A new HDTV has an "aspect ratio" of 16:9, or $\frac{16}{9}$. This means that in terms of unit rates, the ratio between the width and height of the screen is $\frac{16 \text{ inches}}{9 \text{ inches}} = \frac{1.78 \text{ inches}}{1 \text{ inch}}$.

You can use this information to find the height of common HDTV screens. Complete the chart below. To find each screen height, divide the width by 1.78. Round your answers to the nearest whole number.

Screen Width (inches)	Calculation	Screen Height (inches)
16	16 ÷ 1.78	9
41	41 ÷ 1.78	23
44.5		
46.2		

When you want to write a ratio that is expressed verbally, be sure that you put the numbers in the correct places. Let's look at an example.

You and a friend are working on a landscaping project. Working as a team, you know that in 3 hours, you can fill 28 wheelbarrows full of soil and move them to another area. To plan for a future landscaping project, you want to determine the unit rate for this ratio, or wheelbarrows per hour.

You perform the following calculations:

$$\frac{3}{28} = \frac{\frac{3}{28}}{\frac{28}{28}} =$$

$$\frac{0.11}{1} = \frac{0.11 \text{ wheelbarrows}}{\text{hour}}$$

This number does not seem right to you. You know that you can fill more than $\frac{1}{10}$ of a wheelbarrow in one hour! How can you determine where you went wrong? Always include units in your calculations, unlike the calculation above. If you had indicated units, you would have seen that you had set up the ratio incorrectly.

$$\frac{3 \text{ hours}}{28 \text{ wheelbarrows}}$$

Using units will help you determine if your calculations make sense. Recalculate the unit rate in wheelbarrows per hour.

Proportional Relationships

Recall that a ratio is a comparison of two values. You can extend this comparison to sets of value pairs that have the same ratio. These sets of data represent a **proportional relationship**.

Let's look at an example. Suppose you are working as a construction planner and need to order cement mix, which is priced by the pound. You receive the pricing chart on the right from a cement company.

Weight (lbs.)	Cost ($)
1,000	350
2,000	700
3,000	1,050
4,000	1,040
5,000	1,750

You notice that each time the weight increases by 1,000 pounds, the price increases proportionally by $350. So, this means that ratio of cost to weight is:

$$\frac{\$350}{1,000 \text{ lbs}}$$

... and the unit rate is:

$350/1,000$ lbs. $= \$0.35/\text{lb}$

So, the cement costs 35 cents per pound ($0.35/lb). Now look at a graph of cost and weight data.

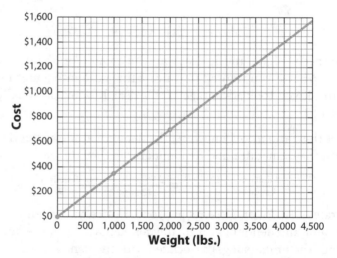

Now, you can determine the equation of this line. The *y*-intercept on the *y*-axis is zero, since zero pounds of cement would cost $0.00. To calculate the slope you can use the points (2,000, 700) and (3,000, 1,050). Recall the formula for slope:

slope = rise/run

So, $\frac{\$1,050 - \$700}{3,000 \text{ lbs.} - 1,000 \text{ lbs.}} = \frac{\$350}{1,000 \text{ lbs.}} = \$0.35/\text{lb}$.

Notice that the slope of the line equals the unit rate that you calculated previously. Therefore, the equation of the line is:

$$y = 0.35x$$

This is an example of a proportional relationship. It is a linear equation that has the form:

$$y = kx$$

The *y*-intercept is zero, and the slope is represented by *k*, which is called the **constant of proportionality**.

Apply Proportional Relationships

You can use the equation for a proportional relationship, $y = kx$, to solve different types of problems using unit rates. Sometimes, it is helpful to rearrange this equation into the form $\frac{y}{x} = k$. Remember that k is the unit rate, so this will help you keep track of the units for y and x.

For example, if a unit rate is expressed in kilometers per hour, you can use the equation $\frac{y}{x} = k$ to tell you the units for y and x.

$$\frac{y}{x} = k$$

$$\frac{y}{x} = \frac{\text{kilometers}}{\text{hours}}$$

Then you know that the units for x are hours and the units for y are kilometers.

Say your family purchased a pool for the summer. The pool holds 12,000 gallons of water when it is full. As you fill up the pool, you record the time and the number of gallons marked on the water meter outside your house.

Time (hours)	Gallons
0.5	600
1.5	1,800
2.5	3,000
3.5	4,200
4.5	5,400

You can write and solve a linear equation for a proportional relationship to determine the rate of the water flow, or unit rate, in gallons per hour. If you let x represent the time and y represent the number of gallons in the pool, the linear equation for the proportional relationship is: $y = 1{,}200x$

It may help you find unit rates if you remember that the constant of proportionality in a proportional relationship equals the slope of the line that represents the proportional relationship. So, the constant of proportionality equals the unit rate.

Look at this example. An airline uses two different models of aircraft to make the long flight between New York City and Istanbul, Turkey. The time and distance data for the first aircraft is recorded in a table below. The time and distance data for the second aircraft is recorded in the graph on the bottom of this page.

Time (hours)	Distance (miles)
1	450
3	1,350
5	2,250
7	3,150
9	4,050
11	4,950

The speed, or unit rate, of the first aircraft is $\frac{450 \text{ miles}}{\text{hour}}$.

The speed, or unit rate, of the second aircraft is $\frac{500 \text{ miles}}{\text{hr.}}$.

The second aircraft travels to Istanbul faster than the first.

THINK ABOUT MATH

Directions: Use the equation $y = 1{,}200x$ to determine how long it will take the pool to fill up to its 12,000 gallon capacity.

_____

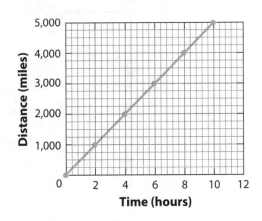

Vocabulary Review

Directions: Use the following terms to complete each sentence. Some words may be used more than once.

constant of proportionality   proportional relationship

1. A _____ exists between two variables if the ratio between them is always the same.

2. The _____ is the value of the ratio between two variables that are proportionally related.

Skill Review

Directions: Choose the best answer to each question.

1. Which of the following is an example of a unit rate?

   A. A car travels 100 miles in two hours.
   B. You spent $1.98 for two pounds of potatoes.
   C. An Olympic sprinter can run 100 meters in 10.5 seconds.
   D. A taxi driver charges $1.75 per mile.

2. You went to a farmer's market where they were offering 3 pints of strawberries for $8.25. What is the unit rate in dollars per pint?

   A. 2.75
   B. 0.36
   C. 3.00
   D. 8.25

3. If a pound of sugar costs $0.65 and you have $3.00, how many pounds of sugar can you buy?

   A. 1.95
   B. 3.65
   C. 4.62
   D. 0.21

4. Which graph represents a proportional relationship?

   A.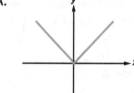

   B.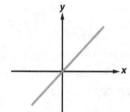

   C.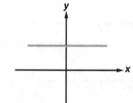

   D.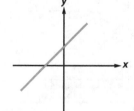

Skill Practice

Directions: Read the problem. Then answer the questions.

1. Imagine that you are looking at a map of the United States, but it doesn't have a scale that indicates the number of miles per inch on the map. You use a ruler to determine the distance between New York City and Washington, DC, on the map. The distance is $2\frac{1}{2}$ inches. You check a website and find out that the actual distance between New York City and Washington, DC, is **200** miles.

   A. What is the scale of the map in miles per inch (unit rate)?

   _____

   _____

   B. Write the equation for the proportional relationship whose constant proportionality is the scale of the map. Let y represent the distance in miles and x represent this distance on the map in inches.

   _____

   C. Using a ruler, you determine that the distance between New York City and Atlanta on the map is 8.75 inches. What is the actual distance between New York City and Atlanta in miles?

   _____

   _____

2. Which of the lines on the graph represents a proportional relationship with a greater constant of proportionality? Explain your answer.

   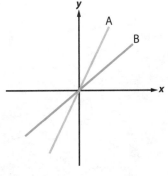

   _____

   _____

3. A car company conducts mileage testing for two new models of cars. For model A, the data for distance traveled and the number of gallons of gasoline consumed is shown in the following table.

Distance (miles)	Gallons
30	0.7
60	1.4
90	2.1
120	2.8
150	3.5

For model B, the distance traveled and the number of gallons of gasoline consumed is given in the following graph.

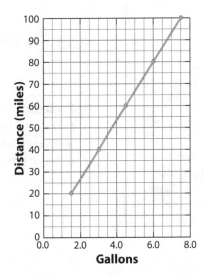

What is the gas mileage (unit rate) of each model car in miles per gallon? Which car gets better gas mileage?

Model A: _____

_____

Model B: _____

_____

Solve Proportions

MATH LINK

Recall that equivalent fractions are fractions that represent the same number, such as $\frac{1}{2}$ and $\frac{3}{6}$.

KEY CONCEPT: Understand how to use proportions to solve problems.

Write each ratio in two other ways.

1. $\frac{3}{5}$ **2.** 9:4   **3.** 5 to 9

Simplify each fraction.

4. $\frac{6}{10}$ **5.** $\frac{24}{60}$ **6.** $\frac{6}{26}$

Understand Proportions

A **proportion** is an equation made up of two **equivalent**, or equal, ratios. If you think of ratios as fractions, then the two ratios in a proportion are equivalent fractions.

Example 1 Write a Proportion

Write a proportion using 3 to 4 as one of the ratios.

Step 1 Write the given ratio as a fraction: $\frac{3}{4}$.

Step 2 Write an equivalent fraction.

$$\frac{3}{4} = \frac{3 \times 2}{4 \times 2} = \frac{6}{8}$$

Step 3 Write the proportion.

$$\frac{3}{4} = \frac{6}{8}$$

Sometimes, you can use the idea of equivalent fractions to decide if two ratios form a proportion or not. You are posing the same question if you ask either, "Do the two ratios form a proportion?" or "Are the two fractions equivalent?"

Example 2 Use Equivalent Fractions to Identify a Proportion

Do the two ratios $\frac{4}{10}$ and $\frac{5}{12}$ form a proportion?

Step 1 Find a common denominator: 60.

Step 2 Write each as a fraction with the denominator 60.

$$\frac{4}{10} = \frac{4 \times 6}{10 \times 6} = \frac{24}{60}$$
$$\frac{5}{12} = \frac{5 \times 5}{12 \times 5} = \frac{25}{60}$$

Step 3 Are the two fractions equivalent? No, since $\frac{24}{60} \neq \frac{25}{60}$.
Do the two ratios form a proportion? No.

Another way to see if two ratios form a proportion or not is to compare their two products using **cross-multiplication**. Each product is the numerator of one ratio multiplied by the denominator of the other ratio. If the products of cross-multiplication are equal, the ratios form a proportion. If they are not equal, the ratios do not form a proportion.

Example 3 Use Cross-Multiplication to Identify a Proportion

Do the two ratios $\frac{2}{15}$ and $\frac{3}{16}$ form a proportion?

Step 1 Perform cross-multiplication:
$2 \times 16 = 32$, and
$3 \times 15 = 45$.

Step 2 Are the products equal? No, because $32 \neq 45$.
Do the two ratios form a proportion? No.

In summary:

- A proportion is made up of two equal ratios.

- A product of cross-multiplication is the numerator of one ratio multiplied by the denominator of the other ratio.

- To see if two ratios form a proportion, check to see if the products of cross-multiplication are equal.

MATH LINK

Two ratios that form a proportion must be equivalent. So the two ratios must be equivalent fractions or equivalent rates. Also, their products of cross-multiplication must be equal.

THINK ABOUT MATH

Directions: Decide if the two ratios form a proportion. If so, write the proportion.

1. 50 to 20, 10 to 4   2. 8:3, 24:9   3. $\frac{1}{4}$, $\frac{2}{9}$

Directions: Form a proportion using the ratio given. There is more than one correct answer.

4. 12 to 13   5. 6:3   6. $\frac{25}{20}$

Core Skill
Represent Real-World Problems

The American Flag has a set of specifications that must be met in order to be considered an official flag and can be flown at a governmental building. These specifications govern not only the length and width of the flag but also the distance between the stars and the size of the stars. The hoist:fly (width:length) ratio is 10:19.

Consider the following problem. Vanessa is visiting her statehouse and wants to buy an official flag for her house. The gift shop has official flags, and Vanessa can tell that the flag is 12 inches wide. How long must the flag be in order for it to be official? In a notebook, write down the correct proportion, and then solve it.

Solve Proportions

When solving a proportion, much like when solving an equation, you find the missing **value**, or amount, of the variable in the proportion. You can use what you have already learned about proportions to solve them.

Example 4 Use Proportions to Solve a Rate Problem

On Monday, Chetan drove 100 miles in 2 hours. On Tuesday, he drove 125 miles in $2\frac{1}{2}$ hours. Did he drive the same rate on Tuesday as on Monday? If so, write a proportion showing that the two ratios (rates) are equal.

Step 1 Divide to find each rate.

100 miles ÷ 2 hours = 50 miles per hour

125 miles ÷ $2\frac{1}{2}$ hours = 50 miles per hour

Step 2 Answer the question.
Yes, Chetan drove the same rate on Tuesday as on Monday (50 miles per hour).

Step 3 Write a proportion showing that the rates are equal.
$\frac{100}{2} = \frac{125}{2.5}$

Step 4 Use cross-multiplication to check your answer.

$100 \times 2.5 = 250; 125 \times 2 = 250$

The products are equal, so the rates are equal.

For many problems, you can write and solve a proportion.

Example 5 Solve a Proportion

Solve the proportion for x. $\frac{5}{8} = \frac{x}{10}$

Step 1 Write the cross-products. Since this is a proportion, they are equivalent.

$$5(10) = 8 \times x$$
$$50 = 8x$$

Step 2 To solve, divide by 8 on both sides.

$$\frac{50}{8} = \frac{8x}{8}$$
$$6.25 = x$$

Step 3 Write the value for x into the proportion, and check using cross-multiplication.

$$\frac{5}{8} = \frac{6.25}{10}; 5 \times 10 = 50; 6.25 \times 8 = 50$$

The products of cross-multiplication are equal, so $x = 6.25$.

Example 6 Write a Proportion to Solve a Problem

A college advertises a 4:5 ratio of male students to female students. If there are about 1,200 male students, how many female students are there?

Step 1 Write a proportion. Make sure the order of the values is the same for each of the two ratios. Let f represent the number of female students.

$$\frac{\text{male}}{\text{female}} = \frac{4}{5} = \frac{1,200}{f}$$

Step 2 Solve the proportion.

$$(5 \times 1,200) = 4f$$
$$\frac{6,000}{4} = \frac{4f}{4}$$
$$1,500 = f$$

Step 3 Check using cross-multiplication in the proportion $\frac{4}{5} = \frac{1,200}{1,500}$.

$$4 \times 1,500 = 6,000; 1,200 \times 5 = 6,000$$

Step 4 State your answer.

There are about 1,500 female students.

Vocabulary Review

Directions: Complete each sentence with the correct word.

cross-multiplication   equivalent   proportion   value

1. If the products of _____ are equal, the proportion is true.

2. Two fractions that name the same value are called _____ fractions.

3. The missing _____ that needs to be found in a proportion is indicated by a variable.

4. A(n) _____ is a relationship of equivalency between two ratios.

Skill Review

Directions: Show four ways a proportion can be written for the data given in each of the following:

1. 5 circles for every 8 triangles and 10 circles for every 16 triangles

2. $2 for 7 miles and $6 for 21 miles

3. 80 seeds every 15 feet and 32 seeds every 6 feet

4. 7 cups of flour for every 5 tablespoons of sugar and 10.5 cups of flour for every 7.5 tablespoons of sugar

5. 72 chairs for 8 tables and 126 chairs for 14 tables

Directions: Solve each problem.

6. Photographs can measure either 3 inches by 5 inches or 4 inches by 6 inches. Do the ratios of $\frac{width}{length}$ form a proportion? Show your work.

7. Guadelupe mixed three parts of blue paint with one part of yellow paint to make green paint. Asura mixed six parts of blue paint with two parts of yellow paint to make green paint. Will the two shades of green paint be the same? Why or why not?

8. Sanaye typed 650 words in 10 minutes. Rashid typed 780 words in 12 minutes. Do the two rates form a proportion? Are their typing speeds the same? Explain.

Skill Practice

Directions: Choose the best answer to each question.

1.

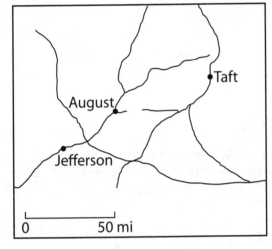

The scale on a highway map is 1 inch:50 miles. What is the distance in miles between Jefferson and Taft, which are $2\frac{1}{2}$ inches apart on the map?

A. 20

B. $52\frac{1}{2}$

C. $100\frac{1}{2}$

D. 125

2. Jakob earned $100 mowing 8 lawns. Which proportion can be solved to determine how much he will earn mowing 10 lawns at this rate?

A. $\frac{100}{10} = \frac{d}{8}$

B. $\frac{10}{d} = \frac{100}{8}$

C. $\frac{8}{10} = \frac{d}{100}$

D. $\frac{100}{8} = \frac{d}{10}$

3. Which statement is true of the proportion $\frac{2}{3} = \frac{10}{15}$?

A. 2:3 and 15:10 are equivalent ratios.

B. $\frac{2}{3}$ and $\frac{10}{15}$ are equivalent fractions.

C. 2×10 and 3×15 are equal.

D. It is not a true proportion.

4. The ratio of rock songs to pop songs on Jason's mp3 player is 3:8. If the number of pop songs he has on his player is 400, how many rock songs does he have?

A. 50

B. 100

C. 150

D. 300

Introduction to Percents

KEY CONCEPT: Percents, like decimals and fractions, represent part of a whole.

Divide.

1. $230 \div 5$ **2.** $23 \div 5$ **3.** $2.3 \div 5$ **4.** $0.23 \div 5$

Understand Percents

A **percent** is another means of expressing a number as part of a whole. The word *percent* means "for each 100." For example, you can write the ratio 3 to 100 as 3% and read it as 3 percent.

Example 1 Use Percents

The 10-by-10 grid below has 100 squares. What percent of the grid is shaded?

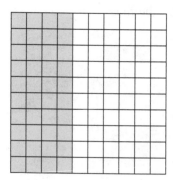

Step 1 Count the number of shaded squares: 40.

Step 2 Write a percent.
40 of the 100 squares are shaded: 40%.

Step 3 Summarize: 40% of the grid is shaded.

COMPARE AND CONTRAST

Writers use a **comparison** when they want to examine a **similarity**, or the way two or more people, things, or ideas are alike. Writers use **contrast** to look at the differences between people, things, or ideas. In comparing items that are alike, writers use terms such as *similar*, *both*, or *like*. When contrasting unlike things, terms such as *different*, *but*, *in contrast*, and *instead of* are often used.

Read the following paragraphs. As you read, look for the similarities and differences between percents and fractions.

Fractions are a way of comparing a part to a whole. The whole can be divided into any number of equal parts. This number is used as the denominator. A certain number of the parts are compared to the whole. This is the numerator. Operations such as adding, subtracting, multiplying, and dividing can be performed on fractions. Fractions are written with a slash (1/2) or with a fraction bar $\left(\frac{1}{2}\right)$.

Percents are another way to compare a part to a whole. *Percent* means "per hundred." Unlike fractions, percents always have the same number of equal parts: 100. The number out of 100 that is being specified is the percent. Percents can also have operations, such as addition, subtraction, multiplication, and division, performed on them. One hundred percent is a whole. Percents are written with a percent sign (%) or the word *percent*.

Similarities	Differences
both compare parts to a whole	fractions: any number of equal parts
operations performed in both	percents: always 100 equal parts
both can be greater or less than one	fractions: written with a slash or fraction bar
	percents: written with a percent sign (%) or the word *percent*

The writer has given the same type of information for both percents and fractions—what they are, whether operations can be performed on them, and how they are written. You must decide when the information about the items is the same or is different.

You have learned that data can be displayed in many different ways: in tables, in number lines, and in diagrams. Percents also can be displayed visually. A circle graph shows parts of a whole. The whole looks like a pizza, and the parts look like the slices. The size of each slice represents the percent part of the whole that is assigned to that slice.

Monica takes a poll of her class to find out how the students get to school every day. Here are the results of her poll: 43 percent take the school bus, 37 percent are driven to school, 17 percent ride a bicycle, and 3 percent walk. Monica represents this information in the following circle graph. If the labels in the circle graph didn't contain percentages, could you tell just by looking at it whether the majority of the polled students took a bus to school? Explain your answer.

TRANSPORTATION TO SCHOOL

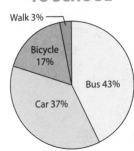

Percents as Fractions

Because a percent is a form of a ratio and because you can write any ratio as a fraction, you can write a percent as a fraction.

Example 2 Write Percents as Fractions

Write 14% as a fraction.

Step 1 Write the percent as a ratio that compares the number to 100 (a fraction with denominator 100).

$$14\% = 14 \text{ of } 100 = \frac{14}{100}$$

Step 2 Write the fraction in lowest terms.

$$\frac{14}{100} = \frac{14 \div 2}{100 \div 2} = \frac{7}{50}$$

Step 3 Summarize: $14\% = \frac{7}{50}$

Percents as Decimals

Notice from Step 1 of Example 2 that 14% is equal to 14 of 100, or 14 hundredths. Recall that 0.14 is read *14 hundredths*. This is helpful information to remember when you want to write a percent as a decimal.

Example 3 Write Percents as Decimals

Write 35% as a decimal.

Step 1 Write the percent as a ratio that compares the number to 100 (a fraction with denominator 100).

$$35\% = 35 \text{ of } 100 = \frac{35}{100}$$

Step 2 Write the fraction as a decimal.

$$\frac{35}{100} = \text{thirty-five hundredths} = 0.35$$

Step 3 Summarize: $35\% = 0.35$

THINK ABOUT MATH

$\sqrt{x}$

1. Write 67% as a decimal.

3. Write 3% as a decimal.

2. Convert 346% to a fraction.

4. Write 40% as a fraction.

Decimals as Percents and Fractions

Just as percents can be written as decimals, decimals can be written as percents. To change a decimal to a percent, simply move the decimal point two places to the right.

Example 4 Write Decimals as Percents

Write 0.6 as a percent.

Step 1 To write 0.6 as a percent, multiply it by $\frac{100}{100}$.

$$0.6 \times \frac{100}{100} = \frac{0.6 \times 100}{100} = \frac{60}{100} = 60\%$$

Step 2 Summarize: $0.6 = 60\%$

Decimals can also be written as fractions or mixed numbers.

Example 5 Write Decimals as Fractions

Write 0.6 as a fraction.

Step 1 To change 0.6 to a fraction, write the decimal as a fraction with a denominator that is a power of 10. (10, 100, 1,000, …) $\frac{6}{10}$

You can also write the decimal as it is read out loud.

$$0.6 \longrightarrow 6 \text{ tenths} \longrightarrow \frac{6}{10}$$

Step 2 Then reduce to lowest terms.

$$0.6 = \frac{6 \div 2}{10 \div 2} = \frac{3}{5}$$

Example 6 Write Decimals as Mixed Numbers

Write 1.76 as a fraction.

$$1.76 = 1\frac{76}{100} = 1\frac{76 \div 4}{100 \div 4} = 1\frac{19}{25}$$

Fractions as Decimals and Percents

Just as a percent can be written as a fraction and a decimal, a fraction can be written as a percent. To do this, you must first write the fraction as a decimal.

Example 7 Write Fractions as Decimals

Write $\frac{5}{8}$ as a decimal.

To change $\frac{5}{8}$ to a decimal, divide the numerator by the denominator until you get a remainder of 0. Add a decimal point and zeros as needed.

$\frac{5}{8} = 0.625$

$$\begin{array}{r} 0.625 \\ 8\overline{)5.000} \\ -4\,8 \\ \hline 20 \\ -16 \\ \hline 40 \\ -40 \\ \hline 0 \end{array}$$

MATH LINK

Remember $\frac{100}{100}$ is equal to 1. So in Example 4, 0.6 is actually being multiplied by 1 in the form of $\frac{100}{100}$. The shortcut for changing a decimal to a percent is to move the decimal point two places to the right (multiply by 100) and to write the percent sign (divide by 100). So, for example, $0.13 = 13\%$.

MATH LINK

The three dots seen at the end of the decimal 0.7777… are the symbol for *continues in this pattern*. The notation $0.\overline{7}$ can also be used. A bar over digits means those digits repeat infinitely.

If two or more digits repeat in the same pattern, place the bar over all of the digits that repeat: $\frac{1}{11} = 0.\overline{09}$, $\frac{1}{12} = 0.8\overline{3}$, and $\frac{1}{7} = 0.\overline{142857}$. When dividing on a calculator, the last digit in the display may be rounded up. $\frac{2}{3}$ will be displayed as 0.6666667 because the 6 rounds to a 7. This does not mean that the decimal stops repeating, only that the space available for displaying the answer is limited.

Do not be confused when you see a percent that already contains a decimal. Remember to move the decimal point two places to the left to change a percent to a decimal.

0.1% is not the same as 0.1.

0.1% = 0.001

10% = 0.1

Example 8 Write Fractions as Percents

Write $\frac{5}{8}$ as a percent.

Step 1 First write $\frac{5}{8}$ as a decimal (see Example 7).

$$\frac{5}{8} = 0.625$$

Step 2 Move the decimal point two places to the right by multiplying by 100. Add a percent sign.

$$\frac{5}{8} = 0.625 = 62.5\%$$

Sometimes when changing a fraction to a decimal, no matter how many zeros you add in the dividend, the answer either does not end or repeats one or more digits over and over. A **repeating decimal** is when digits are repeated over and over.

Example 9 Fractions as Repeating Decimals and Percents

Write $\frac{7}{9}$ as a decimal and as a percent.

Step 1 Divide the numerator by the denominator.

$$7 \div 9 = 0.77\frac{7}{9} \text{ or } 0.777...$$

Step 2 To write as a percent, move the decimal two places to the right and write a percent sign:

$$77\frac{7}{9}\%$$

```
      0.777
  9)7.000
   -6 3
     70
    -63
     70
    -63
      7
```

THINK ABOUT MATH

Directions: Write each number as a percent, decimal, and fraction.

1. $\frac{3}{12}$ 3. 0.068   5. 37%

2. $\frac{2}{3}$ 4. 2.4   6. 0.2%

Vocabulary Review

Directions: Complete each sentence with the correct word.

percent   repeating decimal   similarity

1. _____ is a word that means "of 100."

2. A _____ between two items is the way in which the two items are the same.

3. A _____ has one or more digits that repeat without end.

Directions: The table below shows how a number in each row is written as a percent, decimal, and fraction. Complete each column for the number given in each row. Identify which representations are similar and are easy to change between the two forms.

	Percent	Decimal	Fraction
1.	45%		
2.		0.8	
3.			$\frac{7}{20}$
4.		2.06	
5.	24.1%		
6.			$1\frac{3}{8}$

Directions: Write a sentence that compares and contrasts each of the following pairs of numbers.

7. 0.75 and $\frac{3}{4}$

8. $7\frac{5}{6}$ and $7.8\overline{3}$

9. $\frac{2}{3}$ and 66.667%

10. $\frac{1}{9}$ and $0\overline{1}$

Skill Practice

Directions: Choose the best answer to each question.

1. A band hopes that 2% of the people attending its concert will purchase a CD. Which fraction is the same as 2%?

   A. $\frac{1}{5}$
   B. $\frac{1}{50}$
   C. $\frac{100}{2}$
   D. $\frac{20}{100}$

2. Juan is tiling a hallway in his home. The diagram below shows how much of the floor he has completed. Which percent represents how much of the hallway Juan has completed?

   A. 3%
   B. 7%
   C. 30%
   D. 70%

3. Lishan said that she completed only $\frac{4}{5}$ of her math test. What percent of her math test did she complete?

   A. 45%
   B. 80%
   C. 125%
   D. Not enough information is given.

4. Tomas ran 12 minutes out of the 30 minutes he had hoped to run. What percent of the minutes he had hoped to run did Tomas actually run?

   A. 2.5%
   B. 12%
   C. 18%
   D. 40%

Solve Percent Problems

KEY CONCEPT: Decimals, fractions, and proportions can be used to solve percent problems.

Multiply.

1. 4.3×19 **2.** 0.35×8 **3.** $28 \times \frac{1}{4}$ **4.** $600 \times \frac{1}{5}$

Solve for the missing part of each proportion.

5. $\frac{3}{4} = \frac{\square}{28}$ **6.** $\frac{\square}{10} = \frac{5}{25}$ **7.** $\frac{6}{4} = \frac{9}{\square}$ **8.** $\frac{60}{\square} = \frac{25}{100}$

Percent of a Number

A percent is a **portion** (part) of a number. Often, it is necessary to figure out what that portion is. For example, if you decide to leave a 20% tip at a restaurant, you need to calculate the amount of money that is 20% of the total bill.

Example 1 Use a Percent as a Decimal

What is 20% of $25?

> **Step 1** Write the percent as a decimal.
> 20% = 0.20
>
> **Step 2** Multiply.
> 0.20 × $25 = $5.00
>
> **Step 3** Summarize.
> 20% of $25 is $5.00.

To calculate with percents, you first need to change the percent to either a decimal or a fraction. Example 1 uses a decimal, and Example 2 uses a fraction. You may use whichever form of the number is easier for you to work with, because both forms will give you the same answer.

Example 2 Use a Percent as a Fraction

What is 25% of $400?

Step 1 Write the percent as a fraction.

$25\% = \frac{25}{100} = \frac{1}{4}$

Step 2 Multiply.

$\frac{1}{4} \times \$400 = \$100$

Step 3 Summarize.

25% of $400 is $100.

Some problems require you to find the percent of a number. Be sure to read the problem carefully. With percent problems, look for the word *of* since most percent problems ask you to find the percent *of* a number. When the answer you need is the percent of a number, multiply to find the answer.

Example 3 Percent of a Number

A state official predicts that 70% of the registered voters in a state will vote on election day. If there are 514,000 registered voters in that state, how many are predicted to vote?

Step 1 Understand the question.
You must find how many of the 514,000 registered voters are predicted to vote on election day.

Step 2 Decide what information is needed.
Use the percent of voters predicted to vote: 70%.
Also, use the number of registered voters: 514,000.

Step 3 Choose the most appropriate operation.
This problem asks, "What is 70% of 514,000?"
The operation to use is multiplication.

Step 4 Solve the problem.
70% of 514,000 = 0.70 × 514,000 = 359,800 voters

Step 5 Check your answer.
To check, see if the ratio $\frac{359,800}{514,000}$ equals 70%.
359,800 ÷ 514,000 = 0.7 = 70%
So 359,800 voters are predicted to vote.

THINK ABOUT MATH

Directions: Calculate each of the following.

1. 10% of 870

2. 47% of 1,000

3. 75% of 16

4. 50% of $50

Percents are one of the most important topics to understand because they are used everyday. From shopping sales at a store to tipping at restaurants, percents show up in all places. Being able to calculate percentages is a valuable tool you can have to make sure you are getting the correct price or tipping the correct amount.

Consider the following scenario. Silvia and Sandra are shopping at a local department store. Silvia sees a sign that says 40% off a pair of $60 jeans. She sets up the following proportion and solves it.

$$\frac{x}{60} = \frac{40}{100}$$
$$x = 60 \times \frac{40}{100}$$
$$x = 24$$

She decides that the jeans now cost $24 and decides to buy them. In a notebook, determine if Silvia is correct. If not, explain what she did wrong and what the sale cost would have been.

Use Proportions to Solve Percent Problems

One way to solve percent problems is to use proportions (two equal ratios). Solve a proportion with the percent written as a ratio comparing a number to 100.

$$\frac{part}{whole} = \frac{\%}{100}$$

By solving the proportion, you find the missing value: the part, the whole, or the percent. 100 is always in the proportion. Be careful to write the proportion correctly before you solve it. After solving the proportion, multiply to check your result.

Example 5 Find the Part

What is 30% of 90?

Step 1 Identify the part, the whole, and the percent.
Part: (missing)   Whole: 90   Percent: 30

Step 2 Write the proportion $\frac{part}{whole} = \frac{\%}{100}$.
$$\frac{\square}{90} = \frac{30}{100}$$

Step 3 Solve the proportion.
$(90 \times 30) \div 100 = 27$

Step 4 Summarize.
30% of 90 is 27.

Step 5 Use multiplication to check your answer.
30% of 90 = $0.3 \times 90 = 27$

You are familiar with finding an arithmetic mean, or average. Say, for example, that you take two tests, and your scores are 84 and 92.

(Score 1 + Score 2) $\div$ 2 = the arithmetic mean
$(84 + 92) \div 2 = 176 \div 2 = 88$
Your average test score is 88.

There is another kind of mean, too, but it is not arithmetic. It is geometric, or a **geometric mean**. You can multiply any number of positive values (n) and find the nth root of the product to find the geometric mean.

For example, imagine multiplying two positive values, x and y. Find the geometric mean of x and y, when $x = 9$ and $y = 16$.

$9 \times 16 = 144$
$\sqrt{144} = 12$

When $x = 4$ and $y = 36$, the same geometric mean is also found because $4 \times 36 = 144$.

Now, apply what you know to write a proportion between the two pairs of numbers.

$$\frac{4}{9} = \frac{16}{36}$$

In the proportion, the numbers 9 and 16 represent the **means** because they are closest to the geometric mean (12). The numbers 4 and 36 represent the **extremes**, or numbers farthest from the mean.

Scientists often calculate geometric means to predict exponential growth. Exponential growth occurs more and more quickly over time, as a population of living things increases. For example, scientists can use geometric means to predict global human population growth over decades.

Example 6 Find the Percent

What percent of 90 is 18?

Step 1 Identify the part, the whole, and the percent.
Part: 18   Whole: 90   Percent: (missing)

Step 2 Write the proportion $\frac{part}{whole} = \frac{\%}{100}$.
$\frac{18}{90} = \frac{\square}{100}$

Step 3 Solve the proportion.
$(18 \times 100) \div 90 = 20$

Step 4 Summarize.
20% of 90 is 18.

Step 5 Use multiplication to check your answer.
20% of 90 = 0.2 × 90 = 18

Example 7 Find the Whole

80% of what number is 20?

Step 1 Identify the part, the whole, and the percent.
Part: 20   Whole: (missing)   Percent: 80

Step 2 Write the proportion $\frac{part}{whole} = \frac{\%}{100}$.
$\frac{20}{\square} = \frac{80}{100}$

Step 3 Solve the proportion.
$(20 \times 100) \div 80 = 25$

Step 4 Summarize.
80% of 25 is 20.

Step 5 Use multiplication to check your answer.
80% of 25 = 0.8 × 25 = 20

To solve a percent problem, set up a ratio, $\frac{part}{whole} = \frac{\%}{100}$, comparing a number to 100. Insert each value that you are given into the ratio, and then solve for the missing value.

THINK ABOUT **MATH**

Directions: Answer each question.

1. What is 20% of 45?

2. What percent of 40 is 25?

3. 42% of what number is 14.7?

Core Skill
Use Percents

You use percents everywhere—even when you're not aware of them. The tax you pay on things you buy, for example, is a percentage of the purchase price. The tip you give the server in a restaurant is a percentage of the cost of a meal. Baseball batting averages also are a percentage; you calculate a batting average by dividing the number of hits the player gets by the number of times a player comes to the plate.

The key to finding the percent in every problem is looking for the symbol % or the word *percent*. The whole is right after the word *of*. When attempting to solve such problems, therefore, you should begin by first identifying these three elements. In the problems below, the part is circled, the whole is underlined, and the percent is boxed.

50% of 62 is what number?

3 is 75% of what number?

12 is what percent of 24?

In a notebook, copy the questions below. Circle the part, underline the whole, and put a box around the percent. Then write the proportion $\frac{part}{whole} = \frac{\%}{100}$ for each.

6 is 25% of what number?
17 is what percent of 51?
What is 100% of 49?

Vocabulary Review

Directions: Complete each sentence with the correct word.

extremes   means   portion

1. A percent is one way to represent a(n) _____ of a whole.

2. The two numbers in a proportion closest to the geometric mean are the _____.

3. The two numbers in a proportion furthest from the geometric mean are the _____.

Skill Review

Directions: Circle the *part*, underline the *whole*, and put a box around the *percent* in each problem below. Write the proportion $\frac{part}{whole} = \frac{\%}{100}$ for each. Then answer the question.

1. What number is 25% of 80?

2. 10% of what number is 8?

3. What percent of 44 is 11?

4. 9 is what percent of 100?

5. 16% of what number is 200?

6. 3% of 500 is what number?

7. 7 is 1% of what number?

8. What percent of 16 is 12?

9. Annabelle drank 340 cups of coffee in one year. She drank 73 cups during January alone. What percent of cups of coffee did Annabelle drink during January?

10. Lucio received 52% of the votes to win an election. There were 215,400 voters. How many people voted for Lucio?

11. Panya bought some mittens on sale for $12. She paid only 80% of the original price. What was the original price of the mittens? How much money did she save by buying the mittens on sale?

Skill Practice

Directions: Choose the best answer to each question.

1. The sales tax rate in Marissa's home town is 6%. If she purchased a new car for $12,500, how much will she pay in tax?

    A. $75
    B. $208.33
    C. $750
    D. $2,083.33

2. Delfina is buying an $80 dress for 25% off. What calculation can help her to find how much less she will be paying for the dress?

    A. $\frac{1}{25} \times \$80$
    B. $\frac{1}{4} \times \$80$
    C. $\frac{1}{2} \times \$80$
    D. $\frac{1}{4} \div \$80$

3. To pass an exam, Kamol must get at least 70% of the 30 problems correct. How many problems must he get correct to pass?

    A. 9
    B. 10
    C. 15
    D. 21

4. A newborn baby sleeps an average of 16 hours in a 24-hour period. What percent of its time does a newborn baby spend sleeping?

    A. about 15%
    B. about 38%
    C. about 67%
    D. about 75%

Use Percents in the Real World

KEY CONCEPT: Simple interest can be calculated using a formula and percents.

Write each ratio as a fraction.

1. 28 to 52　　　　2. 55 to 365　　　3. 9 to 12

Multiply.

4. $\$500 \times 0.055$　　5. $0.08 \times \frac{1}{2}$　　6. $20 \times 0.3 \times \frac{5}{12}$

Lesson Objectives

You will be able to

- Understand the interest formula
- Use a formula to find simple interest

Skills

- **Core Skill:** Make Sense of Problems
- **Core Skill:** Solve Real-World Arithmetic Problems

Vocabulary

convert
formula
interest
principal
rate
time

Core Skill
Make Sense of Problems

Because both the interest rate and time in the simple interest formula deal with time, both time units must be the same, whether it is in months, years, or centuries!

Suppose you deposit $100 into a 10% annual savings account, but pays interest monthly. In 12 months, you would have $10, but multiplying 100 × 0.1 × 12 = $120! This happened because the time units were different.

In a notebook, figure out the monthly interest rate for the account.

Simple Interest Problems

Interest is money earned by an investment or paid when money is borrowed. Simple interest is the most basic type of interest. In simple interest, the amount of interest is calculated only on the principal. In other words, interest is not calculated on previous interest. The amount of simple interest depends upon three things: principal, rate, and time. Both the rate (%/time) and time have time units attached to them and must be the same in order for the formula to calculate interest correctly. You can use the formula ...

$$\text{Interest} = \text{Principal} \times \text{Rate} \times \text{Time}$$

The **principal** is the amount of money invested or borrowed. The **rate** is the annual interest rate, usually given as a percent. The **time** is the length of time (in years) the money is invested or borrowed.

Example 1 Calculate Simple Interest with Time in Years

Laurel put $1,000 into a savings account. The bank pays 3% simple interest annually on this account. How much interest will Laurel earn if she leaves the money in the account for 2 years?

Step 1 Identify the principal, the rate, and the time.
　　　　Principal: $1,000　　Rate: 3% = 0.03　　Time: 2 years

Step 2 Multiply.
　　　　Interest = Principal × Rate × Time
　　　　　　　　= $1,000 × 0.03 × 2
　　　　　　　　= $60

Step 3 Summarize.
　　　　Laurel will earn $60 in interest.

When the interest owed or earned is over a time that is not stated in years, you must **convert**, or change, it into years. Remember, there are 365 days (except for leap year), 12 months, or 52 weeks in a year.

Example 2 Convert Time to Years to Calculate Interest

A credit card company charges 12% annual interest on any balance due. If the balance due on Rafael's credit card is $150 and he waits 30 days to pay the bill, how much interest will he owe?

Step 1 Identify the principal, the rate, and the time.
Principal: $150   Rate: 12% = 0.12   Time: 30 days

Step 2 Write a ratio to convert the time to years.
There are 365 days in one year, so 30 out of 365 days
is $\frac{30}{365}$ year.

Step 3 Multiply.
Interest = Principal × Rate × Time
$$= \$150 \times 0.12 \times \frac{30}{365}$$
$$= \$1.479452055 \text{ or } \$1.48$$

Step 4 Summarize.
Rafael will owe $1.48 in interest.

The amount of the interest depends upon three things: principal, rate, and time (in years). Use the formula, Interest = Principal × Rate × Time, to calculate interest. Any time not given in years must be converted to years. This assumes that the the rate given is per year. If the rate is a monthly rate, then time would need to be given in months. The same is true for any type of rate.

THINK ABOUT MATH

Directions: Calculate simple interest. If necessary, round to the nearest cent.

1. principal = $10,000
   rate = 8%
   time = 5 years

2. principal = $2,500
   rate = 4%
   time = 6 months

3. principal = $120,000
   rate = 9%
   time = 30 years

4. principal = $600
   rate = 5.5%
   time = 100 days

Core Skill
Solve Real-World Problems

When your parents or someone you know takes out a loan to buy a car, they pay interest on the money they borrowed. The bank pays you interest on money you put in a savings account. In each case, the interest that is paid is generally a percentage of the money that is borrowed or saved. This is another place where "percent" plays a role in everyday life.

When solving real-world interest problems, you are applying what you have already learned during your studies of ratios, percents, and multiplication with whole numbers, decimals, and fractions. Now, however, you are putting those skills to new use.

Look at the numbers being multiplied in Example 1: $1,000 × 0.03 × 2. In this example, which describes a real-world situation, a percent is changed to a decimal and then multiplied by whole numbers.

In a notebook, determine the simple interest rate on an account that paid $50 on a $1000 principal over 2 years.

MATH LINK

Compound interest can be thought of as earning interest on interest. It is found by adding earned interest to the principal. It is the interest typically used by banks and investment institutions.

Vocabulary Review

Directions: Complete each sentence with the correct word.

convert   formula   interest   principal   rate   time

1. The _____ is the amount of money invested or borrowed.

2. The length of _____ is the number of years the money is being borrowed.

3. One _____ can be used to find the area of a square, while another is used to calculate interest owed.

4. You can _____ a percent to a decimal or a fraction.

5. The _____ of the interest owed is usually given as a percent.

6. _____ can either be owed or earned.

Skill Review

Directions: Solve each problem. Apply information about percents, the interest formula, and problem solving that you learned in this lesson and previous lessons.

1. Find the interest earned in 1 year on a principal of $2,000 that pays 10% annual interest.

2. Find the interest earned in 100 days on a principal of $5,000 that pays 11% annual interest.

3. Jonah bought a car priced at $18,500. He made a 10% down payment and borrowed the rest of the money at a 9% annual interest rate. He will pay the balance due in five years. How much interest will he pay?

4. On her eighteenth birthday, Enriqua put $1,500 into a savings account. The annual interest rate on the account is 6%. How much money will she have in the account on her twenty-first birthday?

Skill Practice

Directions: Choose the best answer to each question.

1. Hye Su put $4,000 into a savings account that pays 4% annual interest. How much interest will she earn in 3 years?

   A. $48
   B. $480
   C. $4,800
   D. $48,000

2. Jamil borrowed $75,000 from a mortgage company to buy a house. He will repay the loan at 8% annual interest in 30 years. How much will he pay the mortgage company in interest?

   A. $180,000
   B. $18,000
   C. $1,800
   D. $180

3. Tanya is comparing two loans. With loan A, she will pay 5% simple interest for 26 weeks. With loan B, she will pay 6.5% simple interest for 18 weeks. She will borrow $12,500. Which expresses the difference in what she will pay in interest on the two loans?

   A. Loan A costs $31.25 more than loan B.
   B. Loan B costs $31.25 more than loan A.
   C. Loan A costs $1,625 more than loan B.
   D. Loan B costs $1,625 more than loan A

4. A credit card company charges 24% annual interest on any balance due. If the balance due on Bette's credit card is $100, and she waits 60 days to pay the bill, about how much interest will she owe?

   A. $1.20
   B. $3.95
   C. $14.40
   D. $39.50

1. There are 2 red, 6 gray, and 12 black cars in a car rental lot. What is the ratio of black cars to the number of cars in the car rental lot?
   A. 4:3
   B. 3:4
   C. 5:3
   D. 3:5

2. Which statement is true about $\frac{6}{24} = \frac{x}{32}$?
   A. $x = (24 \times 32) \div 6$
   B. $x = (24 \times 6) \div 32$
   C. $x = (6 \times 32) \div 24$
   D. $x = (24 \div 6) \times 32$

3. Carlotta is planting 40 tulip bulbs in her garden. She has dug holes for 15% of the bulbs so far. How many holes has she dug?
   A. 60   C. 0.6
   B. 6   D. 0.06

4.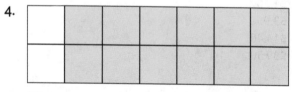

   What percent of the squares are shaded? Round your answer to the nearest percent.
   A. 90%
   B. 86%
   C. 85%
   D. 80%

5. The ratio of teachers to students in one school district is 2 to 47. If the number of teachers in the district is 130, how many students are there?
   A. 6,110   C. 260
   B. 3,055   D. 5.53

6. Han opened a savings account that earns 3% annually with $6,700. He makes no more deposits. Which equation can he use to find how much interest he will have if he leaves the money in the account for 6 years and 3 months?
   A. $I = 6{,}700 \times 0.03 \times 6.25$
   B. $I = 6{,}700 \times 0.03 \times 6.3$
   C. $I = 6{,}700 \times 0.3 \times 6.3$
   D. $I = 6{,}700 \times 0.03 \times 6$

7. In an 8-hour shift in a candy factory, 15,000 boxes of candy can be processed. What proportion can be used to find how many boxes of candy can be processed in 7 hours?
   A. $\frac{15{,}000}{x} = \frac{7}{8}$ C. $\frac{8}{15{,}000} = \frac{7}{x}$
   B. $\frac{8}{15{,}000} = \frac{x}{7}$ D. $\frac{15{,}000}{x} = \frac{7}{8}$

8. Shane has $12,455 in his checking account. He wants to move 73% of the money to a high-yield savings account. How much will he put in the account?
   A. $17,061.64   C. $10,500
   B. $12,000.25   D. $9,092.15

9. Which pair of ratios form a proportion?
   A. 3 to 4, 5 to 8
   B. 1:4, 2:9
   C. $\frac{16}{3}, \frac{8}{1}$
   D. 8:3, 24:9

10. Sales tax is 6.5%. Jay buys a car for $12,500. How much sales tax will he pay?
    A. $812.50   C. $650.50
    B. $750   D. $125

11. Frederick ran 8 miles in 60 minutes. What is the unit rate he ran?
    A. 7.5 miles/minute
    B. .133 miles/minute
    C. 8 miles/minute
    D. 4.2 miles/minute

Review

12. Sharon is painting her house. She needs 5 gallons of brown paint for every 7 gallons of white paint. How many gallons of white paint does she need if she has already bought 14 gallons of brown paint?

    A. 10 gallons   C. 16 gallons
    B. 19.6 gallons   D. 2.5 gallons

13. 68% of 50 is what number?

    _____

14. Daniel's bill at a restaurant is $19. He wants to leave a 15% tip. What is the total amount Daniel should write on his credit card receipt?

    A. $2.85
    B. $15.00
    C. $21.85
    D. $34.00

15. Padmini drives an average of 55 miles per hour. Her destination is 275 miles away. She wants to arrive by 4:00 P.M. and allow an extra hour for stops. By what time should she start driving?

    A. 10:00 P.M.
    B. 12:00 noon
    C. 11:00 A.M.
    D. 10:00 A.M.

16.

    In the pattern, there are 3 circles for every 21 stars. How many stars will there be if there are 8 circles?

    _____

Check Your Understanding

On the following chart, circle the number of any item you answered incorrectly. Near each lesson title, you will see the pages you can review to learn the content covered in the question. Pay particular attention to reviewing those lessons in which you missed half or more of the questions.

Chapter 7: Ratios, Proportions, and Percents	Procedural	Conceptual	Application/ Modeling/ Problem Solving
Ratios and Rates pp. 212–217	1		
Unit Rates and Proportional Relationships pp. 218–223	11		15
Solve Proportions pp. 224–229	2	9, 16	5, 7, 12
Introduction to Percents pp. 230–235	4		3
Solve Percent Problems pp. 236–241		13	8, 10
Use Percents in the Real World pp. 242–245	14		6

Exponents and Roots

If you have free time and space, you may wish to write out an expression such as $7 \times 7 \times 7 \times 7 \times 7 \times 7 \times 7 \times 7 \times 7$. Most people would rather write this multiplication by using 7 as the base and 9 as the exponent in the power, 7^9. This saves space and effort and helps to eliminate errors made from miscounting.

Finding roots of numbers is the reverse of using exponents and has important applications in higher math and science. You may be asked to find or estimate a square root or a cube root. For example, you may know that your bedroom floor is in the shape of a square that measures 100 square feet. If you want to buy molding for only the length of one wall, you will need to find the square root of the area of the floor.

Writing numbers using scientific notation has the advantage of making very large or very small numbers easy to compare. Scientific notation uses the rules of exponents.

The Key Concepts you will study include:

Lesson 8.1: Exponents
Extend understanding of numbers to exponents and arithmetic expressions that contain exponents.

Lesson 8.2: Roots
Develop and extend understanding of numbers to include the concepts of square roots and cube roots.

Lesson 8.3: Scientific Notation
Develop understanding of large numbers to include scientific notation and how to translate between numbers written in scientific notation and standard notation.

Goal Setting

Before starting this chapter, set goals for your learning.

Look at the topics of this chapter.

- To which operations do exponents, roots, and scientific notation have a connection?

  _____

  _____

  _____

- How do you think knowing about exponents will further your understanding of numbers and operations?

  _____

  _____

  _____

- What do you know already that will help you to find the square roots of numbers?

  _____

  _____

  _____

Exponents

Lesson Objectives

You will be able to
- Evaluate exponents
- Evaluate arithmetic expressions with exponents

Skills

- **Core Skill:** Evaluate Expressions
- **Core Skill:** Calculate Area and Volume

Vocabulary

base
exponent
power

KEY CONCEPT: Extend understanding of numbers to exponents and arithmetic expressions that contain exponents.

Evaluate each expression when x = 3 and y = −2.

1. $x + y$
2. $2x − 3$
3. $3y + 4$

4. $6x − 5y$
5. $4(x + 2y)$
6. $−2(3x − y)$

Evaluate Exponents

The expression 2^4 is called a **power**: 2 is the **base**, and 4 is the **exponent**. The expression 2^4 is read *two to the fourth power*. To find the value, use the base as a factor and the exponent as the number of times it is multiplied. The expression 2^4 has the same product as $2 \times 2 \times 2 \times 2$. The product of both expressions is 16.

Example 1 Find the Value of an Exponential Expression

Find the value of 3^5.

Step 1 Identify the base and exponent.
Base: 3
Exponent: 5

Step 2 Use the base as a factor as many times as the exponent indicates.
$3 \times 3 \times 3 \times 3 \times 3$

Step 3 Multiply.
$3 \times 3 \times 3 \times 3 \times 3 = 243$
The value of 3^5 is 243.

Example 2 Find the Value of an Expression with Zero as the Exponent

Find the value of 5^0.

Step 1 Identify the base and exponent.
Base: 5
Exponent: 0

Step 2 The exponent is 0. 5 is used as a factor 0 times, and the value of the expression is 1.
The value of 5^0 is 1.

MATH
LINK

When the exponent of a non-zero base is zero, the value of the expression is always 1.

MATH
LINK

You can use a **mnemonic device** (a memory aid) such as PEMDAS to help recall the order of operations for arithmetic expressions. PEMDAS stands for *Parentheses, Exponents, Multiplication, Division, Addition, Subtraction*.

Example 3 Use a Calculator to Evaluate an
Exponential Expression

Find the value of 9^6.

Press

The display should read

> 9^6 Math ▲
>
>   531441

The value of 9^6 is 531,441.

THINK ABOUT MATH

$\sqrt{x}$

Directions: Find the value of each expression.

1. 4^3 **2.** 2^5 **3.** 5^2 **4.** 3^3

Directions: Use a calculator to find the value of each expression.

5. 8^7 **6.** 24^5 **7.** 43^4 **8.** 12^6

Evaluate Arithmetic Expressions with Exponents

Recall that an arithmetic expression contains numbers and one or more operations. It is evaluated by using the order of operations. When you don't follow the order of operations, you will get the wrong value for an arithmetic expression.

An arithmetic expression can also contain powers with exponents or square roots. Square roots will be introduced in Lesson 8.2. The order of operations that you learned about in Chapter 1 is expanded to include powers with exponents and square roots. Follow this sequence of steps when evaluating expressions:

1) Operations within parentheses
2) Exponents and roots
3) Multiplication and division from left to right
4) Addition and subtraction from left to right

In mathematics, it is important to learn the sequence of steps you follow to solve expressions and equations, use your calculator, and perform operations. As you gain experience in mathematics, you willl begin to recognize one of the singular features of the subject: its regularity. You work your way to the solution of similar problems by following the same sequence of steps over and over again.

When you are evaluating an expression that contains exponents, for example, you must understand the meaning of each part of the expression. You want to identify the base, the number that will be the factor that is multiplied by itself. You also want to identify the exponent, the number that tells you how many times to multiply the base by itself. Then you perform the multiplication to get the product. The steps you take to arrive at a solution never vary.

After reading the section that lists the sequence in the order of operations, describe in a notebook why it is important to use the order of operations to evaluate $12 - (5 + 3)^2 \div 2^3$. Include in your description the value of the expression if you use the order of operations. Use a possible value if you do not use the order of operations.

Two of the things that exponents are used for are finding area and volume. The most basic shapes that have their area and volume calculated are the square and cube. In fact, the units to describe area and volume are square units and cubic units, respectively, and are named after the square and cube.

Suppose that the length of the side of a square is s inches long. What would the area be? The formula for the area of a rectangle is $A = l \times w$. A square is a rectangle with both sides equal, so the area of the square is $A = s^2$. What about a cube with a side length s? The volume of the cube is $V = s^3$.

Finding area and volume will be covered later in the book. But in both examples, exponents are used. When speaking of these two formulas, the squares area would be said "s to the second power, or s squared", while the volume would be "s to the third power, or s cubed". These shortcuts for the two powers (2nd and 3rd), are used because they describe the shape whose area or volume is being calculated.

In a notebook, determine what would happen to a square's area if all of the sides doubled. Make sure to use the properties of exponents correctly.

Example 4 Use the Order of Operations to Find the Value of an Expression

Find the value of $45 - 3 \times 2^2 + (8 \times 5)$.

Step 1 Do operations within parentheses. $\qquad (8 \times 5) = 40$

Step 2 Do exponents and roots. $\qquad\qquad\qquad 2^2 = 4$

Step 3 Do multiplication and division. Work from left to right. $\qquad\qquad\qquad\qquad 3 \times 4 = 12$

Step 4 Do addition and subtraction.
$45 - 12 + 40 = 33 + 40 = 73$
The value of the expression is 73.

THINK ABOUT MATH

$\sqrt{x}$

Directions: Find the value of each expression.

1. $(1 + 2 + 3)^2$
2. $3^2 + 6^2 \div 3$
3. $(2^3 + 3^3) \div 7$
4. $9 \times 8^0 + (6 - 1)$
5. $24 \div (1^5 + 5)$
6. $3 \times (10 - 4) \div 9 + 4^2$

Vocabulary Review

Directions: Match each word to one of the phrases below.

1. _____ base

2. _____ exponent

3. _____ power

A. contains a base and an exponent

B. the number 3 in the expression 3^4

C. the number that indicates how many times a number is multiplied by itself

Directions: Answer the following question.

1. How does understanding sequence help you find the value of an expression that contains two or more operations?

_____

Directions: Describe the sequence you would use to find the value of each of the following expressions. Then find the value of the expression.

2. $4^2 + 3^3 \div 9$

_____

3. $2 \times (14 - 7^0) + 28 \div 2^2$

_____

Skill Practice

Directions: Choose the best answer to each question.

1. Which of the following has the same value as 4^3?

   A. 3^4
   B. 8^2
   C. 43^1
   D. 64^0

2. Which operation in the following expression should be performed first?

   $(3 + 6)^2 - 2^3 \div 4 \times 3$

   A. Evaluate 2^3.
   B. Evaluate 6^2.
   C. Multiply 4×3.
   D. Add $3 + 6$.

3. By what factor would the volume of a cube change if all of the sides doubled?

   A. 1 (the volume would stay the same)
   B. 2 (the volume would double)
   C. 4 (the volume would quadruple)
   D. 8 (the volume would octuple)

4. Tabina sold 2^5 air conditioners last week and 3^3 air conditioners this week. What is the difference in the number of air conditioners she sold?

   A. 1
   B. 2
   C. 5
   D. 8

Roots

KEY CONCEPT: Develop and extend understanding of numbers to include the concepts of square roots and cube roots.

Find the value of each expression.

1. 7^2

2. 2^5

3. 3^4

4. 6^3

Use a calculator to find the value of each expression.

5. 5^6

6. 8^8

7. 12^4

8. 41^5

Lesson Objectives

You will be able to
- Find square roots
- Find cube roots

Skills

- **Core Skill:** Evaluate Reasoning
- **Core Skill:** Interpret Data Displays

Vocabulary

cell
cube root
perfect cube
perfect square
radical sign
square root
squared

Find Square Roots

The expression 7^2 is sometimes called "7 **squared**" or "the square of 7." The exponent 2 indicates that the base is squared. Recall that 7^2 is the same as 7×7, so the value of 7^2 is 49.

The expression $\sqrt{49}$ is read, "the **square root** of 49." The symbol $\sqrt{}$ is called a **radical sign**. Finding the square root of a number is the opposite of finding the square of a number. A number's square root is the number that, multiplied by itself, will yield the original number.

A **perfect square** is a whole number whose square root is a whole number. For example, 16 is a perfect square because 4^2 is 16.

Example 1 Find the Square Root of a Perfect Square

Find the value of $\sqrt{100}$.

Step 1 Think: What number multiplied by itself is 100?
$$n \times n = 100$$
$$10 \times 10 = 100$$

Step 2 Write the square root.
$$\sqrt{100} = 10$$
The value of $\sqrt{100}$ is 10.

Example 2 Use a Calculator to Find a Square Root

Find the value of $\sqrt{2{,}304}$.

Press (2nd) (x²) (2) (3) (0) (4) (enter).

The display should read

$\sqrt{2{,}304}$	Math ▲
	48

The value of $\sqrt{2{,}304}$ is 48.

MATH LINK

Many square roots are memorized as part of the multiplication facts. Look at columns 1 and 2 in the table on the next page. You probably already know the square roots of the perfect squares to 10. You might also know that 11×11 is 121 and 12×12 is 144. Both 121 and 144 are perfect squares.

UNDERSTAND A TABLE

In addition to giving specific data, tables can help show information as a whole. Looking across a row or down a column can tell you certain things about that set of data. You may not have to analyze every cell in the table to understand the data. A **cell** is a place in a table or spread sheet where column and row intersect. Sometimes, trends are noticeable across rows or down columns. Other times, comparing two rows or columns can give information about the data.

Number (x)	Square (x²)	Cube (x³)
1	1	1
2	4	8
3	9	27
4	16	64
5	25	125
6	36	216
7	49	343
8	64	512
9	81	729
10	100	1,000

Study the table above. Without using the definition of squares and cubes, what can you tell about what happens to whole numbers when they are squared and cubed?

Notice that the numbers in the square column increase more rapidly than those in the number column as the list progresses. Likewise, the numbers in the cube column increase at a greater rate than the numbers in the square column. However, the numbers in the row for 1 are all the number 1. So you can infer that for whole numbers, with the exception of 1, cubed numbers will always be greater than squared numbers. Additionally, both squares and cubes increase more rapidly as the whole numbers increase.

Tables will often present data or information in a way that makes it easy to use. First, study the information in the table so that you understand what it is telling you, and then use that information to solve problems. Some ways of using data to solve a problem include performing operations on the data, such as adding or subtracting, using the data to make a graph, or looking for a pattern.

Look again at the data in the table on the previous page. Think about ways in which you could use the data. Now look at Example 1. You are asked to find the value of $\sqrt{100}$. If you look at the first two columns in the table, you can see that the square of 10 is 100. If the square of 10 is 100, then the square root of 100 is 10. So you can use column 2 in the table to find the square root in column 1.

In a notebook, explain how you can use the data in the table to approximate square roots.

If a number is not a perfect square, you can **approximate,** or estimate, the square root by finding the two following, or **consecutive,** whole numbers, such as 8 and 9, between which the square root lies.

Example 3 Approximate a Square Root

Find the two consecutive whole numbers between which $\sqrt{150}$ lies.

Step 1 Think: Which two perfect squares are closest to 150.
Try the squares of 12 and 13.
$12 \times 12 = 144$
$13 \times 13 = 169$

Step 2 Write the perfect squares as square roots, and compare.
Write the square roots using $<$.
$\sqrt{144} < \sqrt{150} < \sqrt{169}$
$\sqrt{150}$ lies between $\sqrt{144}$ and $\sqrt{169}$.

Step 3 Find the square roots of the perfect squares.
$12 < \sqrt{150} < 13$
$\sqrt{150}$ is between 12 and 13.

Example 4 Solve Problems Involving Square Roots

Find the side length of a square with an area of 324 square meters.

Step 1 The area of a square is the length of one side squared, so find the value of $\sqrt{324}$ to find the side length of the square. Try numbers that, when multiplied by themselves, are equal to 324.
Try 17.   Try 18.
$17 \times 17 = 289$ $18 \times 18 = 324$

Step 2 Write the square root.
$\sqrt{324} = 18$
The side length of the square is 18 meters.

Example 5 Do Operations with Square Roots

Find the sum of $\sqrt{81}$ and $\sqrt{144}$.

Step 1 Find each square root.
$9 \times 9 = 81$, so $\sqrt{81} = 9$.
$12 \times 12 = 144$, so $\sqrt{144} = 12$.

Step 2 The operation is addition, so add the square roots.
$9 + 12 = 21$
The sum of $\sqrt{81}$ and $\sqrt{144}$ is 21.

Find Cube Roots

The expression 7^3 is sometimes called "7 cubed" or "the cube of 7." The exponent 3 indicates that the base is cubed. Since 7^3 is the same as $7 \times 7 \times 7$, the value of 7^3 is 343.

The expression $\sqrt[3]{343}$ is read, "the **cube root** of 343." The radical sign has a 3 in the corner to indicate that this is a cube root. The cube root of a number is the one number that multiplied three times will give the cube of the number. A **perfect cube** is a number whose cube root is an integer. For example, 8 is a perfect cube because $2^3 = 8$.

Example 6 Find the Cube Root of a Number

Find the value of $\sqrt[3]{125}$.

Step 1 Think: What number multiplied three times is 125?
$$n \times n \times n = 125$$
$$5 \times 5 \times 5 = 125$$

Step 2 Write the cube root.
$\sqrt[3]{125} = 5$; the value of $\sqrt[3]{125}$ is 5.

Example 7 Use a Calculator to Find a Cube Root

Find the value of $\sqrt[3]{1,728}$.

Press .

The display should read [   ].

The value of $\sqrt[3]{1,728}$ is 12.

Example 8 Solve Problems Involving Cube Roots

Find the side length of a cube with a volume of 729 cubic centimeters.

Step 1 The volume of a cube is the length of one side cubed. Find the value of $\sqrt[3]{729}$ to find the side length of the cube.

Try 8. | Try 9.
$8 \times 8 \times 8 = 512$ | $9 \times 9 \times 9 = 729$.

Step 2 Write the cube root.
$\sqrt[3]{729} = 9$; the side length of the cube is 9 centimeters.

THINK ABOUT MATH

Directions: Find the value of each of the following. Use a calculator if necessary.

1. $\sqrt[3]{8}$ 3. $\sqrt[3]{27}$ 5. $\sqrt[3]{216}$ 7. $\sqrt[3]{64}$

2. $\sqrt[3]{1,000}$ 4. $\sqrt[3]{3,375}$ 6. $\sqrt[3]{15,625}$ 8. $\sqrt[3]{5,832}$

Vocabulary Review

Directions: Fill in the blanks with one of the words or phrases below.

cube root   perfect cube   perfect square   radical sign   square root   squared

1. A _____ is the symbol for the square root of a number.

2. The _____ of 121 is 11.

3. If the square root of a number is a whole number, then it is a _____.

4. A number that is _____ is multiplied by itself.

5. The _____ of 64 is 4.

6. A _____ has a whole number for the cube root.

Skill Review

Directions: Study the table below. Then answer the questions.

x	$\sqrt[3]{x}$	x	$\sqrt[3]{x}$
1	1	1,331	11
8	2	1,728	12
27	3	2,197	13
64	4	2,744	14
125	5	3,375	15
216	6	4,096	16
343	7	4,913	17
512	8	5,832	18
729	9	6,859	19
1,000	10	8,000	20

1. Describe what the columns and rows in the table show.

2. Describe any patterns in the table.

3. Explain the ways in which you could use the data in the table.

4. Explain how you could use the data in the table to approximate the value of $\sqrt[3]{326}$.

Skill Practice

Directions: Choose the best answer to each question.

1. Between which two consecutive whole numbers does $\sqrt{33}$ lie?

   A. 3 and 4
   B. 4 and 5
   C. 5 and 6
   D. 6 and 7

2. What is the value of $\sqrt{289 - 225}$?

   A. 64
   B. 8
   C. 4
   D. 2

Directions: Answer the following questions.

3. A square parking lot has a total area of 6,400 square meters. What is the length, in meters, of one side of the parking lot?

   _____

4. A photo shop makes custom photo cubes. Elliot needs to make a photo cube with a volume of 2,744 cubic centimeters. What is the length, in centimeters, of each side of the cube?

   _____

Scientific Notation

KEY CONCEPT: Develop understanding of large numbers to include scientific notation and how to translate between numbers written in scientific notation and standard notation.

Evaluate each expression.

1. 12^4 3. 4^3 5. 2^6

2. 10^2 4. 10^1 6. 3^5

Lesson Objectives

You will be able to

- Translate standard notation to scientific notation

- Translate scientific notation to standard notation

Skills

- **Core Practice:** Attend to Precision

- **Core Skill:** Perform Operations

Vocabulary

annex zeros
powers of ten
scientific notation
standard notation

Translate Standard Notation to Scientific Notation

Scientific notation is a way to write very large numbers (or very small numbers) using multiplication and **powers of ten**, such as 10^3, 10^8, 10^{12}, and so on. Scientists and others who work with very large numbers, such as the distance from Earth to Saturn, use scientific notation, because the numbers are too great in standard notation. **Standard notation** is the way we generally represent numbers in everyday usage. In standard notation, the distance from Earth to Saturn is about 1,320,000,000 kilometers. In scientific notation, the distance is written as 1.32×10^9 kilometers.

A number written in scientific notation includes a number greater than or equal to 1 and less than 10 multiplied by a power of 10. Some examples are 8×10^4, 2.1×10^{22}, and 5.273×10^5.

Example 1 Write a Number in Scientific Notation

Write 25,500,000 in scientific notation.

Step 1 Move the decimal point to the left so that the number to the left of the decimal point is between 1 and 10. Write the number as a decimal. Drop the zeros.   25,500,000   2.55

Step 2 Count the number of places the decimal point moves to the left, and write the number of places as the exponent of a power of ten.   7 places $= 10^7$

Step 3 Write the number times the power of ten. 25,500,000 written in scientific notation is 2.55×10^7. 2.55×10^7

MATH LINK

A number written in scientific notation will always have one non-zero digit to the left of the decimal point.

MATH LINK

If a number written in standard notation is greater than 1, the exponent of the power of 10 will be positive.

THINK ABOUT MATH

Directions: Write each number in scientific notation.

1. 18,400
2. 453,260,000
3. 20,000,000
4. 870,000
5. 12,650,000,000
6. 9,348,000

Translate Scientific Notation to Standard Notation

When translating standard notation to scientific notation, move the decimal point to the right and **annex zeros,** if necessary. Annex, or add, zeros so that the number of places after the decimal point in the original number is the same as the exponent in the power of ten.

Example 2 Write a Number in Standard Notation

Write 3.9×10^5 in standard notation.

Step 1 Use the power of ten to determine how many places to move the decimal point.
The exponent of 10^5 is 5, so move the decimal point 5 places.

Step 2 Move the decimal point to the right. Annex zeros, if needed.

3.90000

Step 3 Write the number in standard notation.
3.9×10^5 is 390,000 in standard notation.

390,000

Example 3 Use a Calculator to Translate Scientific Notation to Standard Notation

Translate 5.9874×10^8 to standard notation.

Press (2nd) (5) (.) (9) (8) (7) (4) (x10ˣ) (8) (enter).

The display should read

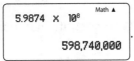

5.9874×10^8 written in standard notation is 598,740,000.

Numbers that are expressed in scientific notation can be thought of as mathematical expressions. Fortunately, the expressions in scientific notation contain only one operation: multiplication. The calculation is simple—in theory, at least.

If you're not careful, however, when converting numbers in scientific notation to numbers in standard notation, you could come up with a number that is much too large or much too small. Make sure you convert to the correct number of decimal places.

Look at the pattern in the table below. Notice what happens to the standard form of the number as the power of ten increases by 1.

Power of Ten	Standard Notation
10^0	1
10^1	10
10^2	100
10^3	1,000
10^4	10,000
10^5	100,000

In a notebook, explain the pattern in the table. Then use the pattern to predict the number of zeros in 10^{14}. Finally, write the number in standard form.

The Voyager Interstellar Mission is a pair of satellites, Voyager I and II, which NASA sent into space in September and August 1977, respectively. As stated on NASA's Voyager website, its mission is "to extend the NASA exploration of the solar system beyond the neighborhood of the outer planets to the outer limits of the Sun's sphere of influence, and possibly beyond".

Voyager I is approximately 18.467 billion kilometers and Voyager II is approximately 15.171 billion kilometers from Earth. In a notebook, write down both distances in scientific notation.

THINK ABOUT MATH

Directions: Write each number in standard notation.

1. 3.1×10^5

2. 7×10^{12}

3. 4.06×10^6

4. 2.913×10^8

Directions: Use a calculator to find each number in standard notation.

5. 6.641×10^8

6. 1.002×10^7

7. 5.9×10^9

8. 8.22×10^4

Vocabulary Review

Directions: Write the word next to the statement it matches.

annex zeros   powers of ten   scientific notation   standard notation

1. _____ a way of writing numbers in which a number between 1 and 10 is multiplied by a power of ten

2. _____ expressions, such as 10^9, in which 10 is written with an exponent

3. _____ a way of writing numbers, such as 4,445

4. _____ adding zeros to a number so that a number has the correct number of places

Skill Review

Directions: Apply what you have learned about converting numbers from standard notation to scientific notation or from scientific notation to standard notation to answer the questions below.

1. Translate the numbers 3,786,000,000 and 92,433 to scientific notation.

_____

2. Translate 4.0×10^{12} and 1.9236×10^7 to standard notation.

_____

Skill Practice

Directions: Choose the best answer to each question.

1. What is 853,491 written in scientific notation?

    A. 0.853491×10^6
    B. 8.53491×10^5
    C. 85.3491×10^4
    D. 853.491×10^3

2. What is 3.4587×10^7 written in standard notation?

    A. 34,587,000
    B. 34,587
    C. 3,458,700
    D. 345,870

3. Which of the following is true of scientific notation?

    A. The number must be positive.
    B. The exponent of 10 must be zero.
    C. The number must be between 1 and 9.
    D. The exponent of 10 must be greater than zero.

4. What must be done to convert from scientific to standard notation?

    A. Annex the same amount of zeros as the exponent of 10 at the end of the number.
    B. Move the decimal point to the right past the last digit of the number.
    C. Rewrite the number without a decimal point.
    D. Move the decimal point to the right the same number as the exponent of ten and annex enough zeros (if needed).

5. The distance of Earth to the Sun is approximately 93,000,000 miles. What is that number using scientific notation?

    A. 93×10^6
    B. 9.3×10^6
    C. 9.3×10^7
    D. $.93 \times 10^7$

Directions: Choose the best answer to each question.

1. Which of the following has the greatest value?

   A. 200^1 C. 1^{200}
   B. 5^4 D. 2^{10}

2. What is 32,450 written in scientific notation?

   A. $3,245 \times 10^2$
   B. 324.5×10^3
   C. 3.245×10^3
   D. 3.245×10^4

3. How can $\sqrt{90}$ be approximated?

   A. Divide 90 by 2.
   B. Find the two perfect squares closest to 90. $\sqrt{90}$ is between the square roots of those perfect squares.
   C. $\sqrt{90}$ is between 81 and 100.
   D. Add 10 to 90 and subtract 10 from 90 and find the square roots of those two numbers.

4. What is the value of $\sqrt[3]{64}$?

   A. 4   C. 16
   B. 8   D. 262,144

5. Which two expressions are equal?

   A. $3^2 + 4^2$ and 5^2
   B. $\sqrt[3]{124}$ and 24^2
   C. $13^2 - 8^2$ and 5^2
   D. $4^2 + 5^2$ and 9^2

6. Which has the greatest value?

   A. 8.3×10^4
   B. 6.7×10^3
   C. 4.3×10^3
   D. 1.2×10^5

7. An Astronomical Unit (AU) is the mean distance from Earth to the Sun. One unit is about 93,000,000 miles. What is this distance written in scientific notation?

   A. $93 \times 1,000,000$
   B. 9.3×10^6
   C. $9.3 \times 10,000,000$
   D. 9.3×10^7

8. How do you determine how 6.04×10^5 is written in standard form?

   A. Drop the decimal point and annex 5 zeros to the right of the 4.
   B. Move the decimal point five places to the right and annex 3 zeros to the right of the 4.
   C. Move the decimal point 5 places to the left and annex 4 zeros to the left of the 6.
   D. Move the decimal point 2 places to the right and annex 3 zeros to the right of the decimal point.

Review

9. There are 25 members of a soccer club. Each member pays $25 in monthly dues. How much does the soccer club collect in dues each month?

    A. $2,500   C. $500
    B. $625   D. $50

10. What does the expression 6^3 mean?

    A. $3 \times 3 \times 3 \times 3 \times 3 \times 3$
    B. 6×3
    C. $6 + 6 + 6$
    D. $6 \times 6 \times 6$

11. What is the value of $3^2 + 6 \times 2 - 15$?

    A. 6
    B. 29
    C. 15
    D. −69

12. Which is the best approximation for $\sqrt[3]{145}$?

    A. $12 < \sqrt[3]{145} < 13$
    B. $10 < \sqrt[3]{145} < 11$
    C. $8 < \sqrt[3]{145} < 9$
    D. $5 < \sqrt[3]{145} < 6$

13. The volume of a cube is 3,375 cubic centimeters. What is the length of one edge of the cube?

    A. 15 cm
    B. 58.1 cm
    C. 125 cm
    D. 1125 cm

14. Jiao is tiling her bathroom with square tiles. The sides of the tiles are 12 inches long. What is the area of each tile in square inches?

    A. 24
    B. 48
    C. 72
    D. 144

Check Your Understanding

On the following chart, circle the number of any item you answered incorrectly. Near each lesson title, you will see the pages you can review to learn the content covered in the question. Pay particular attention to reviewing those lessons in which you missed half or more of the questions.

Chapter 8: Exponents and Roots	Procedural	Conceptual	Application/ Modeling/ Problem Solving
Exponents pp. 250–253	11	1, 5, 10	9, 14
Roots pp. 254–259	4, 12	3	13
Scientific Notation pp. 260–263	2	6, 8	7

UNIT 4

Data Analysis and Probability

Data

Chapter 9 is an introduction to data. Data are groups of information. The media show representations of data in the form of graphs, plots, and tables. You might see a line graph showing stock prices or a table listing bus or train times. An accurate graph can make it easier for readers to understand the information that is being presented. A misleading graph can lead readers to draw incorrect conclusions. Knowing how to read graphs and knowing the difference between accurate and misleading graphs help you to make informed decisions in many choices in life.

Data are described, in part, by measures of central tendency and range. The measures of central tendency are mean, median, and mode. Range is the difference between the greatest and least values in a data set or the spread of the data. Each of these values has importance in how we interpret data.

The Key Concepts you will study include:

Lesson 9.1: Measures of Central Tendency and Range
Understand how data are collected and then analyzed using measures of central tendency and range.

Lesson 9.2: Graphs and Line Plots
Understand how to analyze data presented in a bar graph, line graph, circle graph, or line plot.

Lesson 9.3: Plots and Misleading Graphs
Understand how to analyze stem-and-leaf plots and to recognize misleading graphs.

Goal Setting

Before reading this chapter, set goals for your learning. Think about how knowing more about data will help you in your daily life.

- What do you hope to learn about reading graphs and tables?

  _____

  _____

  _____

- How do you think learning about data will help you make decisions?

  _____

  _____

  _____

- How do you think finding the mean, median, and mode are related to previous mathematics you have learned?

  _____

  _____

  _____

Measures of Central Tendency and Range

Lesson Objectives

You will be able to

- Find the mean, median, and mode
- Find the range
- Understand measures of central tendency

Skills

- **Core Practice:** Model with Mathematics
- **Core Skill:** Calculate Mean, Median, and Mode

Vocabulary

data
mean
measures of central tendency
median
mode
range

KEY CONCEPT: Understand how data are collected and then analyzed using measures of central tendency and range.

Add, subtract, or divide.

1. $28 + 35 + 17 + 24$
2. $39 - 2$
3. $180 \div 4$
4. $22 + 6 + 0 + 33 + 9$
5. $64 - 46$
6. $46 \div 5$

Data

Data are information that is collected and analyzed. It is often, but not always, numerical. Statisticians use many different methods for collecting and analyzing data. Two important characteristics of a set of data are its center (measures of central tendency) and its spread (range).

Measures of Central Tendency

The mean, median, and mode are called the **measures of central tendency** because they are measures that describe the center of a data set. One of these measures might be more appropriate than another for a given data set. Here are some definitions of these measures.

The **mean** is the average value of a data set.

The **median** is the middle value of a data set listed in order from least to greatest. If there are an even number of values in the data set, then the median is the average of the two middle values.

The **mode** is the item that occurs most often in a data set.

For the examples, use Robin's running data to find these measures of central tendency.

Robin's Running Data

Robin is training for a 5-kilometer race. Each day, she runs 5 kilometers and records her time to the nearest minute. Here are the data she collected one week: 20, 24, 22, 22, 21, 20, 25.

Example 1 Find the Mean

Refer to Robin's running data. Find the mean.

Step 1 Find the sum of all the items in the data set.
$20 + 24 + 22 + 22 + 21 + 20 + 25 = 154$

Step 2 Count the number of items in the data set: 7 items.

Step 3 Divide the sum by the number of items in the data set.
$154 \div 7 = 22$. The mean is 22 minutes.

Example 2 Find the Median of an Odd Number of Items

Refer to Robin's running data. Find the median.

Step 1 List items in order from least to greatest.
20, 20, 21, 22, 22, 24, 25

Step 2 Count the number of items in the data set: 7 items.

Step 3 Since the number of items is odd, identify the middle value
in the ordered list.
20, 20, 21, <u>22</u>, 22, 24, 25
The middle value is 22. The median is 22 minutes.

Example 3 Find the Median of an Even Number of Items

Find the median of this data: 20, 24, 22, 22, 21, 20.

Step 1 List items in order from least to greatest.
20, 20, 21, 22, 22, 24

Step 2 Count the number of items in the data set: 6 items.

Step 3 If the number of items is even, identify the two middle values
in the ordered list.
20, 20, <u>21</u>, <u>22</u>, 22, 24
The two middle values are 21 and 22.

Step 4 Find the average (mean) of the two middle values.
$21 + 22 = 43$; $43 \div 2 = 21.5$. The median is 21.5.

Example 4 Find the Mode

Refer to Robin's running data. Find the mode(s), if any.

Step 1 Group items in the data set that are the same.
20, 20   21   22, 22   24   25

Step 2 Find the item or items that occur most often. A set of data
might have one mode, more than one mode, or no modes.
The items 20 and 22 both occur most often (twice).
The modes are 20 minutes and 22 minutes.

MATH LINK

When the number of
items in a data set is
even, the median is
the average of the two
middle values.

The term *central tendency* suggests that these three ways of looking at data all focus on the middle range of a set of numbers. This is certainly true of mean and median, since they both name numbers that are close to the middle. Mode, like median and mean, is certainly a measure of tendency, since it weighs the mean towards it. But even though the repeating number(s) could be close to the middle, the mode could also be at the high or low end of the range of numbers.

In a notebook, find the mean, median, and mode of the following set of numbers. Would you describe the mode as a measure of *central* tendency in this case? Explain your answer.

91, 40, 2, 78, 26, 51, 53, 35, 68, 22, 8, 87, 34, 54, 91, 43

Range

The **range** is the difference between the greatest and least items of a data set. The spread of a data set can be described by its range.

Example 5 Find the Range

Refer to Robin's running data. Find the range.

Step 1 Identify the items with the greatest value and the least value.
Greatest value: 25   Least value: 20

Step 2 Subtract the item with least value from the item with greatest value.
25 − 20 = 5. The range is 5 minutes.

THINK ABOUT MATH

$\sqrt{x}$

Directions: Calculate the mean, median, and mode of each data set.

1. Number of siblings in families: 5, 1, 4, 3, 4, 3, 2, 1, 3, 3, 2, 4, 4.

2. Ages (in years) of employees in a department: 24, 35, 58, 22, 33, 35, 29, 28, 64, 48.

Directions: Find the range for each set of data.

3. Wages for office staff: $340, $478, $370, $370, $865

4. Hours worked by sales department: 35, 48, 29, 35, 35, 50

Vocabulary Review

Directions: Complete each sentence with the correct word.

data   mean   measures of central tendency   median   mode   range

1. The _____ is the average value in a data set.

2. _____ can be collected and analyzed.

3. The _____ is the item that occurs most often in a data set.

4. If the data set has an odd number of items, then the _____ is the middle value when the data is listed in order from least to greatest.

5. The mean, mode, and median are called the _____.

6. The _____ is the difference between the greatest and least value in the data set.

Directions: Use the method for sorting numbers that you learned in order to find the median and mode of the following data sets. Then determine the median and mode of each data set.

1. 21, 78, 69, 71, 31, 92, 67, 16, 27, 74, 43, 67, 63, 33, 28, 30, 13, 92, 72, 81, 70, 86, 34, 48

2. 95, 36, 70, 37, 99, 70, 37, 74, 62, 67, 96, 59, 42, 95, 74, 12, 17, 37, 95, 14, 88, 22, 43, 29

3. 47, 51, 84, 33, 20, 17, 83, 23, 88, 96, 35, 54, 21, 19, 81, 63, 76, 5, 16, 9, 42, 38, 92, 77, 3

Directions: Think of a scenario for which the following might be possible.

4. Can the mean, median, and mode of a data set ever be equal (all the same number)? Explain your answer.

Skill Practice

Directions: Choose the best answer to each problem.

1. Houses have sold in Parwana's neighborhood for $85,000; $108,000; $95,500; $120,000; $105,000; $99,900; and $124,000. What is the median price for the houses sold?

    A. $39,000
    B. $105,000
    C. $120,000
    D. $122,900

2. Arnan's test scores for this grading period are 75%, 72%, 88%, 90%, 85%, 100%, 77%, and 86%. What is the mean of Arnan's test scores to the nearest percent?

    A. 88%
    B. 86%
    C. 84%
    D. 28%

3. The age range for players on a soccer team is 10 years. The youngest player is 15 years old. What is the age of the oldest player?

    A. 25 years
    B. 20 years
    C. 12.5 years
    D. 5 years

4. Which sentence can help you determine the mode in the following data set? Favorite water sport: swimming, skiing, scuba diving, swimming, skiing, fishing, rafting, skiing, sailing

    A. Since swimming occurs twice in the list, swimming is the mode of the data set.
    B. Since swimming and skiing occur more than once in the data list, swimming and skiing are the modes of the data set.
    C. Because fishing and rafting are the only sports that do not begin with the letter s, fishing and rafting are the modes of the data set.
    D. Since skiing occurs more often than any other sport, skiing is the mode of the data set.

Graphs and Line Plots

KEY CONCEPT: Understand how to analyze data presented in a bar graph, line graph, circle graph, or line plot.

Find the mean, median, mode, and range of each set of data.

1. Price of books on sale: $14, $19, $20, $15, $14, $16, $19, $22, $14

2. Temperatures in degrees Fahrenheit: 35, 63, 44, 54, 77, 93, 35, 63

Bar Graphs

Data can be displayed in many different ways. A **graph** gives a visual picture of the data. With a graph, you can often see things about the data that are difficult to see by looking only at the numbers.

A **bar graph** can help you make visual comparisons among numerical data. A bar graph is made up of rectangular bars that extend upward or lengthwise. The height of each bar corresponds to one number in the data. The higher the bar, the greater the number. The bar graph below displays the data given in the table.

AVERAGE MONTHLY TEMPERATURES IN ANCHORAGE, ALASKA

Month	Temperature (°F)
January	13
April	36
July	58
October	36

AVERAGE MONTHLY TEMPERATURES IN ANCHORAGE, ALASKA

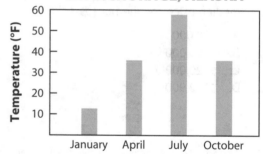

The table and the graph provide the same information, but the graph makes comparing quick and easy. For example, you can see that April and October have the same average monthly temperature because their bars are the same height.

Here are some things to notice about the bar graph. This information will help you to read, interpret, and draw bar graphs.

• Along the **horizontal axis**, the bottom of the graph, the months being compared are listed: January, April, July, and October.

• Along the **vertical axis**, the left side of the graph, numbers representing temperatures from 10°F to 60°F are listed.

• The top of each bar aligns with the correct number on the vertical axis. In this graph, a bar for each of the four months gives the average temperature for the month.

MAKE PREDICTIONS

A **prediction** is an attempt to answer the question, "What will happen next?" Predicting gets you involved in what you are reading and can keep you interested. When predicting, readers use clues in the text, prior knowledge, and experience to make reasonable guesses—or guesses that make sense—about what will happen next.

Here are some key points to remember when predicting:

- Start by looking at the author's name and the title to help predict what the passage is about.

- Make predictions before and during your reading. Predicting during reading means thinking ahead about how things might turn out.

- Use prior knowledge. Ask yourself: *Have I studied something like this before? What happened next in that case?*

- **Adjust**, or change, your prediction as you read. Remember that predictions should make sense, but they don't have to be right.

Read the following passage. Predict what will happen on Thursday if the weather is 80°F and sunny.

> Karen has a vegetable booth at the farmer's market. On Monday, it was 75°F and sunny, and she took in $150. On Tuesday, it was 68°F and cloudy, and she took in $53. On Wednesday, there was a steady downpour, and she took in $28.

Karen's sales are correlated to the weather. When the weather is cold or wet, her sales decrease. Because Thursday is warm and sunny, she will likely make good sales.

Core Skill
Interpret Data Displays

The tables and lists that you have already worked with are not the only sorts of visual tools that can be used to display data. You can also picture data in graphs. In fact, graphs often reveal a **trend**, or developing pattern, that you might not have spotted otherwise. Thus, by revealing trends that exist over time, graphed data enables you to make predictions. You just need to know how to interpret the data.

A relationship between two quantities can be shown in a graph. For example, look at the bar graph about the temperatures in Anchorage, Alaska, on page 274. The graph shows the relationship between the month and the temperature. From the shape of the graph, you can make a prediction about the average temperatures in Anchorage.

Imagine you are planning a trip to Alaska. Write one prediction about temperatures in Anchorage that can be made by analyzing the graph.

The next three examples show how to read the bar graph below to answer questions about the data displayed.

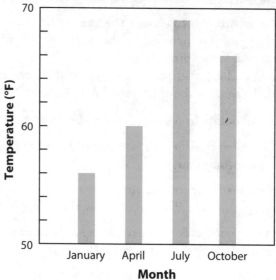

AVERAGE MONTHLY TEMPERATURES IN LOS ANGELES, CALIFORNIA

Example 1 Read the Horizontal Scale on a Bar Graph

What is the average monthly temperature in Los Angeles in April?

Find the bar for April. Trace a horizontal line from the top of this bar to the vertical axis. Read the number on the axis: 60.

The average temperature in Los Angeles in April is 60°F.

Example 2 Read the Vertical Scale on a Bar Graph

In what month is the average temperature in Los Angeles 69°F?

Since 69 does not appear on the vertical axis, determine that each mark represents 2 degrees. Trace a horizontal line from where 69 would be (halfway between 70 and the mark directly below it) to the top of a bar. Read the month for that bar: July.

The average temperature in Los Angeles is 69°F in July.

Example 3 Find the Range of Data Displayed on a Bar Graph

What is the range of average monthly temperatures in Los Angeles?

Step 1 Read from the graph the lowest temperature, shown by the lowest bar: 56°F. Read from the graph the highest temperature, shown by the highest bar: 69°F.

Step 2 Subtract to find the range: 69°F − 56°F = 13°F.

To read a bar graph, find the bar that displays the information you seek, trace a line from the top of the bar to the vertical axis, and read the number on the axis. In some cases, do the opposite. Find the number first, then the bar.

MATH LINK

A **point** is a mark made on a graph to represent the position of a data value. The line you draw to connect two points on a graph is called a **line segment**.

Line Graphs

A **line graph** can help you see patterns and trends in data. A line graph is made up of points that are connected by line segments. Line graphs are often used to display data over a period of time.

The line graph below displays the data given in the table.

STOCK FUND A, NOVEMBER 21, 1999–2009

Year	Price of One Share
1999	$12.82
2000	$14.74
2001	$16.37
2002	$15.84
2003	$14.89
2004	$12.37
2005	$14.30
2006	$8.14
2007	$22.19
2008	$26.80
2009	$24.86

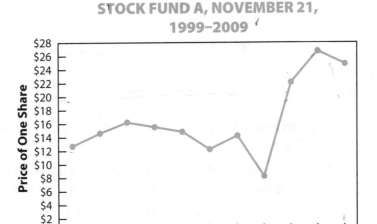

STOCK FUND A, NOVEMBER 21, 1999–2009

The table and the graph provide the same information, but the graph visually shows the dramatic decrease in the price in 2009.

Like bar graphs, line graphs have a horizontal and a vertical axis.

Along the horizontal axis, the years refer to annual dates from November 21, 1999 to November 21, 2009.

Along the vertical axis, numbers that fall within the range of the data represent the price of one share of Stock Fund A. Prices range from about $8 to about $27.

In the body of the graph, the points display the data. Points connected by line segments show how the price changes (increases or decreases) from year to year.

The next two examples show how to read the line graph below.

INVESTMENT INCOME

Example 4 Read the Horizontal Scale on a Line Graph

If you invest $10,000 at an 8% annual interest rate, what will be its value in 25 years?

Find 25 years on the horizontal axis. Find the point directly above 25. Trace a line across to the vertical axis: about $68,000.

In 25 years, the value of the money will be about $68,000.

Example 5 Read the Vertical Scale on a Line Graph

If you invest $10,000 at an 8% annual interest rate, how long will it take for your money to be worth $40,000?

Find $40,000 on the vertical axis. Trace directly across to the graph. Where you intersect the line graph, draw a point. From this point, trace down to the horizontal axis: about 18.

It will take about 18 years for the money to be worth $40,000.

Circle Graphs

A **circle graph** shows parts of a whole. The circle is the whole, or 100%. The circle is divided into parts, and all the parts add up to 100%. So, circle graphs are often used to display data given as percents.

The circle graph on the next page displays the data given in the table. The graph shows that air exhaled by the human body contains parts of nitrogen, oxygen, and carbon dioxide. It also shows the size of each part. For example, nitrogen has the largest part of the circle (78%) because nitrogen is the largest part of exhaled air (78%).

EXHALED AIR

Gas	Percent
Carbon dioxide	4%
Oxygen	18%
Nitrogen	78%

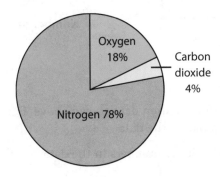

The examples are based on the circle graph above.

Example 6 Find a Percent on a Circle Graph

What percent of exhaled air is oxygen?

Find oxygen in the circle graph. Read the percent: 18%.
Exhaled air is 18% oxygen.

Example 7 Find a Category on a Circle Graph

What makes up 4% of exhaled air?

Find 4% in the circle graph. Read the label: carbon dioxide.
Carbon dioxide makes up 4% of exhaled air.

Line Plots

A **line plot** is a data display that uses a number line with X's (or other marks) to show how often each data value occurs.

THINK ABOUT MATH

Directions: Refer to the circle graph above to answer the following questions.

1. What two gases make up 82% of exhaled air?

2. What percent of exhaled air is nitrogen?

Using graphs to interpret data is one way of understanding what the data means. Sometimes the numbers can be so large or small (or even too many numbers) that it can be hard to see what information can be gathered from the data. Creating graphs from data is actually rather simple. First, pick a topic that you find interesting and that would have multiple categories to compare, for example, "The Population of Six US States". Second, determine the categories you are going to compare, such as Arizona, Delaware, Kansas, Nebraska, Texas, and Virginia. Third, determine the data necessary to compare the categories and choose a specific type of graph you want to use. Lastly, create the graph, whether it is a bar graph, circle graph, line graph, or a line plot.

In a notebook, create a circle graph using data you collect from research on a topic you find interesting and want to know more about. Then determine the percentages in each category and interpret what the data means.

Twelve people were asked how many children they have. The data was recorded in the table below. The line plot displays the data in the table.

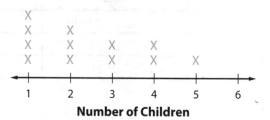

Number of Children					
2	2	3	1	1	4
5	4	2	1	1	3

Number of Children

Example 8 Use a Line Plot to Find the Mean, Median, Mode, and Range

Refer to the Number of Children data shown above. Find the mean, median, mode, and range of the data.

To find the mean, add all the numbers in the data set, then divide by the number of items.
$(1 + 1 + 1 + 1) + (2 + 2 + 2) + (3 + 3) + (4 + 4) + 5 = 29$
$29 \div 12$ is about 2.4. The mean is about 2.4.

To find the median, find the center X on the plot. There are 12 X's on the plot, so the median is the average of the 6th and 7th X from the left. The 6th and 7th X's are both 2's. Their average is 2. The median is 2.

To find the mode, find the data column with the greatest number of X's. The mode is 1. Most people asked have 1 child.

To find the range, subtract the lowest value in the data set from the highest value in the data set. $5 - 1 = 4$. The range is 4.

Vocabulary Review

Directions: Complete each sentence with the correct word.

bar graph   circle graph   graph   horizontal axis   line graph
line plot   trend   vertical axis

1. A _____ displays data given in percents.

2. A _____ shows data over a period of time.

3. The _____ is the scale across the bottom of a graph.

4. The _____ is the scale across the side of a graph.

5. A _____ is made up of rectangular bars that extend vertically or horizontally.

6. A _____ uses X's on a number line to show how often each data value occurs.

7. A _____ can come in many forms. It is a visual picture of data.

8. A _____ is a pattern in the way data relates to other data.

Directions: Refer to the line graph on page 278 to answer each question.

1. If you invest $10,000 at 8% annual interest rate, estimate the value of the money after 20 years.

2. Predict how many times greater the estimated value of the money will be after 30 years than it will be before any interest is calculated.

Skill Practice

Directions: Choose the best answer to each question.

1. The entire nontidal coastline of the United States is 12,383 miles long. The circle graph below shows the percent of the coastline that is on each of the four coasts. How many miles long is the coastline of the United States on the Gulf Coast?

U.S. COASTLINE

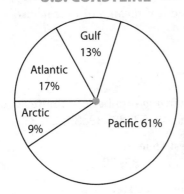

A. about 7,554
B. about 4,829
C. about 2,105
D. about 1,610

3. The bar graph below shows the number of each type of medal won by the United States in the 1994 Winter Olympics. How many more silver medals were won than bronze medals?

MEDALS WON BY THE U.S. IN THE 1994 WINTER OLYMPICS

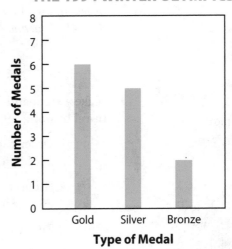

A. 2   C. 4
B. 3   D. 5

2. Refer to the line graph on page 277. During which of the years from 1999 to 2009 was the price of the stock the closest to $20?

A. 2003
B. 2006
C. 2009
D. 2007

4. How can you find the mode of the data displayed in a line plot?

A. Count all the X's and divide by 2.
B. Look at the item with the fewest number of X's above it.
C. Look at item with the greatest number of X's above it.
D. Start from left or right and count X's until you reach the middle one. Read the item below it.

Plots and Misleading Graphs

Lesson Objectives

You will be able to

- Understand stem-and-leaf plots
- Identify misleading displays of data

Skills

- **Core Practice:** Critique the Reasoning of Others
- **Core Skill:** Interpret Data Displays

Vocabulary

key
leaf
mislead
outlier
stem
stem-and-leaf plot

MATH LINK

In a stem-and-leaf plot, use a stem of 0 and a leaf of 8 to show a number such as 8.

Answer the following.

1. Find the mean, median, mode, and range of these data values: 28, 32, 23, 24, 36, 41, 28, 56, 55, and 44.

2. What is the next number in this pattern: 10, 20, 30, 40, _____?

3. What is the next number in this pattern: 0, 6, 12, 18, _____?

Stem-and-Leaf Plots

A **stem-and-leaf plot** is used to organize data. It shows how often data values occur. The plot has two columns. The left column is labeled **stem** and can have one or more digits indicating numbers with place values. The right column is labeled **leaf** and can have only one digit: the last digit of a piece of data. To determine the number of data values, count each leaf. A stem-and-leaf plot will also have a **key**, or legend, to show what the stem and leaf mean. For example: *Key: 45|8 means 458.*

Example 1 Make a Stem-and-Leaf Plot

Make a stem-and-leaf plot of the following number of minutes people travel to work: 29, 25, 60, 38, 53, 55, 38, 53, 39, 39, 35, 27, 37.

Step 1 Order the data from greatest to least.
60, 55, 53, 53, 39, 39, 38, 38, 37, 35, 29, 27, 25

Step 2 Group the numbers with the same stems.
60   55, 53, 53   39, 39, 38, 38, 37, 35   29, 27, 25

Step 3 Title the plot and make a key for the plot.

Minutes Spent Traveling

Stem	Leaf

Key: 3|7 means 37 miles.

Step 4 Complete the plot for the data values given.

Minutes Spent Traveling

Stem	Leaf
6	0
5	3 3 5
4	
3	5 7 8 8 9 9
2	5 7 9

Key: 3|7 means 37 miles.

Core Practice
Critique the
Reasoning of Others

UNDERSTAND PERSUASIVE TECHNIQUES

Writers use many techniques when they are trying to **persuade** readers. Persuading is causing someone to believe something.

One technique writers use is persuasive language. The words that a writer chooses may reflect his or her feelings or beliefs about the topic. The author is trying to persuade the reader to agree with the point of view presented and get the reader to act. Not all writing uses persuasive language. When you read, look for strong words, descriptions, and the details that the author uses.

A second persuasive technique is biased questions, which are worded in such a way that a particular answer is favored over others.

Read the questions below. Decide which response the questioner is hoping to get.

> (1) Would you support the opening of another government office, creating 214 new jobs?
>
> (2) Do you consider the governor's plan a failure because it has not met all of its objectives?
>
> (3) Would you vote for someone with so little experience?

1. The question highlights the creation of jobs, which many people would support. However, it ignores the costs associated with the opening and operating of an office. The questioner wants the answer to be in favor of opening the new office.

2. The respondent is faced with either agreeing that the plan is a failure or trying to justify the plan as a success even though it has not met all of its objectives. The questioner wants the plan to be considered a failure.

3. The phrase *so little experience* is a clue that the questioner does not think the candidate is qualified and wants the respondent to agree.

This lesson talks about the ways in which data can be presented to make you believe something. When studying data that are displayed in a graph, therefore, don't assume that you are looking at a report that accurately represents a certain situation. Consider the blue-card/red-card bar graph in Example 3. Are there really twice as many blue cards as red cards? No, there are only two more blue cards than red cards.

It is always a good idea to ask questions when data have been presented visually. For example, "Can I trust the scales in the graph, or are they set up in a way that may misrepresent the data? Do I recognize signs of bias in the visual presentation? In other words, is this diagram trying to persuade me to adopt a certain point of view, or is it trying to present the information in an accurate and bias-free manner?"

Example 2 Analyze a Stem-and Leaf-Plot

The plot below shows the ages of 12 people taking a survey. Find the median and range of the ages. Do not include any outliers.

Age of Survey Taker

Stem	Leaf
7	0 2 6
6	5 5
5	5
4	8 9
3	4 4 7
2	
1	1

Key: 3|4 means 34 years old.

Step 1 Find the median.
Start at the ends, 76 and 11. 11 is an outlier, so do not count it. Count to the middle beginning with 34. The middle is 55.

Step 2 Find the range. Remember that 11 is an outlier.
76 − 34 = 42. The range is 42 years

Misleading Graphs

The following examples will demonstrate a few ways in which graphs can be made to **mislead,** or lead the reader to make a wrong conclusion.

Example 3 The Vertical Scale Does Not Start at 0

The graphs below show the number of blue, red, and green game cards.

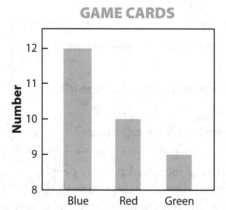

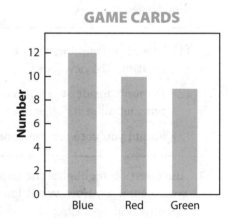

Why do the graphs look different?
The scales are different. The graph on the left does not start at 0.

How does the graph on the left mislead you?
The graph looks as if there are twice as many blue cards as red cards and also twice as many red cards as green cards. However, there are only two more blue cards than red cards, and there is only one more red card than green card.

Example 4 The Vertical Scale Has Irregular Intervals

The graphs below both show the sales of pet fish.

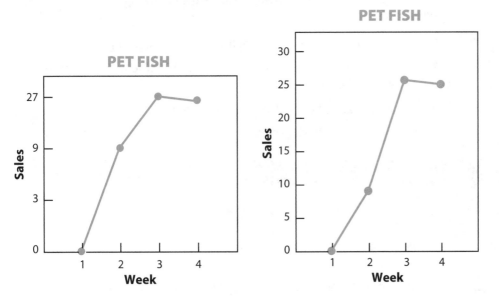

Why do the graphs look different?
The scales are different. The graph on the left has a scale that does not go up in even steps.

How does the graph on the left mislead you?
The graph distorts the data. It makes it look like the biggest increase in sales was from Week 1 to Week 2 rather than from Week 2 to Week 3.

Example 5 Percents Do Not Add to 100%

The following graph was made to show how Mia's workday is spent.

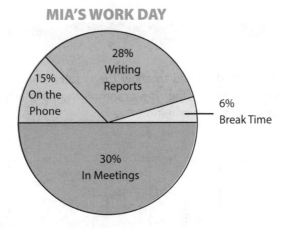

Why is this graph misleading?
The percents do not add up to 100%. 30% + 15% + 28% + 6% = 79%

Why might Mia have chosen to display her work day this way?
She may want it to seem that she spends most of her day in meetings. She may not know what she does in the missing 21% of her day.

Authors who are trying to persuade and not merely inform you can try to reinforce their arguments with misleading statistics or misleading graphs. These are graphs or statistics that technically may be accurate but are presented in a certain way to distort the facts. Therefore, think carefully when evaluating the data. It is important to consider what information is left out of a statistic or graph when you are examining it.

For example, consider the statistic, "86% of respondents consider private school better than public school." Now consider the statistic again knowing that the only people questioned were students currently attending a private preparatory school. The statistic takes on a new meaning. It tells us that 86% of students at this particular school prefer private to public school, but it has no bearing on what the general population thinks about schooling.

It is important to examine statistics and graphs carefully as you read. In a notebook, keep a list of statistics and graphs you encounter in this lesson and in your everyday life. Note whether they are trying to convince the reader of something.

THINK ABOUT MATH

Directions: Answer each of the following.

1. What is one way a circle graph could be misleading?

2. Explain why a vertical scale of 0, 1, 2, 10, 50 could make a bar graph misleading.

3. Use *Key: 25|8 means 25.8* to determine what *16|9* means.

Vocabulary Review

Directions: Complete each sentence with the correct word.

key   leaf   mislead   outlier   stem   stem-and-leaf plot

1. A(n) _____ is a type of graph that shows how often data values occur.

2. The _____ for the number 125 is 12.

3. The _____ can only be one digit, the last digit in the data value.

4. A(n) _____ is a data value that falls well outside the range of the other values.

5. The _____ to a graph is an explanation of the symbols or numbers used in the graph.

6. If a graph's scale does not start at 0, the graph could distort the data and _____ the person reading the graph.

Skill Review

Directions: The two bar graphs below were designed to show the results of a walking competition. The graph on the left was created by Ian. The graph on the right was created by Dora. Refer to these graphs when answering each question.

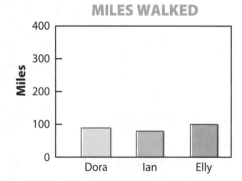

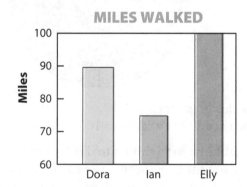

1. What do you think Ian was trying to persuade you of when he made the graph on the left? How has Ian distorted his graph to mislead you?

2. What do you think Dora was trying to persuade you of when she made the graph on the right? How has Dora distorted her graph to mislead you?

Skill Practice

Directions: Choose the best answer to each problem.

1. Which statement best explains why the bar graph below is misleading?

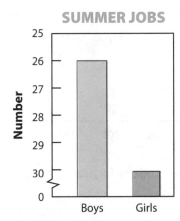

SUMMER JOBS

A. The data should be displayed in a circle graph.
B. The scale does not begin at 0.
C. The intervals in the scale are not equal.
D. The scale is numbered in reverse order.

2. Marcos wants to make a stem-and-leaf plot of the following data values: 1,314; 1,315; 1,320; 1,322; 1,329; 1,331; 1,335; and 1,336. Which key would be most appropriate for Marcos to put with his plot?

A. *Key: 13|1 means 131.*
B. *Key: 131|4 means 1,314.*
C. *Key: 1|314 means 1,314.*
D. *Key: 13|14 means 1,314.*

3. Which best explains why the circle graph below is misleading?

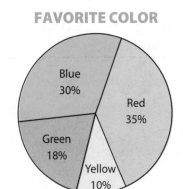

FAVORITE COLOR

A. Percents do not add up to 100%.
B. Each section should be the same color as its label.
C. There are more colors than the four shown in the graph.
D. It does not say the age of the people who voted for the colors.

4. Which best explains why the line graph below is misleading?

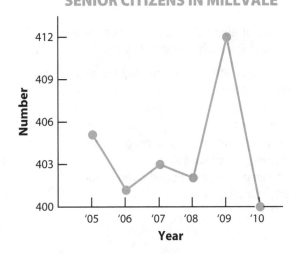

SENIOR CITIZENS IN MILLVALE

A. The graph does not say why the number of senior citizens changes each year.
B. The scale does not begin at 0.
C. The scale does not go up to 500.
D. The scale does not increase by ones, twos, or tens.

Directions: Choose the best answer to each question.

Questions 1 and 2 refer to the following graph.

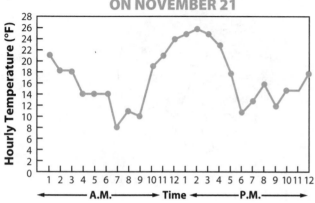

**HOURLY TEMPERATURES
ON NOVEMBER 21**

1. During which hour was the temperature change greatest?

    A. 6 A.M. to 7 A.M.
    B. 9 A.M. to 10 A.M.
    C. 5 P.M. to 6 P.M.
    D. 7 P.M. to 8 P.M.

2. Sunset was at about 5:00 p.m., and sunrise was about 7:30 a.m. Which statement is true?

    A. The temperature rose steadily from sunrise until midafternoon.
    B. The temperatures at sunset and midnight were about the same.
    C. The temperature did not drop in the afternoon until sunset.
    D. The temperature dropped overall during the early morning hours until sunrise.

Use the following line plot to answer Questions 3 through 5.

MONTHLY ELECTRIC BILLS

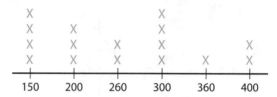

3. What is the mode(s) for the monthly electric bills?

    A. $300
    B. $200 and $150
    C. $150
    D. $150 and $300

4. How many households paid less than $300 for the electric bills for the month?

    A. 3   C. 9
    B. 4   D. 13

5. Which is the mean for the monthly electric bills?

    A. $275   C. $250
    B. $253.13   D. $103.13

6. Faysal found that the numbers of hours worked by members of his department were 35, 37, 41, 34, 18, 35, 35, 35, 32, 40, 38. If he drops the outlier from the data, which of the statements is true?

    A. The range will not change.
    B. The mean will not change.
    C. The mode will change.
    D. The median will not change.

Review

Use the following bar graph to answer Questions 7 and 8.

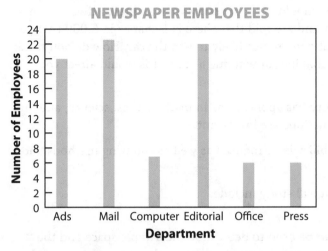

NEWSPAPER EMPLOYEES

8. How can the range of the number of employees in each department be found?

    A. Subtract the number of employees in the Office from the number in the Mail department.

    B. Subtract the number of employees in the Computer department from the number in the Mail department.

    C. Subtract the number of employees in the Press department from the number in the Ads department.

    D. Subtract the number of employees in the Editorial department from the number in the Ads department.

7. Which three departments have a combined total of 34 employees?

    A. Editorial, Computer, Office
    B. Computer, Press, Ads
    C. Office, Computer, Mail
    D. Office, Press, Editorial

Use the bar graph on the right to answer Question 9.

9. Sabrina took a poll of her 10 classmates asking which type of juice they liked better: apple or orange. She placed the results in the bar graph shown. What can be said of her findings?

    A. The number of students who prefer Apple is double that of Orange.
    B. The number of students who prefer Orange is double that of Apple.
    C. The graph is misleading because the totals don't add up to 12.
    D. The graph is misleading because the scale does not start at 0.

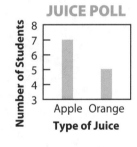

JUICE POLL

Check Your Understanding

On the following chart, circle the number of any item you answered incorrectly. Near each lesson title, you will see the pages you can review to learn the content covered in the question. Pay particular attention to reviewing those lessons in which you missed half or more of the questions.

Chapter 9: Data	Procedural	Conceptual	Application/ Modeling/ Problem Solving
Measures of Central Tendency and Range pp. 270–273	6	5	7
Graphs and Line Plots pp. 274–281	8	1, 2, 3, 4	
Plots and Misleading Graphs pp. 282–287	9		

Probability

Probability is the likelihood that an event will occur. The weather forecaster says that there is a 75% chance of rain. This means that it is likely to rain, and you should probably carry your umbrella. If you enter a drawing for a new car and are told that there is 1 chance in 5,000 that you will win, it means that you are not likely to win the car. How do you think knowing what the probability of winning a contest is would affect your decision to enter?

The study of probability has applications in mathematics, science, and any activity in which predictions are important.

In Chapter 10, probability is introduced as well as counting methods and compound events.

The Key Concepts you will study include:

Lesson 10.1: Counting Methods
Counting methods can be used to determine the sample space and the number of possible outcomes in experiments.

Lesson 10.2: Introduction to Probability
Understand and use concepts of probability to find probabilities and make predictions.

Lesson 10.3: Compound Events
Extend your understanding of probability to finding the probability of compound events.

Goal Setting

Before starting this chapter, set goals for your learning. Think about how learning more about probability could be useful in your daily life.

- What do you know already about probability?

  _____

  _____

  _____

- What do you hope to learn about probability?

  _____

  _____

  _____

- How do you think learning about probability will help you make decisions?

  _____

  _____

  _____

  _____

Counting Methods

Lesson Objectives

You will be able to

- Count possible outcomes

- Understand and use tree diagrams

Skills

- **Core Skill:** Utilize Counting Techniques

- **Core Practice:** Reason Abstractly

Vocabulary

compound event
Counting Principle
event
outcome
sample space
tree diagram

MATH LINK

A tree diagram or a table gives the sample space and the number of possible outcomes in a compound event. The Counting Principle gives *only* the number of possible outcomes.

KEY CONCEPT: Counting methods can be used to determine the sample space and the number of possible outcomes in experiments.

Find each product.

1. 8×14 **2.** 15×12 **3.** $6 \times 4 \times 7$ **4.** $3 \times 5 \times 8$

Count Possible Outcomes

An **outcome** is the result of an experiment, such as flipping a coin and having it land on tails. An **event** is one or more outcomes of an experiment. A **compound event** is the result of two or more events, such as rolling two sixes when rolling two number cubes. A **sample space** is a list of all possible outcomes.

Example 1 Use a Table

Find the sample space and possible outcomes for flipping two coins.

Step 1 Make a table. Let H = head, and let T = tail. The two possibilities for one coin are given in the top row: H and T. The two possibilities for the other coin are given in the left column: H and T. Complete the table. This gives all the possible outcomes.

	H	T
H	HH	HT
T	TH	TT

Step 2 Find the sample space and the total number of possible outcomes. The sample space is HH, HT, TH, and TT. There are four possible outcomes.

A **tree diagram** is a special type of diagram that shows the sample space and the number of possible outcomes.

Example 2 Use a Tree Diagram

Find the sample space and total possible outcomes for choosing a cone or cup; vanilla, chocolate or strawberry; and sprinkles or hot fudge.

Step 1 Draw a tree diagram. Each branch of the tree leads to a possible outcome. See the diagram on the next page.

Step 2 Find the sample space and the total number of possible outcomes. Let O stand for cone, U stand for cup, V stand for vanilla, C stand for chocolate, S stand for strawberry, P stand for sprinkles, and H stand for hot fudge. The sample space is OVP, OVH, OCP, OCH, OSP, OSH, UVP, UVH, UCP, UCH, USP, USH. There are 12 possible outcomes.

UNDERSTAND A DIAGRAM

A diagram is a drawing or a graphic illustration that presents information. It may show how something is organized, how something works, or other features. It can stand alone or with a passage.

A diagram may give additional information related to the passage or restate the information in a different way. Diagrams have **labels** that give more information about the drawing. Labels are words or numbers written on the diagram to identify what the images represent.

Describe how the diagram below relates to the passage.

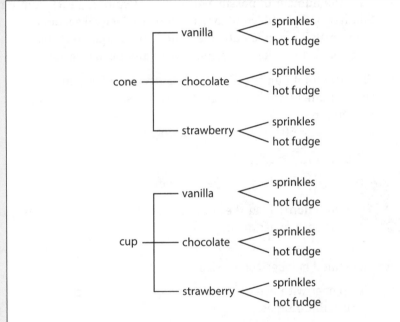

An ice cream stand offers 3 flavors, vanilla, chocolate, and strawberry, in a cone or a cup. There are 2 toppings: hot fudge or sprinkles. If a customer chooses 1 container, 1 flavor, and 1 topping, there are 12 possible combinations.

The diagram is a tree diagram. It reinforces and adds to the information in the passage. The passage states that there are 12 possible combinations. The tree diagram proves there are 12 combinations and shows what each possible combination is.

Core Skill
Utilize Counting Techniques

Tree diagrams are a valuable tool for counting all possible outcomes because they visually show all possible combinations of outcomes. Study the tree diagram. Use it as a model to solve the following problem. The spinner below is spun twice.

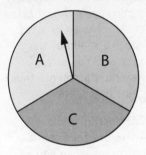

Copy and complete the tree diagram below to count the possible outcomes of spinning the spinner.

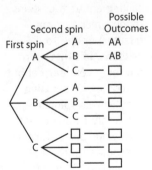

Earlier, you learned about the range of a set of numbers, which is one way to describe a set of numbers. Probability, which you will learn about in the next lesson, is a way to describe events. In probability, tree diagrams are used to help show possible outcomes. You then use this information that is represented visually to reach certain conclusions about the data. In other words, you begin with the concrete: occurrences of particular events. Then you move into the realm of the abstract. You arrive at general statements that describe events.

Consider Example 3. There are four possible outcomes when you spin Spinner C: A, B, C, D. Do you have to use letters for the four possible outcomes? Not really. You could be talking about four coins (penny, nickel, dime, quarter) or the four seasons. In other words, anything you learn about outcomes by spinning the 4-letter spinner will apply to any event involving four different outcomes that are equally possible.

The **Counting Principle** uses multiplication to get the total number of possible outcomes in a compound event.

Example 3 Use the Counting Principle

Find the total number of possible outcomes of spinning the three spinners below.

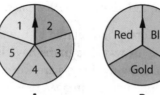

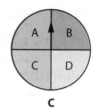

A   B   C

Step 1 Find the number of possible outcomes of spinning each spinner.
Spinner A: 5 possible outcomes; there are 5 equal sections.
Spinner B: 3 possible outcomes; there are 3 equal sections.
Spinner C: 4 possible outcomes; there are 4 equal sections.

Step 2 Find the total number of possible outcomes of spinning the three spinners. Multiply together the possible outcomes of each spinner.
$5 \times 3 \times 4 = 60$. There are 60 possible outcomes.

THINK ABOUT MATH

Directions: Find the total number of possible outcomes for each experiment.

1. Roll a 6-sided number cube twice.

2. Spin a spinner with five equal sections of red, blue, yellow, green, and purple three times.

Vocabulary Review

Directions: Match each word to one of the phrases below.

1. _____ compound event

2. _____ Counting Principle

3. _____ event

4. _____ outcome

5. _____ sample space

6. _____ tree diagram

A. one or more outcomes of an experiment

B. a diagram with branches that gives the sample space and total number of possible outcomes

C. one result of an experiment

D. a list of all possible outcomes

E. two or more events

F. uses multiplication to find the total number of possible outcomes

Directions: Answer the following questions.

1. How does using tree diagrams help you determine possible outcomes for an event?

_____

2. Describe the relationship between the passage and tree diagram below.

   A flower shop offers 2 choices of vase and 4 choices of flowers. The vases are glass or stoneware. The flowers are lilies, roses, gerbera daisies, or a combination of all three. If a customer chooses 1 vase and 1 selection of flowers, there are 8 possible combinations.

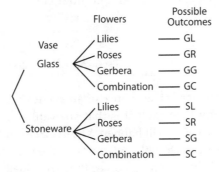

_____

3. Draw a tree diagram that shows the sample space for spinning a spinner with five equal sections of A, B, C, D, and E and flipping a coin. Then state the total number of possible outcomes.

   Total possible outcomes: _____

Skill Practice

Directions: Choose the best answer to each question.

1. Bomani wants to know the sample space for spinning a spinner with 3 equal sections numbered 1, 2, and 3 twice. Which of the following is the sample space?

   A. 1, 1; 2, 2; 3, 3
   B. 1, 1; 1, 2; 1, 3; 2, 1; 2, 2; 2, 3; 3, 1; 3, 2; 3, 3
   C. 6
   D. 9

2. Jacinda drew a tree diagram to find the total number of possible outcomes of boys and girls in a family of three children. How many possible outcomes are there?

   A. 3   C. 8
   B. 6   D. 9

3. A number cube is rolled once, and a spinner with four equal sections labeled A, B, C, D is spun once. What is the total number of possible outcomes?

   A. 24
   B. 48
   C. 96
   D. 144

4. Hiroko is choosing a 4-digit PIN number for her bank account. Each digit can be a number from 0 to 9, and the numbers can repeat. What is the total number of possible PIN numbers?

   A. 100   C. 10,000
   B. 1,000   D. 100,000

Introduction to Probability

Lesson Objectives

You will be able to

- Find theoretical probability

- Find experimental probability

- Make predictions

Skills

- **Core Skill:** Evaluate Reasoning

- **Core Skill:** Determine Probabilities

Vocabulary

certain event
combination
experimental probability
impossible event
permutation
probability
support
theoretical probability
trials

MATH LINK

The closer the probability of an event is to 0, the less likely it is to happen. The closer the probability of an event is to 1, the more likely it is to happen. A probability of $\frac{1}{2}$ means the event is equally likely to happen or not to happen.

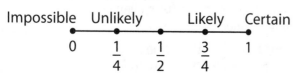

Impossible	Unlikely		Likely	Certain
0	$\frac{1}{4}$	$\frac{1}{2}$	$\frac{3}{4}$	1

KEY CONCEPT: Understand and use concepts of probability to find probabilities and make predictions.

Find the total number of possible outcomes.

1. Roll a number cube twice.

2. Spin a spinner with three equal sections, A, B, and C, three times.

3. Flip a coin three times.

4. Flip a coin and roll a number cube.

Theoretical Probability

Probability is the chance of an event occurring. Probability only measures how likely it is that an event will occur; it does not guarantee that the event will happen. **Theoretical probability** is the ratio of the number of favorable outcomes to the total number of possible outcomes, and it is based on outcomes that are equally likely. **Favorable outcomes** are the outcomes that are specified by a problem. If a question asks about the probability of drawing a blue marble, drawing the blue marble is considered a favorable outcome. You can calculate the theoretical probability of an event using the ratio below.

$$\text{Theoretical probability of an event} = \frac{\text{number of favorable outcomes}}{\text{total number of possible outcomes}}$$

You can express a probability as a fraction, decimal, or percent. Write the fraction in lowest terms. The probability of an event can range from 0 to 1. A probability of 0 is an **impossible event**. It can never happen. A probability of 1 is a **certain event**. It always happens.

Example 1 Find Theoretical Probability

A number cube has faces numbered 1 to 6. Find the probability of rolling an even number.

Step 1 Find the total number of possible outcomes.
There are 6 different faces, so the total number of possible outcomes is 6.

Step 2 Find the number of favorable outcomes.
The favorable outcome is that the cube lands on an even number (2, 4, or 6). There are 3 different favorable outcomes.

Step 3 Calculate the probability.
$\frac{\text{favorable outcomes}}{\text{total possible outcomes}} = \frac{3}{6} = \frac{1}{2}$, 0.5, or 50%

In Example 1, if the number cube had the numbers 2, 4, 6, 8, 10, 12 on its sides, the probability of rolling an even number would be 1 or 100%. The event would be certain.

EVALUATE SUPPORT FOR CONCLUSIONS

An author may draw a conclusion based on information presented in his or her writing. It is important for the reader to **evaluate**, or closely study, support an author provides for conclusions. A good conclusion should be logical and should be supported. Details that **support** a conclusion give it backup and strength. Ask yourself:

- Is this information factual? What is the source of this information?

- Is the author biased, with strong positive or negative feelings?

- Is the conclusion logical? Does it make sense based on the data?

Read the following passage and the conclusions that follow.

---

(1) Cintia doesn't like to sit in the sun because her skin is sensitive. Her aunt has two tickets to a baseball game, but the seats are not in the shade. (2) Cintia's favorite basketball team is the Chicago Bulls, so she must love all sports. (3) The tickets are expensive, and Cintia hates to waste money. She will definitely go to the baseball game.

---

(1) Cintia will sit in the sun for hours if she goes to the baseball game. Because she has sensitive skin, she doesn't like to sit in the sun. This detail does not support the conclusion.

(2) Cintia likes a certain basketball team, but the author gives no evidence that she likes sports in general or that she likes baseball at all. This detail neither supports nor denies the conclusion. It is neutral.

(3) Cintia hates to waste money. While this detail appears to support the conclusion that Cintia will go to the game, it is not as strong as it seems. The author does not explicitly say that Cintia has been invited. Also, there is no evidence that if she declines, the ticket will go to waste. The evidence does not support the conclusion.

21st Century Skill
Ethics and Probability

Decisions based on probability sometimes can be linked to ethics. Ethics are the moral guiding principles that govern a person's or a group's behavior. Suppose a certain medical procedure is successful in only 1 case out of 50. How would one determine who gets the treatment and who is denied it? Should every individual be given the treatment who needs it? Discuss with a partner how probability and ethics may intersect in real-life decision making.

Consider the following scenario. You and a friend were bored one day and decided to flip a quarter to see how many times it would land on its head. After landing heads 6 times, your friend says "the next time, it has to be tails because it's landing heads so often". Is this line of reasoning correct? Is the probability the coin will land on its tails greater because of its past behavior? Assuming it is a fair coin (both sides are equally likely), does this mean that both sides must fall an equal number of times?

In a notebook, describe why your friend's reasoning is incorrect. Also, describe what it means for a coin to have a 50/50 chance of landing on its head.

MATH LINK

In Example 2, Sam chooses the photos without replacing them. When he has picked 1 of the 5 for the first position, he then has only 4 choices for the second position. When he has picked 1 of the 5 for the first position and 1 of 4 for the second position, he has only 3 choices for the third position.

Permutations and Combinations

A **permutation** is a selection of items or events in which order is important. To find the total possible outcomes of a permutation, multiply together the number of choices for each position in the order.

Example 2 Find Theoretical Probability Using Permutations

Sam will hang 3 out of 5 photographs of animals on a wall in a row. The photographs include a zebra, a gorilla, a chimpanzee, an elephant, and a giraffe. Find the probability that the first photograph in the row is an elephant, the second is a gorilla, and the third is a zebra.

Step 1 Find the total number of possible outcomes.
The arrangement of the photographs is a permutation.
There are 5 choices for the first position in the row, 4 choices for the second position, and 3 choices for the third position.
$5 \times 4 \times 3 = 60$
There are 60 possible outcomes.

Step 2 Find the number of favorable outcomes.
There is only 1 favorable outcome in which the photographs can be hung in the order of elephant, gorilla, and zebra.

Step 3 Calculate the probability.
$\frac{\text{favorable outcomes}}{\text{total possible outcomes}} = \frac{1}{60}$, about 0.017, or about 1.7%

A **combination** is a selection of items in which order does not matter.

Example 3 Find Theoretical Probability Using Combinations

A bag contains 12 different letter tiles, including P and S. Two letter tiles are randomly drawn from the bag. Find the probability that the two letter tiles are P and S.

Step 1 Find the total number of possible outcomes. Since you are drawing 2 tiles out of 12, there are $12 \times 11 = 132$ possible outcomes.

Step 2 Find the number of favorable outcomes. Since P and S is what needs to be drawn, in any order, there are two ways for that to happen: PS or SP.

Step 3 Calculate the probability.
$\frac{\text{favorable outcomes}}{\text{total possible outcomes}} = \frac{2}{132} = \frac{1}{66}$, about 0.015, or about 1.5%

THINK ABOUT MATH

Directions: Find each probability.

1. The probability that you spin a 9 on a spinner with 8 equal sections numbered 1 to 8.

2. The probability that you draw a blue tile from a bag of 5 blue tiles, 9 red tiles, and 1 green tile.

3. The probability that Eduardo and Julietta are the two names randomly drawn from a bag that contains 8 different names, including Eduardo and Julietta.

Suppose you want to predict the number of times a yellow marble will be drawn from a bag in 500 draws. First, determine probabilities. Theoretical probability is a good choice if you already know the number of each color marble in the bag. If half the marbles were yellow, you would expect to pick a yellow marble 250 times in 500 draws. However, suppose you actually drew a marble from the bag 100 times and then returned it each time. If you found that you drew yellow 55% of the time, your prediction would change to 275.

Suppose a bag has an equal number of A, B, C, D, and E letter tiles in it. Use theoretical probability to predict the number of times a vowel will be drawn in 100 draws (replacing the tile after each draw). Then predict using the experimental probability results shown in the table below. Show your work.

Letter	Number of Times
A	16
B	22
C	25
D	19
E	18

Experimental Probability

Experimental probability is based on the ratio of the number of **successes,** or favorable outcomes, to the total number of trials. **Trials** are the number of times an experiment is repeated. You can use the following ratio to find experimental probability.

Experimental probability $= \frac{\text{number of favorable outcomes or successes}}{\text{total number of trials}}$

Example 4 Find Experimental Probability

Nina got a hit 15 out of 65 times that she was at bat. Find the probability that she will get a hit the next time she is at bat.

Step 1 Find the total number of trials.
Nina was at bat 65 times, so there are 65 trials.

Step 2 Find the number of successes.
She got a hit 15 times, so there are 15 successes.

Step 3 Calculate the probability.
$\frac{\text{favorable outcomes}}{\text{total trials}} = \frac{15}{65} = \frac{3}{13}$, about 0.23, or about 23%

Example 5 Make a Prediction

Yasir's favorite basketball team has won 20 out of 28 games. Predict the number of games the team will win out of 42 games.

Step 1 Calculate the probability of winning based on the team's record.
$\frac{\text{favorable outcomes}}{\text{total trials}} = \frac{20}{28} = \frac{5}{7}$, about 0.71, or about 71%

Step 2 Use the probability to make the prediction. The probability that the team will win a game is about 71%. When making predictions, use the exact probability and round afterward to avoid any rounding errors.
Multiply the probability times the number of games.
$\frac{5}{7} \times 42 = 30$
The prediction is that the basketball team will win 30 games out of 42.

Vocabulary Review

Directions: Fill in the blanks with one of the words or phrases below.

certain event combination experimental probability impossible event
permutation probability theoretical probability trials

1. _____ are the number of times an experiment is performed.

2. A(n) _____ has a probability of 0.

3. In a(n) _____, order is important.

4. _____ is the chance of something happening.

5. A selection in which order does not matter is called a(n) _____.

6. If the probability is 1, the event is a(n) _____.

7. Use the ratio of favorable outcomes to the number of possible outcomes to
   find _____.

8. _____ is based on the ratio of successes to trials.

Skill Review

Directions: Answer the following.

1. Kwami says that the theoretical probability of drawing a red marble from a bag of 8 red
   marbles, 4 blue marbles, and 6 green marbles supports his conclusion that a red marble
   will be picked about 67 times out of 150 random drawings. Does the evidence Kwami cites
   support his conclusion? Explain.

   _____

2. Jin is arranging 5 brochures at the check-in desk at a hotel. There are 5 different slots in
   which he can arrange the brochures, and he has been instructed to place one brochure per
   slot. Find the probability of placing the brochures in this order in the slots: restaurants,
   parks, shopping, tours, sightseeing. Then write any details that support your conclusion.

   _____

3. How are predicting outcomes based on theoretical probability and on experimental
   probability the same and different?

   _____

Skill Practice

Directions: Choose the best answer to each question.

1. One hundred raffle tickets are sold. The holder of the winning ticket will win a new car. If Jaylen buys five tickets for $3, what is the probability that she will win the car?

   A. 5%
   B. 6%
   C. 50%
   D. 60%

2. An association of 10 committee members chooses the president and vice-president of the committee by drawing all of their names out of a sack. The president is chosen first, then the vice-president. If Patrice and Miguel place their names in the sack, what is the probability that Patrice will be president and Miguel will be vice-president?

   A. $\frac{1}{5}$
   B. $\frac{1}{10}$
   C. $\frac{1}{45}$
   D. $\frac{1}{90}$

Directions: Answer the following questions.

3. Out of every 20 boxes of cereal, a company places in 8 boxes coupons for a free box. What is the probability that a box will contain a coupon? Write your answer as a fraction.

   _____

4. At a pet store, 42 out of 150 customers bought organic pet food. How many customers out of the next 50 can be predicted to buy organic pet food?

   _____

Compound Events

KEY CONCEPT: Extend your understanding of probability to finding the probability of compound events.

Find the sum, difference, product, or quotient.

1. $\frac{3}{5} + \frac{1}{2}$ 2. $\frac{1}{6} \div \frac{2}{3}$ 3. $\frac{5}{8} \times \frac{4}{7}$ 4. $\frac{7}{8} - \frac{5}{6}$

Find each probability.

5. Find the probability of spinning a consonant on a spinner with 8 equal sections labeled A, B, C, D, E, F, G, and H.

6. Find the probability of rolling a number greater than 2 on a number cube with faces 1, 2, 3, 4, 5, and 6.

Lesson Objectives

You will be able to

- Find the probability of mutually exclusive events

- Find the probability of overlapping events

- Find the probability of independent and dependent events

Skills

- **Core Skill:** Represent Real-World Problems

- **Core Practice:** Reason Abstractly

Vocabulary

dependent events
independent events
mutually exclusive events
overlapping events
replacement

MATH LINK

You can use the formula below for finding the probability of mutually exclusive events:
$P(A \text{ or } B) = P(A) + P(B)$

You can use the formula below for finding the probability of overlapping events:
$P(A \text{ or } B) = P(A) + P(B) - P(A \text{ and } B)$

Mutually Exclusive and Overlapping Events

A compound event is an event consisting of two or more simple events, such as rolling a number cube and spinning a spinner. Before you can find the probability of a compound event, you need to know the type of compound event it is.

Mutually exclusive events are events that cannot happen at the same time. For example, if you are drawing one card from a standard deck of 52 playing cards, you cannot draw a queen and a jack at the same time.

Example 1 Find the Probability of Mutually Exclusive Events

A bag contains a set of 6 letter tiles. The tiles are B, F, I, J, K, and U. Find the probability of drawing a vowel or a K.

Step 1 Determine whether the events are mutually exclusive. Since K is a consonant and cannot be a vowel, the events are mutually exclusive.

Step 2 Find the probability of each event.
$P(\text{vowel}) = \frac{2}{6}$, or $\frac{1}{3}$ I and U
$P(K) = \frac{1}{6}$ K

Step 3 Calculate the probability. Since these are mutually exclusive events, add the probabilities.

$\frac{2}{6} + \frac{1}{6} = \frac{3}{6}$ or $\frac{1}{2}$

The probability of drawing a vowel or K is $\frac{1}{2}$.

Overlapping events can occur at the same time because they are not mutually exclusive. For example, say a school has a band and a yearbook team. It is possible for a student to be in either one of those groups. Because the groups are not mutually exclusive, it is also possible that some students are in both groups. In other words, their membership overlaps.

Example 2 Find the Probability of Overlapping Events

A number cube with 6 faces numbered 1 through 6 is rolled. Find the probability of rolling a number greater than 2 or an even number.

Step 1 Determine whether the events are overlapping.
Since some numbers greater than 2 are even numbers, the events are overlapping.

Step 2 Find the probability that each event occurs and the probability that both events occur.
$P(\text{number greater than 2}) = \frac{4}{6}$, or $\frac{2}{3}$ 3, 4, 5, and 6
$P(\text{even number}) = \frac{3}{6}$, or $\frac{1}{2}$ 2, 4, and 6
$P(\text{number greater than 2 and even number})$
$= \frac{2}{6}$, or $\frac{1}{3}$ 4 and 6

Step 3 Calculate the probability. Since these are overlapping events, add the probabilities of either event happening and subtract the probability of both events happening.
$$\frac{4}{6} + \frac{3}{6} - \frac{2}{6} = \frac{7}{6} - \frac{2}{6} = \frac{5}{6}$$
The probability of rolling a number greater than 2 or an even number is $\frac{5}{6}$.

Some students may have noticed that the set of even numbers were part of the set of numbers greater than 1 in the previous example. Therefore, a quick calculation of 5 favorable out of 6 possible outcomes gives a probability of $\frac{5}{6}$. This is also a valid way to think about the problem. However, the process shown in Example 2 can be used for any problem that appears, especially when the number of possible outcomes is extremely large.

Core Skill
Represent Real-World Problems

The equations for mutually exclusive events and overlapping events are very similar. How can you tell which one you should use? Look at Example 2. The number "2" occurs only in the outcome "even number." The numbers "3," "4," "5," and "6" occur in the outcome "numbers greater than 2." But notice that there is some overlap in the two groups. Because "4" and "6" are possible outcomes for both kinds of events, you don't want to count them twice. This is why you must subtract their probability from the sum of the first two probabilities you calculated.

In a notebook, write the equation that you would use to find the following probability. Then calculate the probability.

You have tiles that are numbered from 1 to 100. Find the probability of randomly picking a tile with an odd number or perfect square on it.

An understanding of probability comes from your experience with real-world situations involving trials and outcomes. This is why the lesson begins by describing events such as rolling die, flipping coins, and drawing cards from decks. Eventually, however, you no longer have to record each single event in order to determine the probability of compound events. The formulas for mutually exclusive events, overlapping events, dependent events, and independent events are a shorthand method for determining probabilities.

Consider Example 4. How many branches would the tree diagram have to have in order to represent each one of the possible outcomes? How many of those branches would represent outcomes in which the first marker was blue and the second was red? Would you want to draw the tree diagram and then try to locate and count the few successful outcomes in the vast forest of data?

Independent and Dependent Events

Compound events can also be independent or dependent. **Independent events** do not affect each other. Examples include flipping a coin and then flipping the coin again, or rolling a number cube and then spinning a spinner. If you randomly draw a bead from a bag, replace it, and then draw another bead, the events are independent because the number of beads in the bag the second time you draw is the same as the number of beads that were in the bag for the first draw. Drawing an object from a container and then returning it before drawing again is often referred to as **replacement**.

Example 3 Find the Probability of Independent Events

A bag contains 3 red tiles, 8 blue tiles, 5 green tiles, and 2 yellow tiles. You draw a tile with replacement and then draw another tile. Find the probability that the first tile is red and the second tile is yellow.

> **Step 1** Determine whether the events are independent.
> The tile from the first draw is replaced, so the events are independent.

> **Step 2** Find the probability of each event.
> 3 out of 18 tiles are red. $P(\text{red tile}) = \frac{3}{18}$ or $\frac{1}{6}$
>
> 2 out of 18 tiles are yellow. $P(\text{yellow tile}) = \frac{2}{18}$ or $\frac{1}{9}$

> **Step 3** Calculate the probability. If two events were independent, multiply the probabilities of the two events to find the probability of the compound event. $\frac{1}{6} \times \frac{1}{9} = \frac{1}{54}$
>
> The probability of drawing the first tile red and the second tile yellow is $\frac{1}{54}$.

THINK ABOUT MATH

Directions: Use the spinner below for each exercise. Identify whether events are mutually exclusive or overlapping. Then find each probability.

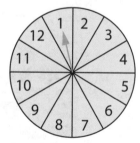

1. $P(\text{number greater than 5 or odd number})$

2. $P(11 \text{ or even number})$

3. $P(\text{number greater than 9 or number less than 2})$

4. $P(\text{odd number or number greater than 7})$

Dependent events do affect each other. If you draw a bead from a bag without replacement (you do not return the bead to the bag before drawing a second bead), then the events are dependent because the number of beads in the bag for the second draw is not the same as the number of beads that were in the bag for the first draw.

MATH
LINK

Example 4 Find the Probability of Dependent Events

A drawer in your desk contains 5 blue markers, 4 green markers, and 3 red markers. You reach in the drawer and randomly choose a marker. Then you choose another marker. Find the probability that the first marker is blue and the second marker is red.

You can use the formula below for finding the probability of independent events.

$P(A \text{ and } B) = P(A) \times P(B)$

You can use the formula below for finding the probability of dependent events.

$P(A \text{ and } B) = P(A) \times P(B \text{ after } A)$

> **Step 1** Determine whether the events are dependent.
> You do not replace the marker after the first draw, so the events are dependent.
>
> **Step 2** Find the probability that each event occurs.
>
> $P(\text{blue marker}) = \frac{5}{12}$ 5 out of 12 markers are blue.
>
> $P(\text{red marker}) = \frac{3}{11}$ 3 out of the remaining 11 markers are red.
>
> **Step 3** Calculate the probability. Since these are dependent events, multiply the probability of the first event times the probability of the second event after the first event occurs.
>
> $\frac{5}{12} \times \frac{3}{11} = \frac{15}{132} = \frac{5}{44}$
>
> The probability of choosing a red marker following a blue marker is $\frac{5}{44}$.

Vocabulary Review

Directions: Match each word to one of the phrases below.

1. _____ dependent events

2. _____ independent events

3. _____ mutually exclusive events

4. _____ overlapping events

5. _____ replacement

A. two or more events that cannot happen at the same time

B. in an experiment, when an object is drawn from a bag, box, jar, or other container and is then placed back into the container

C. compound event in which the result of one event does not affect the other event

D. compound event in which the events could happen at the same time

E. compound event in which the result of one event affects the result of the other event

Directions: Answer the following.

1. With dependent events involving colored tiles, after every pick, does the likelihood of picking a particular color go up or down? Explain your answer.

_____

_____

2. Find the probability in the following problem.

Ernesto created a 4-digit PIN number for an online account. The PIN number consists of numbers from 1 through 9, and the numbers can repeat. What is the probability that the first two numbers are 2?

_____

_____

Skill Practice

Directions: Choose the best answer to each question.

1. At a raffle for charity, people place tickets into jars for prizes they wish to win. Two names will be drawn at random without replacement from each jar at the end of the night. In one jar to win tickets to a movie theater, Gracia's name occurs 7 times, Bryan's 5 times, and Azmera's 6 times. If these are the only names in the drawing, what is the probability that one name is drawn and then another name and that both names are Azmera?

   A. $\frac{5}{51}$ C. $\frac{1}{3}$

   B. $\frac{1}{9}$ D. $\frac{2}{3}$

2. A box contains 3 pink paper clips, 4 yellow paper clips, and 7 green paper clips. Two paper clips are drawn with replacement. What is the probability that the first paper clip is pink and the second is green?

   A. $\frac{5}{7}$ C. $\frac{3}{28}$

   B. $\frac{3}{26}$ D. $\frac{9}{91}$

3. Chantou wants to find the probability that the number spun on a spinner is either odd or a number greater than 6. The spinner has 8 equal sections labeled 1, 2, 3, 4, 5, 6, 7, and 8. Which shows how to find this probability?

   A. Multiply the probabilities of the two events to find the probability of the compound event.
   B. Multiply the probability of the first event times the probability of the second event after the first event occurs.
   C. Add the probabilities of both events
   D. Add the probabilities of either event happening and subtract the probability of both events happening.

4. A number cube with faces labeled 1 through 6 is rolled. What is the probability that the number is less than 4 or equal to 6?

   A. $\frac{5}{6}$ C. $\frac{1}{6}$

   B. $\frac{2}{3}$ D. $\frac{1}{12}$

Review

Directions: Choose the best answer to each question.

Questions 1 through 4 refer to the following information.

> A company sends out packets of discount coupons for local businesses. In each packet are the following coupons: 12 restaurants, 15 clothing stores, 12 home-improvement contractors, and 18 car repairs. The coupons are put into each packet in random order. Lujayn receives one of these packets in the mail.

1. What is the probability that without looking, Lujayn will select a coupon for a clothing store?

   A. $\frac{5}{14}$ C. $\frac{15}{42}$

   B. $\frac{5}{19}$ D. $\frac{4}{19}$

2. Which coupons are equally likely to be selected from the packet?

   A. restaurant and clothing store
   B. car repair and restaurant
   C. clothing store and car repair
   D. restaurant and home improvement

3. Lujayn says the probability of pulling out a coupon for car repair, not replacing it, and then pulling out a coupon for home improvement is found by multiplying $\frac{18}{57} \times \frac{12}{57}$. Why is Lujayn incorrect?

   A. She did not take into account that the events are dependent.
   B. She did not take into account that the events are independent.
   C. She did not take into account that the events are mutually exclusive.
   D. She did not take into account that the events are overlapping.

4. What is the probability of selecting a shoe-store coupon from the packet?

   A. 2   C. $\frac{1}{2}$

   B. 1   D. 0

5. Two number cubes, each with the numbers 1 through 6, are rolled. Which outcome is most likely?

   A. The sum of the numbers is equal to 12.
   B. The same number is rolled on each cube.
   C. The sum of the numbers is equal to 7.
   D. At least one of the numbers is a 6.

Questions 6 and 7 refer to the following information.

> A bag contains 8 letter tiles:
> S, T, S, U, E, X, A, Q.

6. Corey predicts that he will draw a vowel or an S from the bag on his first try. What percent chance does he have that his prediction is correct?

   A. 12.5%   C. 37.5%

   B. 25%   D. 62.5%

7. Tevin draws a tile from the bag, does not replace it, and then chooses another tile. Which will give the probability that Tevin will chose a Q and then an S?

   A. $\frac{1}{8} + \frac{2}{7}$

   B. $\frac{1}{8} \times \frac{2}{7}$

   C. $\frac{1}{8} \times \frac{2}{8}$

   D. $\frac{1}{8} \times \frac{1}{7}$

Questions 8 and 9 refer to the two spinners below.

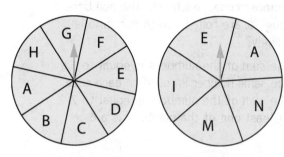

8. The spinners are spun at the same time. What is the probability of spinning H on the first spinner and E on the second spinner?

   A. $\frac{1}{40}$ **C.** $\frac{1}{8}$

   B. $\frac{1}{20}$ **D.** $\frac{1}{5}$

9. The spinners are spun at the same time. What percent of the time should both spinners land on vowels?

   A. 3%   **C.** 15%

   B. 10%   **D.** 38%

10. Which could have a maximum of 90 different combinations?

   A. 50 cars, 10 trucks, 30 SUVs

   B. 5 cars, 6 trucks, 3 SUVs

   C. 18 cars, 10 trucks

   D. 10 cars, 3 trucks, 6 SUVs

11. All of Jane's work clothes can be mixed and matched. She has 7 pairs of pants, 5 skirts, 9 blouses, and 4 sweaters. If she always wears a blouse and a sweater and either a pair of pants or a skirt, how many different outfits can she create?

   A. 25

   B. 315

   C. 432

   D. 1,260

12. Which event has a probability of 1?

   A. Tuesday will immediately follow Wednesday.

   B. Tuesday will be a rainy day.

   C. Wednesday will immediately follow Tuesday

   D. Tuesday and Wednesday will be rainy days.

Check Your Understanding

On the following chart, circle the number of any item you answered incorrectly. Near each lesson title, you will see the pages you can review to learn the content covered in the question. Pay particular attention to reviewing those lessons in which you missed half or more of the questions.

Chapter 10: Probability	Procedural	Conceptual	Application/ Modeling/ Problem Solving
Counting Methods pp. 292–295		10	11
Introduction to Probability pp. 296–301	1, 4	2, 12	
Compound Events pp. 302–306		3, 5	6, 7, 8, 9

UNIT 5

Measurement and Geometry

Measurement

Measurement has many real-world applications. You need to know how much space you have in a room when you buy furniture. You need to know if it will fit through the door or window to get it into the house. Which units would you use to measure to make sure that a piece of furniture fits, yards or inches?

Precision in measurement is an issue in many aspects of life. It is essential whether you are making a cake, building a house, or dispensing medicine. In what ways can your life be affected by an inaccurate measurement?

The two types of measurement in Chapter 11 are customary units and metric units.

The Key Concepts you will study include:

Lesson 11.1: Customary Units
Understand how to use division and multiplication to change from one customary unit of length, capacity, weight, or time to another, and how to change to and from mixed units.

Lesson 11.2: Metric Units
Understand how to change from one metric unit of length, capacity, or mass to another.

Goal Setting

Before starting this chapter, set goals for your learning.

What types of measuring do you do in your daily work or home life?

_____

_____

_____

As you study Chapter 11, practice converting measurements you see or use in your daily life. Copy the following chart to keep a record of the measurements you come across. Indicate another unit to which you have converted the measurements.

Measure		Where Seen or Used	Converted Measure
Customary	Metric		

Which type of unit did you find easier to convert?

_____

Explain why you found it easier.

_____

_____

_____

Customary Units

Lesson Objectives

You will be able to

- Change from one customary unit to another

- Change from mixed units

- Change to mixed units

Skills

- **Core Skill:** Evaluate Expressions

- **Core Skill:** Represent Real-World Problems

Vocabulary

abbreviation
capacity
length
time
unit
weight

KEY CONCEPT: Understand how to use division and multiplication to change from one customary unit of length, capacity, weight, or time to another, and how to change to and from mixed units.

Multiply or divide.

1. 161 ÷ 7   **3.** 8 × 60   **5.** 1,825 ÷ 365

2. 45 × 4   **4.** 224 ÷ 16   **6.** 5,280 × 13

Change Customary Units of Measure

We use different **units** to measure length, capacity, weight, and time. Customary units of measure and their relationships are given in the chart below. Use this information to change from one unit to another.

Customary Units of Measure	
Length	**Weight**
1 foot (ft) = 12 inches (in.)	1 pound (lb) = 16 ounces (oz)
1 yard (yd) = 3 feet = 36 inches	1 ton (T) = 2,000 pounds
1 mile (mi) = 1,760 yards = 5,280 feet	
Capacity	**Time**
1 cup (c) = 8 fluid ounces (fl oz)	1 minute (min) = 60 seconds (sec)
1 pint (pt) = 2 cups	1 hour (hr) = 60 minutes
1 quart (qt) = 2 pints	1 day = 24 hours
1 gallon (gal) = 4 quarts	1 week (wk) = 7 days
	1 year (yr) = 365 days

MATH LINK

Units of measure are often abbreviated. **Abbreviations** are short forms of a word, that are used to save space. Some common abbreviations are inch = in., foot = ft, and yard = yd. It is important to note that the measurement abbreviation does not change if the unit is singular or plural. 1 foot = 1 ft, and 100 feet = 100 ft.

Example 1 Change a Smaller Unit to a Larger Unit

Change 105 days to weeks.

Step 1 Find the relationship between the two units.
There are 7 days in 1 week.

Step 2 To change from a smaller unit (days) to a larger unit (weeks),
divide.
105 days ÷ 7 = 15 weeks
105 days = 15 weeks

Example 2 Change a Larger Unit to a Smaller Unit

Change 3 gallons to cups.

Step 1 Find the relationship between the two units.

The information in the metric table at the beginning of
the lesson does not tell you directly how many cups are in
a gallon. However, the table does tell you that there are 4
quarts in 1 gallon, 2 pints in 1 quart, and 2 cups in 1 pint.

Step 2 To change from a larger unit to a smaller unit, multiply.

3 gallons × 4 quarts = 12 quarts
12 quarts × 2 pints = 24 pints
24 pints × 2 cups = 48 cups

3 gallons = 48 cups

Change *To* and *From* Mixed Units

Some measurements are given in mixed units. For example, a person's
height is given as 5 feet 4 inches. Changing *from* mixed units or *to*
mixed units is an important skill.

Example 3 Change from a Mixed Unit to a Single Unit

Change 5 feet 4 inches to inches.

Step 1 Find the relationship between the two units.
12 inches = 1 foot

Step 2 Change the larger unit to the smaller unit.
5 feet 4 inches = (5 feet) + (4 inches)
= (5 × 12 inches) + (4 inches)
= 60 inches + 4 inches
= 64 inches

5 feet 4 inches = 64 inches

- When changing from a larger unit to a smaller unit, the number of the new unit will be greater. For example, 2 ft = 24 in. (Note that 24 is greater than 2.)

- When changing from a smaller unit to a larger unit, the number of the new unit will be smaller. For example, 36 in. = 3 ft. (Note that 3 is smaller than 36.)

Core Skill
Represent Real-World Problems

Amanda wants to place a fence around her rectangular garden. She knows that the length of her garden is 66 inches long; the width is 2 yards. The store only sells fencing in feet. How many feet of fencing will Amanda need? Write your conversions and answer in a notebook.

Example 4 Change from a Single Unit to a Mixed Unit

Change 200 minutes to hours and minutes.

Step 1 Find the relationship between the two units.
60 minutes = 1 hour

Step 2 Change the smaller unit to the larger unit.
200 minutes = 200 ÷ 60
= 3 R20 or 3 hours and 20 minutes

200 minutes = 3 hours and 20 minutes

THINK ABOUT MATH
Directions: Complete each of the following.

1. 12 gallons = _____ quarts

2. 30 feet = _____ yards

3. 300 minutes = _____ hours

4. 5 tons = _____ pounds

5. 6 feet 2 inches = _____ inches

6. 7 minutes 35 seconds = _____ seconds

7. 11 cups = _____ quarts _____ cups

8. 50 ounces = _____ pounds _____ ounces

Vocabulary Review

Directions: Complete each sentence with the correct word.

capacity   length   time   unit   weight

1. The _____ of the side of a rectangle can be measured with a ruler.

2. The _____ of a bowling ball can be measured in pounds.

3. The mile is a customary _____ of measure for measuring the distance between two cities.

4. A drinking glass with a _____ of one pint is large enough to hold 8 fluid ounces.

5. A clock is a tool for measuring _____ in minutes and hours.

Directions: Use the table at the beginning of the lesson to help you choose the larger unit in each pair.

1. cups or pints

2. pounds or ounces

3. inches or feet

4. gallons or quarts

5. minutes or seconds

6. yards or miles

7. days or hours

8. quarts or pints

9. days or weeks

10. yards or feet

Directions: Use the table at the beginning of the lesson to help you solve each problem.

11. How many gallons are in 32 cups?

12. How many seconds are in a day?

13. Enrique is 6 feet 1 inch tall. Simon is 70 inches tall. Who is taller?

14. Ría worked 8 hours 45 minutes. Sue worked 600 minutes. Who worked longer?

Skill Practice

Directions: Choose the best answer to each question.

1. Which unit is smaller than a minute?

   A. day
   B. week
   C. second
   D. hour

2. Kalume made 20 cups of punch and filled a 1-gallon container. How many cups of punch are left?

   A. 16
   B. 8
   C. 4
   D. 2

3. Which equation shows how to change 8 pounds to ounces?

   A. $8 \times 16 = 128$ ounces
   B. $16 \div 8 = 2$ ounces
   C. $8 \times 8 = 64$ ounces
   D. $2{,}000 \div 8 = 250$ ounces

4. Panya measured the length of her garden to be $60\frac{2}{3}$ yards long. Which of these is another way of saying how long her garden is?

   A. 84 ft.
   B. 180 ft. 9 in.
   C. 182 ft.
   D. 183 ft.

Metric Units

KEY CONCEPT: Understand how to change from one metric unit of length, capacity, or mass to another.

Find each product or quotient.

1. 2.8 × 100   **3.** 50 ÷ 10   **5.** 90 × 1,000

2. 3.5 × 1,000   **4.** 45 ÷ 100   **6.** 56.5 × 100

Lesson Objectives

You will be able to

- Change from one metric unit to another
- Understand the basic metric unit for length
- Understand the basic metric unit for capacity
- Understand the basic metric unit for mass

Skills

- **Core Skill:** Build Solution Pathways
- **Core Skill:** Use Ratio Reasoning

Vocabulary

gram
liter
meter
power of 10
prefix

The Metric System

The metric system of measurement is used widely throughout the world. It is based on the decimal system of numbers. The basic units of metric measure are **meter** for length, **liter** for liquid capacity, and **gram** for mass. Prefixes used with these units form other units. A **prefix** is a letter or a group of letters placed before a word that changes the word's meaning.

The common prefixes are *milli-* for $\frac{1}{1,000}$ or 0.001, *centi-* for $\frac{1}{100}$ or 0.01, and *kilo-* for 1,000 times the basic unit.

Metric units of measure and their relationships are given in the chart below. You can use this information to change from one unit to another.

Metric Units of Measure
Length
1 meter (m) = 100 centimeters (cm) 1 meter = 1,000 millimeters (mm) 1 centimeter = 10 millimeters 1 kilometer (km) = 1,000 meters
Liquid Capacity
1 liter (L) = 1,000 milliliters (mL) 1 kiloliter (kL) = 1,000 liters
Mass
1 gram (g) = 1,000 milligrams (mg) 1 kilogram (kg) = 1,000 grams

MATH LINK

The **powers of** 10 are

$10^1 = 10$
$10^2 = 100$
$10^3 = 1,000$
$10^4 = 10,000$

and so on.

USE PREFIXES

Prefixes are syllables that are added to the beginning of a word. Prefixes change the meaning of a word. Below is a table of some common prefixes and their meanings.

Prefix	Meaning	Example	Example meaning
Pre-	Before	Preread	Read before
Bi-	Two	Biannual	Two times per year
Tri-	Three	Triathlon	Competition with three parts
Un-	Not	Unnecessary	Not necessary
Re-	Again	Rewind	Wind again
Anti-	Against	Antifreeze	Material that prevents freezing

An understanding of prefixes can help you decode unfamiliar words. You can often break a word apart to find the meaning of each part.

For each word below, write its prefix, the prefix meaning, and the meaning of the whole word.

> **1.** unclear   prefix _____ prefix meaning _____
> word meaning _____
>
> **2.** rerun   prefix _____ prefix meaning _____
> word meaning _____

1. prefix <u>un</u>   prefix meaning <u>not</u>   word meaning <u>not clear</u>
Something that is unclear is difficult to understand.

2. prefix <u>re</u>   prefix meaning <u>again</u>   word meaning <u>run again</u>
Reruns are television episodes that have previously aired, or run.

Your work with decimals has taught you that a decimal point moves right and left as you multiply and divide by powers of 10. Multiply by 0.1, and the decimal point moves left one place. Multiply by 100, and it moves right two places—as many places right as the number of zeroes.

As you work your way through this lesson, you will come to realize that you know more about the topic than you thought you did. You have already built the solution pathways you will need in order to convert metric units of measure from one unit to another. Based on your previous knowledge of working with decimals and of changing between the customary units, plus the information given in the metric table, state the rules for changing between units in the metric system. Then in a notebook, use these rules to determine which of these is equal in length to 100 millimeters: 10 meters, 10 centimeters, or 10 kilometers. Then after reading this lesson, go back and see if your answer was correct.

Converting between units of measure uses the exact same processes as converting ratios to unit ratios or converting fractions to their common denominators; all three use the process of multiplying by 1. However, while converting between units, you may need to multiply by 1 more than once.

For example, suppose you want to convert from 1 mile to inches. Is there an easy conversion number to use? No. Instead, use the conversions from miles to feet and then feet to inches (1 mile = 5,280 feet and 1 foot = 12 inches). Therefore the conversion ratios (the "1" to multiply) are:

$\frac{1 \text{ mile}}{5,280 \text{ feet}}$ and $\frac{1 \text{ foot}}{12 \text{ inches}}$.

Both ratios above equal 1 and can be multiplied or divided to any appropriate number. Because we want to convert miles, we must multiply by a ratio of 1 that has miles in the denominator so the miles units will cancel. Therefore, to convert between miles to inches, we make the computation:

1 mile; $\frac{5,280 \text{ feet}}{1 \text{ mile}} \times \frac{12 \text{ inches}}{1 \text{ foot}} =$ 63,360 inches.

In a notebook, convert between 3,567 cm to km.

The diagram below shows the **powers of 10** (numbers formed by multiplying 10 by itself a certain number of times) to multiply by or divide by when changing between units.

$$10^1 = 10 \qquad 10^2 = 100 \qquad 10^3 = 1,000$$

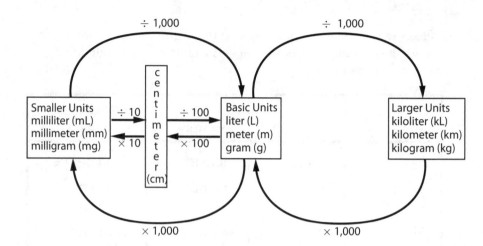

Example 1 Use the Prefix to Change from a Smaller Unit to a Larger Unit

Use the prefixes to find the number of millimeters in 1 meter.

Millimeter is the smaller unit, and 1 millimeter is $\frac{1}{1,000}$ or 0.001 of a meter.

There are 1,000 millimeters in 1 meter.

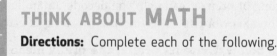

THINK ABOUT MATH

Directions: Complete each of the following.

1. 1 m = _____ km
2. 1 m = _____ cm
3. 1 m = _____ mm
4. 1 km = _____ m
5. 1 km = _____ cm
6. 1 km = _____ mm

7. 1 cm = _____ km
8. 1 cm = _____ m
9. 1 cm = _____ mm
10. 1 mm = _____ km
11. 1 mm = _____ m
12. 1 mm = _____ cm

Example 2 Use the Prefix to Change from a Larger Unit to a Smaller Unit

Use the prefix to change 0.189 kilograms to grams.

Kilogram is the larger unit, and there are 1,000 grams in 1 kilogram.

In 0.189 kilograms there are 0.189 times 1,000 grams.

0.189 kg × 1,000 = 189 g

0.189 kilograms = 189 grams

Example 3 Change from a Smaller Unit to a Larger Unit

Change 70 centimeters to meters.

As with customary units of measure, divide to change from a smaller unit to a larger unit.

70 cm ÷ 100 = 0.70 m or 0.7 m

70 centimeters = 0.7 meter

Example 4 Change from a Larger Unit to a Smaller Unit

Change 5.2 liters to milliliters.

Multiply to change from a larger unit to a smaller unit. Since 1,000 milliliters are in 1 liter, multiply by 1,000.

5.2 L × 1,000 = 5,200 mL

5.2 liters = 5,200 milliliters

To change units of metric measure, first find the relationship between the two units or between the prefixes. If you are changing from a smaller unit to a larger unit, divide by a power of 10. If you are changing from a larger unit to a smaller unit, multiply by a power of 10.

MATH LINK

When multiplying or dividing by a power of 10, move the decimal point the same number of places as the number of zeros in the power of 10.

THINK ABOUT **MATH**

Directions: Complete each of the following.

1. 320 cm = _____ m

2. 0.4 kL = _____ L

3. 5,500 g = _____ kg

4. 89 mm = _____ cm

5. 1.4 kL = _____ L

6. 2,000 L = _____ kL

7. 240 m = _____ km

8. 342,000 mg = _____ g

9. 6.5 L = _____ mL

Vocabulary Review

Directions: Complete each sentence with the correct word.

gram   liter   meter   power of 10

1. The _____ is the basic unit for length in the metric system.

2. The _____ is the basic unit for liquid capacity in the metric system.

3. The _____ is the basic unit for mass in the metric system.

4. Multiplying 10 by itself a certain number of times results in a _____.

Skill Review

Directions: Answer the following questions.

1. The mass of one milliliter of water is approximately 1 gram. What is the approximate mass of 10 milliliters of water?

2. A penny weighs about 2.5 grams. What is the approximate weight of 50 pennies?

3. A DVD is 120 millimeters across. How many centimeters across is a DVD?

4. One liter of water weighs 1 kilogram. How many liters of water would weigh 1,000 kilograms?

5. The symbol for *millimeter* is *mm*, *centimeter* is *cm*, *meter* is *m*, and *kilometer* is *km*. What do you think the symbol for decimeter would be?

Directions: Match each prefix with the correct power of 10.

_____ 6. centi   A. 1,000

_____ 7. milli   B. 0.01

_____ 8. kilo   C. 0.001

Skill Practice

Directions: Choose the best answer to each question.

1. An orange contains about 70 milligrams of protein. How many grams is this?

   A. 0.07
   B. 0.7
   C. 7
   D. 70,000

2. Which of these shows the units of length written from smallest to largest?

   A. millimeter, meter, centimeter, kilometer
   B. centimeter, meter, millimeter, kilometer
   C. meter, centimeter, millimeter, kilometer
   D. millimeter, centimeter, meter, kilometer

3. A glass holds 0.25 liters of water. How many liters of water will 1,000 glasses hold?

   A. 0.0025
   B. 0.25
   C. 250
   D. 2,500

4. The length of a piece of yarn is 800 centimeters. How many meters of yarn is this?

   A. 8
   B. 80
   C. 8,000
   D. 80,000

Directions: Choose the best answer to each question.

1. Which statement shows how to change 12 yards to inches?

   A. Multiply 12 by 3.
   B. Divide 36 by 12.
   C. Divide 12 by 36.
   D. Multiply 12 by 36.

2. Which list shows units of measure arranged from greatest to least?

   A. gram, centigram, kilogram, milligram
   B. kiloliter, liter, milliliter, centiliter
   C. kilometer, meter, centimeter, millimeter
   D. millimeter, centimeter, meter, kilometer

3. Mavis has lived in Tucson for 136 days. Betina has lived there for 5 months. Hector moved to Tucson 18 weeks ago. Jumah has lived there for $\frac{1}{4}$ of a year. Who has lived in Tucson for the longest time?

   A. Mavis
   B. Betina
   C. Hector
   D. Jumah

4. Kiet's coffeemaker brews a one-quart pot of coffee. How many cups of coffee can Kiet pour from 2 pots of coffee?

   A. 4   C. 16
   B. 8   D. 32

5. Winona is 75 inches tall. How many feet and inches is this?

   A. 7 feet 3 inches
   B. 6 feet 3 inches
   C. 6 feet
   D. 5 feet 9 inches

6. There are 300 milliliters of cough medicine in a bottle. How many liters is this?

   A. 0.003
   B. 0.03
   C. 0.3
   D. 3.0

7. Omari runs 5 miles every other day. How many feet does he run?
   (One mile = 5,280 feet)

   A. 26,400
   B. 13,200
   C. 5,280
   D. 5,000

8. Valeria buys 4 pounds of oranges. How many ounces is this?

   _____

Review

9. How many liters are equal to 15 kiloliters?

A. 1.5   C. 15,000
B. 1,500   D. 150,000

12. Farran drives 35 kilometers to work each day. How many meters is this?

A. 0.35   C. 35,000
B. 3,500   D. 350,000

10. Tanh rented a movie that takes 140 minutes to watch. She has 2 hours before she has to leave to pick up her children at school. How can Tanh find out if she has enough time to watch the entire movie before she has to leave?

A. Write both times in terms of minutes. Then divide.
B. Write both times in terms of hours. Then add.
C. Write both times in terms of minutes. Then add.
D. Write both times in terms of minutes. Then compare the times.

13. The mass of a male African elephant is 6,500 kilograms. Which of the following shows how to find how many grams the elephant weighs?

A. $6,500 \times 1,000$
B. $6,500 \div 100$
C. $6,500 \div 1,000$
D. $6,500 \times 0.0001$

11. Yamuna measured a length of wool equal to 12 yards to make a coat. After she cuts 5 feet from the length, how many feet of the original piece of wool does she have left?

A. 36   C. 7
B. 31   D. 2.5

14. The parking lot for the mall is 450 yards long. How many feet is this?

A. 45
B. 150
C. 900
D. 1,350

Check Your Understanding

On the following chart, circle the number of any item you answered incorrectly. Near each lesson title you will see the pages you can review to learn the content covered in the question. Pay particular attention to reviewing those lessons in which you missed half or more of the questions.

Chapter 11: Measurement	Procedural	Conceptual	Application/ Modeling/ Problem Solving
Customary Units pp. 312–315		1, 10	3, 4, 5, 7, 8, 11, 14
Metric Units pp. 316–321	13	2, 9	6, 12

Geometry

You can see geometry in everyday items. The table or counter you eat breakfast on is a geometric shape or combination of shapes. Any buildings or rooms you enter during your day were built using basic geometric ideas, such as right angles and parallel lines. You need to find the areas of the walls each time you decide how much paint to buy to paint a room. What are some instances in your daily life in which you observe or use some of the principals of geometry?

The Key Concepts you will study include:

Lesson 12.1: Geometric Figures
Learn how to identify and classify two-dimensional shapes by their properties.

Lesson 12.2: Perimeter and Circumference
Understand and apply concepts of perimeter and circumference.

Lesson 12.3: Scale Drawings and Measurement
Learn how to read scale drawings, produce real-life measurements from the scale, and draw geometric objects with a certain scale.

Lesson 12.4: Area
Develop and apply the concept of area to find the areas of simple and complex shapes.

Lesson 12.5: Pythagorean Theorem
Understand the Pythagorean Theorem, its proof, and apply it to real-life scenarios.

Lesson 12.6: Geometric Solids and Volume
Extend understanding of geometric figures to include solids and the concept of volume.

Lesson 12.7: Volume of Cones, Cylinders, and Spheres
Calculate the volume of complex 3-D objects by recognizing they are made from solids whose volume formulas are known.

Goal Setting

Before starting this chapter, set goals for your learning. Think about the places and times you use geometry in your home and work life.

How does geometry fit into your life?

_____

_____

_____

Finding examples of geometric figures is as easy as looking into a room. The surface of the door of the room is a rectangle. Walls and windows are also rectangles.

Find examples of geometric figures that you see in objects you pass by or use every day. Copy the following table to record the figure, its dimensions, and its area or volume (whichever is applicable).

Geometric Figure	Dimensions	Area	Volume

Geometric Figures

KEY CONCEPT: Learn how to identify and classify two-dimensional shapes by their properties.

Identify the following shapes by writing circle, rectangle, square, *or* triangle.

1.

2.

3.

4.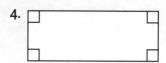

Basic Geometric Terms

In the study of geometry, there are many definitions to learn. Skills of geometry include identifying shapes and understanding relationships among shapes. Carefully study the definitions and figures below.

Line: a straight figure having no thickness that continues forever in both directions

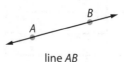

line *AB*

Segment: a finite portion of a line connecting two points, called the endpoints

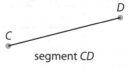

segment *CD*

Ray: a portion of a line continuing in one direction forever, starting at a vertex

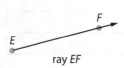

ray *EF*

Angle: the amount of rotation between two intersecting lines or ray

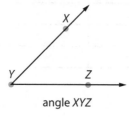

angle *XYZ*

Right angle: an angle that measures 90°

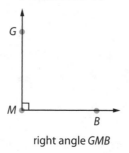

right angle *GMB*

Parallel lines: two (or more) lines that never intersect

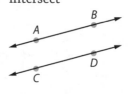

parallel lines *AB* and *CD*

Perpendicular lines: two lines that intersect to form a right angle at the intersection point

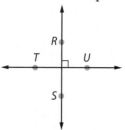

perpendicular lines *RS* and *TU*

Notice that all of the lines and segments are named with two points, and all of the angles are labeled with three points. When an angle is named, the middle point is the **vertex**, the point from which the two rays extend.

THINK ABOUT MATH

Directions: Draw and label each figure described.

1. ray *PQ*   **2.** segment *MN*   **3.** right angle *RST*

CONTRIBUTIONS OF ANCIENT CIVILIZATIONS

You may have heard of Pythagoras, Euclid, and other mathematicians who centuries ago developed and proved the geometry ideas that influenced the modern world. Even longer ago, ancient civilizations learned and used geometry concepts that served as a foundation for later mathematicians. Without it, our modern society would not be possible.

The Babylonians knew how to measure volume and area of shapes. They could estimate the circumference of circles. The Babylonians recorded math problems on clay tablets more than 4,000 years ago.

Ancient Egyptians also calculated areas of shapes and volumes of solids, developing the same formulas that are used today. One example of this is the Moscow Papyrus, written around 1850 BC. The Moscow Papyrus contains hieroglyphics that record and solve word problems. These include finding area, surface area, and volume. You may be surprised how similar the math problems are to the ones you will solve.

The ancient Greeks learned and used the knowledge of the ancient Egyptians, Babylonians, and other civilizations. They also discovered and proved many more geometric concepts.

You will learn basic geometry in this chapter, including different types of geometric figures. You will also learn how to calculate perimeter, circumference, area, and volume. These concepts and skills will provide a foundation as you learn more about geometry and other areas of math, including trigonometry and calculus.

Core Skill
Analyze Events and Ideas

Many consider Archimedes to be the greatest mathematician of antiquity, or Ancient Greece. One of Archimedes' contributions to geometry is finding an approximation of pi, (also known by the symbol π) that was only off by .008%. These calculations, although geometric, used methods still used in calculus today.

On your own, research what Archimedes did to find better approximations of π and write what those approximations were.

MATH LINK

Polygons are named by starting at a vertex and, by moving either clockwise or counter-clockwise, listing the labels of the vertices (plural of vertex) in that order.

Like other areas of math, it is important to check your answers to geometry problems to make sure they make sense.

One way to check for reasonableness in geometry is to study the diagram. For example, if you measure an angle and get 150°, but you see that the angle is acute, you know that the angle cannot be greater than 90°, so your answer is not reasonable.

Be careful, though, not to make assumptions. Diagrams may not be drawn precisely. For example, lines may appear to be parallel or perpendicular in a diagram. Only draw conclusions if they are given or if you can otherwise prove them.

For example, suppose you have a triangle that has two incredibly long sides, with a short side as the third side. Both of the angles next to the short side look like right angles. Can this be the case? Why or why not?

Angles

Angles are classified by their measurement. To **classify** is to sort into groups by some chosen attribute. Angles are measured in **degrees** (°).

CLASSIFICATION OF ANGLES

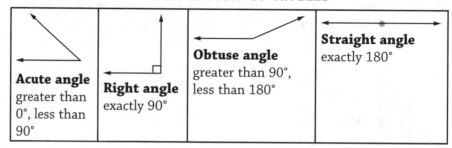

Acute angle
greater than 0°, less than 90°

Right angle
exactly 90°

Obtuse angle
greater than 90°, less than 180°

Straight angle
exactly 180°

Triangles and Quadrilaterals

Triangles

A **polygon** is a closed plane figure made up of three or more non-intersecting segments that are joined together. A polygon that has three sides is a **triangle**. Triangles can be classified by their sides or their angles. The sum of the angles of a triangle is 180°.

Classification by Angles			Classified by Sides		
Acute All angles measure < 90 degrees.	**Right** One angle measures 90 degrees.	**Obtuse** One angle measures > 90 degrees.	**Equilateral** All sides are of equal length.	**Isosceles** Two side lengths are equal.	**Scalene** No side lengths are equal.

THINK ABOUT MATH

Directions: Identify each triangle by its angles and its sides.

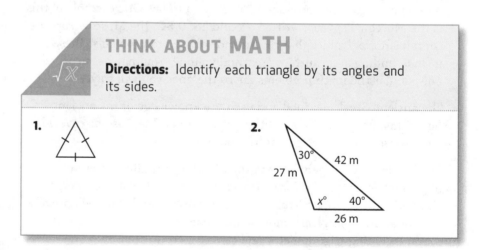

1.

2.
30° 42 m
27 m
$x°$ 40°
26 m

Example 1 Classify a Triangle Given the Angles and Sides

Classify the triangle by its angles and its sides.

Step 1 Identify the angles in the triangle.

The triangle has a right angle symbol, so it is a right triangle.

Step 2 Identify the sides of the triangle.

The triangle has matching tick marks for two sides, which means the two sides have equal length. It is an isosceles triangle, or a triangle that has two sides of equal length.

The triangle is classified as a right isosceles triangle.

Quadrilaterals

Polygons that have four sides are called **quadrilaterals**. Quadrilaterals are classified by their sides or their angles. The sum of the angles of a quadrilateral is 360°. If both pairs of opposite sides of a quadrilateral are parallel, the quadrilateral is a **parallelogram**. The opposite sides of a parallelogram are equal in measure. Two types of parallelograms are **rectangles** and **squares**. A rectangle has all of its angle measures equal to 90°, while a square has all of its side measures equal, and its angle measures equal 90°.

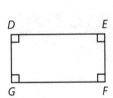

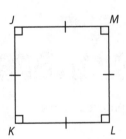

Example 2 Find a Measure of an Opposite Side of a Rectangle

Find the measure of side *EF*.

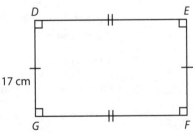

Step 1 Look at the side parallel to side *EF*.

Side *DG* is parallel to side *EF*.

Step 2 Find the measure of side *DG*.

The measure of side *DG* is 17 cm.

Step 3 Since figure *DEFG* is a rectangle, opposite sides have equal measure. Since side *DG* is opposite to side *EF*, it also has an equal measure to side *EF*. Therefore, side *EF* has a measure of 17 cm.

Circles

A **circle** is a plane figure in which every point on it is the same distance from the **center**. The **diameter** is the distance across a circle through its center. It is twice as long as the **radius**, which is the distance from the center to any point on the circle. In circle *R* shown, *R* is the center, line segment *AB* is the diameter, and line segment *RC* is a radius. In this circle, *RA* and *RB* are also radii (plural of radius).

Example 3 Identify the Parts of a Circle

Identify a radius and diameter of circle *A*.

Step 1 Identify a radius.

The radius is the distance from the center of a circle to any point on the circle, so line segments *AB* and *AC* are both radii of the circle.

Step 2 Identify a diameter.

The diameter is the distance across a circle through its center, so line segment *BC* is the diameter.

Vocabulary Review

Directions: Match each word to one of the phrases below.

angle   circle   parallel lines   perpendicular lines   quadrilateral
rectangle   segment   square

_____ A.   a portion of a line continuing in one direction forever, starting at a vertex

_____ B.   a plane figure in which every point on it is the same distance from the center

_____ C.   a four-sided figure

_____ D.   the amount of rotation between two intersecting lines or ray

_____ E.   a figure with four right angles

_____ F.   two lines that intersect to form a right angle at the intersection point

_____ G.   two (or more) lines that never intersect

_____ H.   a figure with all four sides of equal length and four right angles

Directions: Complete the following.

1. How did Archimedes approximate the value of π?

_____

2. Compare and contrast rectangles and squares.

_____

Directions: Answer the following question.

3. Can a triangle have both a right angle and an obtuse angle? Explain.

_____

Skill Practice

Directions: Choose the best answer to each question.

1. Which line segment is the diameter of the circle?

   A. line segment *MP*
   B. line segment *PT*
   C. line segment *QR*
   D. line segment *ST*

2. Which term best describes the angle?

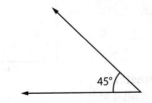

   A. right   C. acute
   B. straight   D. obtuse

3. Which terms best describe the triangle?

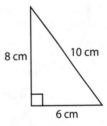

   8 cm   10 cm
   6 cm

   A. acute, scalene
   B. obtuse, equilateral
   C. right, scalene
   D. right, isosceles

4. Which term best describes the figure?

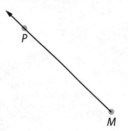

   A. ray *MP*   C. segment *MP*
   B. line *MP*   D. angle *MP*

Perimeter and Circumference

KEY CONCEPT: Understand and apply concepts of perimeter and circumference.

Identify each polygon.

1.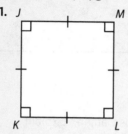

2.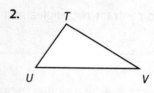

Perimeter

The distance around a figure, such as a triangle, rectangle, or square, is its **perimeter**. To find the perimeter of a figure, add the lengths of all the sides or use a formula. Sometimes not all of the side lengths of a figure are given. In those cases, you will need to find the unknown side lengths and then find the perimeter.

Example 1 Find Perimeter by Adding Lengths of Sides

Find the perimeter of the triangle.

Step 1 Determine the number of sides of the figure and the side lengths.

The triangle has 3 sides, and each side is 32 inches.

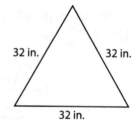

Step 2 Add the side lengths.

$P = 32 + 32 + 32 = 96$ inches

The perimeter is 96 inches.

Example 2 Find Perimeter Using the Formula for a Square

Find the perimeter of the square.

Step 1 Determine the number of sides of the figure and the side lengths.

The square has 4 sides, each with a length of 5 feet.

Step 2 Use a formula to find the perimeter. In this case, you could also add all 4 sides.

The formula for the perimeter of a square is $P = 4s$.
$P = 4 \times 5$ Substitute 5 for s.
$P = 20$

The perimeter of the square is 20 feet.

MATHEMATICAL PRACTICES

Some passages contain many details. If you are reading for a specific purpose, you might find that not all of the details are necessary. Finding the details that you need is an important skill.

When looking at a passage, diagram, or chart, ask: *Do I need all of this information to solve the problem?*

Find the perimeter of the base of the building represented by the diagram below.

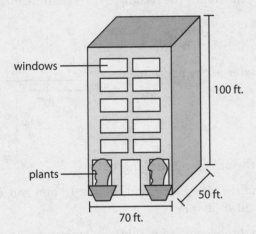

The base of the building is the bottom, and the perimeter is the distance around it. The details that you need are the length and width of the bottom of the building: 50 feet and 70 feet. You can ignore the height, the label for the windows, and the plants.

The perimeter is $50 + 70 + 50 + 70 = 240$ ft.

Core Practice
Model with
Mathematics

Sometimes, problems important to the community can be solved by using interpersonal skills. It is hoped that using individual strengths will support the group.

Suppose a town has a plot of land designated as a community garden, and it plans to paint the fence surrounding the garden. The community group plans to let the 25 families who live nearby paint the fence. How can the amount of fence be divided evenly among the 25 families?

To answer these questions, the work can be divided up among volunteers who have the necessary math skills. The tasks may include:

• Measuring the plot of land to determine its perimeter.

• Using the perimeter to determine the size of each family's fence.

In a notebook, write down the steps to determine the amount of fencing each family would get.

Example 3 Find Perimeter Using the Formula for a Rectangle

Find the perimeter of the rectangle.

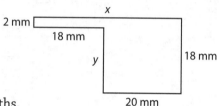

10 cm

28 cm

Core Skill
Build Lines
of Reasoning

Often you will be asked to explain your solution to a math problem in writing. Be sure to write your explanations logically.

First, state your end goal. Then list each step you take as you solve the problem. Explain each step by showing your work.

You may follow the format of the examples in the lesson to clearly show your solution process.

In a notebook, list the steps and determine the perimeter of a square that has one of its side lengths as 3 yards.

Step 1 Determine the number of sides and the side lengths.

The rectangle has 4 sides: two 28 cm lengths and two 10 cm widths.

Step 2 Use a formula to find the perimeter.

The formula for the perimeter of a rectangle is $P = 2l + 2w$.

$P = 2l + 2w$ $l =$ length and $w =$ width

$P = 2(28) + 2(10)$ Substitute 28 for length and 10 for width.

$P = 56 + 20$ $P = 76$

The perimeter of the rectangle is 76 centimeters.

Example 4 Find Perimeter with Unknown Side Lengths

Find the perimeter of the polygon.

x

2 mm

18 mm

y

18 mm

20 mm

Step 1 Find the unknown side lengths.

To find the length of x, add lengths 18 mm and 20 mm.
$18 + 20 = 38$ mm

To find the length of y, subtract 2 mm from 18 mm.
$18 - 2 = 16$ mm

Step 2 Add the side lengths. In this polygon, there are 6 sides.

$P = 38 + 18 + 20 + 16 + 18 + 2 = 112$

The perimeter of the polygon is 112 millimeters.

THINK ABOUT MATH

Directions: Find the perimeter of each figure.

1.

20 yd.

30 yd.

2.

8 cm

17 cm

15 cm

3.

x

4 m

8 m

8 m

12 m

11 m

Circumference

The distance around a circle is its **circumference**. This measure depends on the **radius** or the **diameter** of the circle. Recall that the radius of a circle is the distance from the center to any point on the edge of the circle. The diameter is a line segment that crosses the circle through its center from one side to the other.

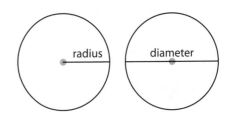

If you know the radius of a circle, multiply by 2 to find its diameter. If you know the diameter of a circle, divide by 2 to find its radius.

To find the circumference of a circle, multiply its diameter by **pi** (π). Pi is the ratio of the circumference of a circle to its diameter. The value of pi is 3.14159.... You can use either 3.14 or $\frac{22}{7}$ as approximations for pi.

Example 5 Find Circumference Using Diameter

Find the circumference of a circle with a diameter of 28 meters.

Step 1 Decide whether to use 3.14 or $\frac{22}{7}$ for π.
The diameter is 28 meters. 28 is a multiple of 7, so use $\frac{22}{7}$.

Step 2 Find the circumference of the circle.
$C = \pi d$
$C = \frac{22}{7} \times 28 = \frac{22}{1\cancel{7}} \times \frac{\cancel{28}^4}{1} = 88$

$C = 88$
The circumference of the circle is about 88 meters.

MATH LINK

Use $\frac{22}{7}$ for π when the diameter or radius is a multiple of 7. Otherwise, use 3.14 for π.

Example 6 Find Circumference Using Radius

Find the circumference of the circle.

Step 1 Find the diameter.

The radius is labeled 10 inches.
The diameter is twice the radius:
10 in. × 2 = 20 in.

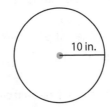

10 in.

Step 2 Find the circumference of the circle. Use the formula $C = \pi d$.
Use 3.14 for π, since 20 is not a multiple of 7.
$C = \pi d$
$C = 3.14 \times 20$
$C = 62.8$
The circumference of the circle is about 62.8 inches.

Example 7 Use a Calculator to Find Circumference

Find the circumference of a circle with a diameter of 12 millimeters.

Press ⓞ ③ ⚬ ① ④ ⓧ ① ② ⓔⁿᵗᵉʳ

The display will read

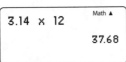

3.14 × 12   Math ▲

37.68

The circumference of the circle is about 37.68 millimeters.

THINK ABOUT MATH

Directions: Find the circumference of each figure.

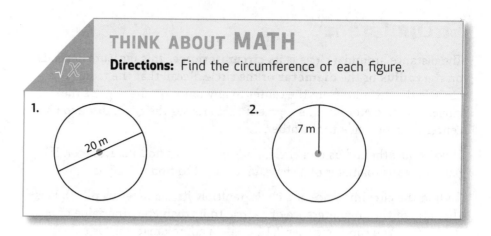

1.

2.

Vocabulary Review

Directions: Fill in the blanks with one of the words below.

circumference   diameter   perimeter   pi   radius

1. _____ is the distance around a polygon.

2. The _____ of a circle is the distance from the center to any point on the circle.

3. The distance around a circle is its _____.

4. _____ is the ratio of the circumference of a circle to its diameter, and has an approximate value of 3.14.

5. A line segment that crosses a circle through its center from one side to the other is the _____.

Skill Review

Directions: Describe the details you need to find the perimeter or circumference of each figure. Then find each perimeter or circumference.

1.

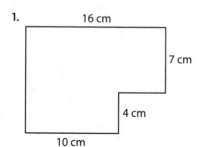

2.

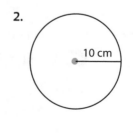

Skill Practice

Directions: Choose the best answer to each question.

1. The diagram shows the design of a rug for a front porch. If the manufacturer puts rubber edging around the rug, how many meters of rubber will be needed?

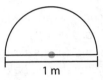

1 m

   A. 1.57
   B. 2.57
   C. 3.14
   D. 4.14

2. What is the perimeter in millimeters, of the figure shown below?

8 mm

   A. 48   C. 72
   B. 64   D. 80

3. Mr. Ruiz wants to put new molding around the ceiling in his living room. How many feet of molding should he buy?

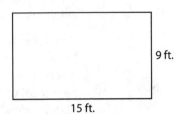

9 ft.

15 ft.

   A. 6
   B. 24
   C. 48
   D. 135

4. The perimeter of a rectangle is 24 feet. If the width is 7 feet, what is the length in feet?

   A. 3
   B. 5
   C. 10
   D. 17

Scale Drawings and Measurement

KEY CONCEPT: You can use scale drawings to discover information about the actual objects those drawings represent.

A ratio is a comparison of two values. Ratios can be expressed in words, such as 3 dogs to 2 cats. They can also be expressed with a colon, as in 3:2, and as a fraction, or $\frac{3}{2}$.

Regardless of how a ratio is written, it is usually expressed in simplest form. For example, say the ratio of girls to boys in a class is 5:4. This means that there could be a total of 5 girls and 4 boys. Or there could be a total of 10 girls and 8 boys, or 10:8. Or there could be a total of 15 girls and 12 boys, or 15:12. In each case, the simplest form of each ratio is the same: 5:4.

Scale Factor

Recall that a ratio is a comparison of two values. The ratio you find on a map is called a **scale factor**. It is a ratio of lengths. In other words, it compares the measure of one length to the measure of another.

You have probably seen or used a variety of maps. Each map includes a scale factor, which you can use to find out how far one place is from another. On the map, the scale factor might be 1 inch = 50 miles. This scale factor tells you that if the distance between two objects on a map is 1 inch, the real-world distance between these objects is 50 miles.

Sometimes you can use mental math to figure out actual distances. At other times, it is helpful to write a **proportion**, or equation. In a proportion, the ratios on either side of the equation are equivalent.

$$\frac{1 \text{ inch}}{50 \text{ miles}} \underset{\diagdown}{\overset{\diagup}{\bowtie}} \frac{2 \text{ inches}}{x \text{ miles}}$$

$1x = 2 (50)$

$x = 100$ miles

Scale Drawings

Scale drawings help you to see real things, like buildings and parks, on paper. Blueprints and maps are examples of **scale drawings**. A scale drawing of a place or object is either a smaller or a larger representation of the real place or object.

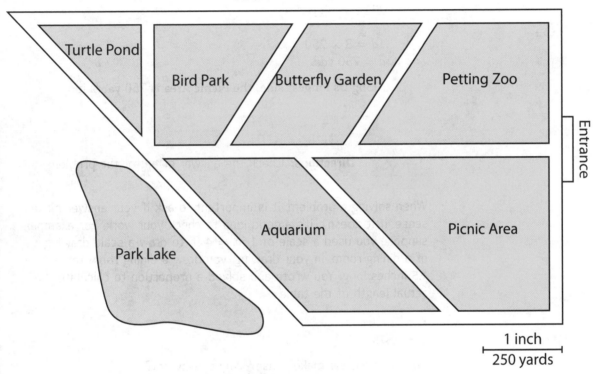

NATURE PARK

Computing Distances on a Scale Drawing

Look at the map of Nature Park. The Picnic Area stretches across the southeastern corner of the park. The map is not to exact scale, but suppose that at its widest point on the map, the Picnic Area is 3 inches long. You can use this information and the scale factor to find the length of the picnic area's longest side.

Step 1 Write a ratio.

Use the scale.

The scale factor is 1 in. = 250 yds. You can rewrite this scale factor as a ratio:

$\dfrac{1 \text{ in.}}{250 \text{ yds.}}$

Look at the numerator in this ratio. It represents the distance on the map. Now look at the denominator. It represents the distance in the real park.

Step 2 Write a proportion.

You used the scale factor to write a ratio. Now write a proportion to find the length of the Picnic Area's longest side. Let the variable d represent the unknown distance. The proportion is:

$$\frac{1 \text{ in.}}{250 \text{ yds.}} = \frac{3 \text{ in.}}{d \text{ yds.}}$$

Look closely at the numerators in the proportion. They both represent lengths on the map. Now look closely at the denominators. They represent real distances in the park.

Step 3 Solve the proportion.

To solve a proportion, use mental math or cross-multiply.

$$\frac{1 \text{ in.}}{250 \text{ yds.}} \bowtie \frac{3 \text{ in.}}{d \text{ yds.}}$$

$1d = 3 \times 250$
$d = 750 \text{ yds.}$

Along its longest side, the Picnic Area is 750 yards long.

THINK ABOUT MATH

$\sqrt{x}$ **Directions:** Check the answer given for the problem.

When solving a problem, it is important to ask if your answer makes sense. If it doesn't, it's a good idea to check your work. For example, suppose you used a scale of 1 in. = 4 ft. to draw a scale drawing of a dining room. In your drawing, you drew a dining room table 1.5 inches long. You wrote and solved a proportion to calculate the actual length of the table.

$\frac{1}{4} = \frac{x}{1.5}$

$x = 0.375$ ft.

Does this answer make sense? Why or why not?

Draw Geometric Shapes with Given Conditions

Imagine you are an architect who wants to draw plans for an office building. In your drawing, each office will have a rectangular door that measures 36 inches wide by 84 inches tall. You want to use a scale factor of 1 in. = 12 in. in your scale drawing. Follow the steps below to determine the correct dimensions of each door in your scale drawing.

Step 1 Write a ratio.

Use the scale factor $\frac{1 \text{ in.}}{12 \text{ in.}}$

Step 2 Write a proportion.
You are calculating each door's height and width, meaning you need to find two dimensions. So, you need two proportions.

Height: $\frac{1 \text{ in.}}{12 \text{ in.}} = \frac{h \text{ in.}}{84 \text{ in.}}$

Width: $\frac{1 \text{ in.}}{12 \text{ in.}} = \frac{w \text{ in.}}{36 \text{ in.}}$

Step 3 Solve the proportions.

Now you can cross-multiply to solve each proportion. All of the measurement units are in inches, so you leave them out of the solution.

Solve for height:

$$\frac{1 \text{ in.}}{12 \text{ in.}} \diagdown \frac{h \text{ in.}}{84 \text{ in.}}$$

$$1 \times 84 = 12 \times h$$

$$84 = 12h$$

$$\frac{84}{12} = \frac{12h}{12}$$

$$h = 7$$

In your scale drawing, you will draw each door 7 inches high.

Solve for width:

$$\frac{1 \text{ in.}}{12 \text{ in.}} = \frac{w \text{ in.}}{36 \text{ in.}}$$

$$1 \times 36 = 12 \times w$$

$$36 = 12w$$

$$\frac{36}{12} = \frac{12w}{12}$$

$$w = 3$$

In the scale drawing, you will draw each door 3 inches wide.

Step 4 Now you have the information you need to draw your scale drawing. You will make each rectangular door 7 inches tall and 3 inches wide. Because each door is a rectangle, you use a protractor (a tool for measuring angles) to be sure each angle measures 90°.

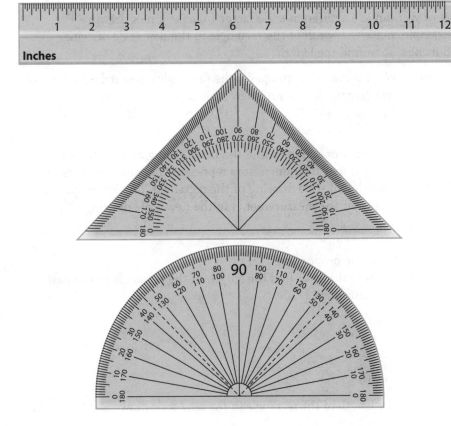

Core Practice
Use Appropriate
Math Tools

When you make scale drawings, it is necessary to make precise measurements. That's because the lengths in your drawings are proportional to real-world lengths. There are a variety of rulers you can use for the job.

Straight rulers measure shorter lengths, such as inches, feet, millimeters, and centimeters. Straight rulers, called yardsticks, measure lengths up to three feet, and meter sticks measure lengths up to 100 centimeters.

A machinist who makes precise tools might use a micrometer, a tool that measures to the nearest 0.01 inch or 0.25 millimeter. Vernier calipers make even more precise measurements possible, up to 0.001 inch or 0.02 millimeter. For long distances, some engineers use laser distance measuring tools with LCD screens. The tools emit a light that strikes a distant object, calculates the distance, and shows the measurement on the LCD screen.

What measurement tools are you used to using? Think of something you would like to build or an area you would like to measure. What measurement tools would you need to create a scale drawing? Research your project to view actual scale drawings and learn about the measurement tools that were used to create them.

Often a problem
involving measurements
of scale requires you to
use both algebra and
geometry to find the
solution.

For example, suppose
that the length of a
rectangle is twice as
long as the width. A
scale drawing of the
rectangle is 3 inches
wide. What is its length
in the scale drawing?

Write an equation to
compare the rectangle's
length to its width:

length = 2 width,
or $l = 2w$

This equation has two
variables, but you
know the value of the
variable w.

$l = 2(3)$
$l = 6$

The length of the
rectangle in the scale
drawing is 6 inches.
So, the rectangle in
the drawing measures
3 inches wide and
6 inches tall.

Imagine you have been
hired as a set designer
for a theater, and you
have been given a scale
drawing of a living room
set for a play. In the
drawing, the back wall
is 7 inches wide and
3 inches high. The actual
back wall of the set is
25 feet wide. How can
you find the height of
the actual back wall?
What is the height of
the actual back wall?

Reproduce a Scale Drawing of a Different Scale

You can find the measurements of an actual object if you have a scale
drawing and a scale factor. Say that a rectangle in a scale drawing is
3 inches tall and 5 inches wide. The scale factor is 1 in. = 3 ft. This gives
you enough information to write and solve a proportion to find the object's
actual measurements.

Begin with finding the width of the actual
rectangle. Write and solve a proportion.

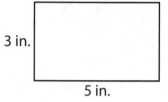

3 in.

5 in.

Step 1 Write a ratio.
Use the information from the
scale:
$\frac{1 \text{ in.}}{3 \text{ ft.}}$

Step 2 Write a proportion.
Use the scale and the measurements of the rectangle.
$\frac{1 \text{ in.}}{3 \text{ ft.}} = \frac{3 \text{ in.}}{w \text{ ft.}}$

Step 3 Solve the proportion. Cross-multiply.
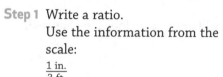
$1 \times w = 3 \times 3$
$w = 9$

The width of the actual rectangle is 9 feet.

You can follow the same steps to find the actual rectangle's length.

Now you can use a different scale factor to make a new drawing. For
example, make the width of the rectangle in the new scale drawing
6 inches. Then find the length.

Step 1 Write a ratio that compares the rectangle's original length to its
new length. This is your new scale factor.

$\frac{3 \text{ in. (old length)}}{6 \text{ in. (new length)}}$

Step 2 Write a proportion to find the new rectangle's length. Keep
in mind that the numerators represent the measurements in
the original scale drawing of the rectangle. The denominators
represent the measurements in the new scale drawing.

$\frac{3 \text{ in.}}{6 \text{ in.}} = \frac{5 \text{ in.}}{x \text{ in.}}$

Step 3 Solve the proportion.
Cross-multiply to find the length of the rectangle in the new
scale drawing.

$3 \times x = 5 \times 6$
$3x = 30$
$\frac{3x}{3} = \frac{30}{3}$
$x = 10$

The length of the rectangle in the new scale drawing is 10 inches.

Vocabulary Review

Directions: Match each term to its definition.

1. _____ proportion

2. _____ scale drawing

3. _____ scale factor

A. a ratio of lengths

B. a reduced or enlarged drawing of a real place or object

C. an equation showing two equivalent ratios

MATH LINK

Sometimes you can use mental math to solve proportions. Look at the numerators in this proportion. Notice that you can multiply the 3 in the first ratio by 4 to get the 12 in the second ratio.

$\frac{3}{5} = \frac{12}{x}$

$\frac{3}{5} \times \frac{4}{4} = \frac{12}{?}$

This means that you can also multiply the denominator by 4 to find the value of x.

$5 \times 4 = 20$, so $x = 20$.

Using mental math when you observe this type of relationship between the numerators or denominators helps you solve proportions quickly.

Skill Review

Directions: Complete each activity. The pairs of shapes in questions 1 and 2 are scale drawings of each other.

1. Find the perimeter of triangle *ABC*.

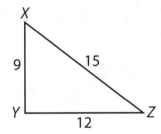

2. Find the perimeter of quadrilateral *ABCD*.

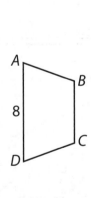

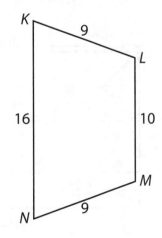

3. The height of a triangular monument is triple the length of the base. If a scale drawing of the monument has a height of 4.5 inches, what is the length of the base in the drawing?

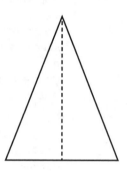

_____

5. A floor plan is also a scale drawing. If the length of one square on the graph paper shown is $\frac{1}{4}$ inch, what is the scale of the drawing?

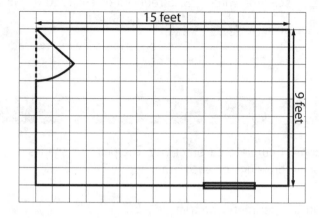

A. $\frac{1}{4}$ inch = 1 foot

B. $\frac{1}{2}$ inch = 1 foot

C. 1 inch = 1 foot

D. 2 inches = 1 foot

4. The length of a rectangular bench is 5 feet longer than the height. A scale drawing was made of the bench with a height of 2 inches and a length of 7 inches. What are the height and length of the actual bench?

_____ , _____

6. The designer of the following floor plan wants to add a 7-foot-long couch. How long will the couch be in the scale drawing?

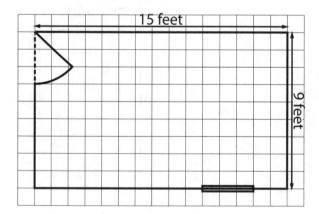

A. 1 inch

B. 1.75 inches

C. 2.33 inches

D. 4 inches

Skill Practice

Directions: Complete each activity.

1. Suppose that a picture of a shopping cart needs to be drawn to scale for a sign that a grocery store is making. The basket is a rectangle with an actual height of 2 feet and width of 3 feet. The diameter of each wheel is $\frac{1}{2}$ foot. The scale of the drawing is 3 in. = 2 ft.

   Note: drawing is not to scale.

   a. What is the height of the cart in the scale drawing?

   _____

   b. What is the width of the cart in the scale drawing?

   _____

   c. What is the diameter of each wheel in the drawing?

   _____

2. An amoeba is a single-celled organism. A scale drawing of an amoeba is shown. This drawing is 75 times larger than an actual amoeba. The scale is 3 in. = 1 mm. The diameter of the nucleus in this scale drawing is $\frac{1}{2}$ inch. What is the actual diameter of the nucleus?

   A. $\frac{1}{6}$ millimeter

   B. $\frac{1}{2}$ millimeter

   C. $\frac{2}{3}$ millimeter

   D. 2 millimeters

   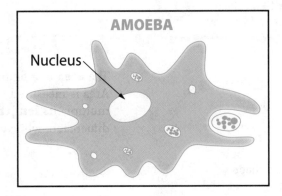

3. A scale drawing of a decorative window includes a triangle that has side lengths of 6, 7, and 10 centimeters. If the longest side of the actual window is 12 inches, what is that scale of the drawing?

   _____

4. Describe how to draw a triangle with side lengths 3 inches, 4 inches, and 5 inches using a compass.

   _____

   _____

5. You are given a scale drawing of a floorplan for a house. The length of the front door in the scale is 2 inches, but the foreman wants the scale plans to have the front door with a length of 3 inches. Write what must be done to accomplish this.

   _____

KEY CONCEPT: Develop and apply the concept of area to find the areas of simple and complex shapes.

Find the perimeter or circumference of each figure.

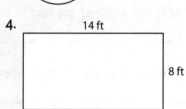

1. 11 m

2. 4 cm 5 cm 3 cm

3. 6 mm

4. 14 ft 8 ft

Area of Rectangles and Squares

The **area** of a figure is the amount of surface covered by the figure. Area is measured in square units. To find the area of a rectangle, multiply its **length** times its **width**. Length in a rectangle is the longer dimension, while width is the shorter dimension.

Example 1 Area of Rectangles

Find the area of the rectangle.

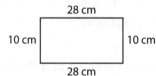

28 cm
10 cm   10 cm
28 cm

Step 1 Identify the measures of length and width.

length: 28 cm   width: 10 cm

Step 2 Find the area.

$A = l \times w$ Area = length × width
$A = 28 \times 10$
$A = 280 \text{ cm}^2$

The area is 280 square centimeters (cm²).

Example 2 Area of Squares

Find the area of a square with sides 5 feet long.

Step 1 Identify the measures of length and width.

The figure is a square; length and width are both 5 feet.

Step 2 Find the area of the square.

$A = l \times w$
$A = 5 \times 5$
$A = 25 \text{ ft}^2$ The area is 25 square feet (ft²).

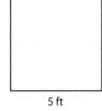
5 ft

MATH LINK

You can also use the formula for the area of a square: $A = s^2$, where s is one side of the square.

Area of Triangles

To find the area of a triangle, first identify the **base** and **height** of the triangle. The base is always the side that the triangle rests on. The height is the length of the segment perpendicular to the base and extending to the "top" of the triangle. In each triangle shown below, the base is labeled b and the height is labeled h.

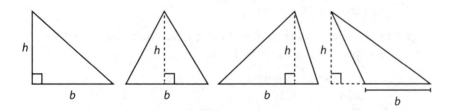

Example 3 Area of a Triangle

Find the area of the triangle.

Step 1 Identify the base and height of the triangle and their measures.

The base is segment BC, and the measure is 4 meters.
The height is segment AB, and the measure is 3 meters.

Step 2 Find the area of the triangle.

The area of a triangle is $\frac{1}{2}$ the total of its base (b) times its height (h).

$A = \frac{1}{2}(b \times h)$

$A = \frac{1}{2}(4 \times 3)$

$A = \frac{1}{2}(12)$

$A = 6 \text{ m}^2$ The area of the triangle is 6 square meters (m²).

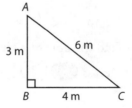
A
3 m   6 m
B   4 m   C

Core Skill
Build Lines
of Reasoning

When trying to determine the area of a triangle, you can first look at right triangles. If you place two identical triangles together by rotating the second and connecting them by their hypotenuses, you have created a rectangle. The area of the rectangle is just $A = bh$, where b and h are the base and height of the original triangle. Therefore, the area of the triangle is half that! This is where the formula came from. Of course, it takes more work to guarantee that this is the formula for all triangles, but it can be done.

Try drawing your own triangle. It can be any kind as long as it isn't a right triangle. Complete the rectangle by "adding in" right triangles. Then compute the area of the rectangle, and subtract the area of each of the right triangles. What is your result? It should be the area of the original triangle, using the formula you already know!

Core Skill
Build Solution Pathways

Finding the area of a shape is simpler when you have a plan.

1) Identify whether it is a shape for which you know how to find the area or if it is complex. If it is complex, draw lines to divide it into familiar shapes and find the area of each.

2) Determine what you need. Depending on the shape, you may need the length and width, base and height, or radius.

3) Use the correct formula for your shape, substituting in the values you found in step 2.

4) Check your answer for reasonableness. One way is to imagine square units covering the figure.

In a notebook, determine the area of a circle that has a circumference of 25.12 cm.

Example 4 Use a Calculator to Find the Area of a Triangle

Find the area of a triangle with base 4.26 inches and height 2.48 inches.

Press  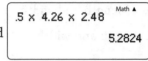.

The display will read

.5 × 4.26 × 2.48   Math ▲

5.2824

The area of the triangle is 5.2824 square inches (in.²).

Area of Circles

The area of a circle is π times the radius times the radius again. Use either $\frac{22}{7}$ or 3.14 for π. The formula for the area of a circle is $A = \pi r^2$.

Example 5 Area of a Circle

Find the area of the circle shown.

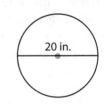
20 in.

Step 1 Find the radius.

Diameter ÷ 2 = 20 ÷ 2 = 10
The radius is 10 inches.

Step 2 Find the area of the circle. In this case, the radius is not a multiple of 7, so use 3.14 for π.

$A = \pi \times$ radius $\times$ radius
$A = 3.14 \times 10 \times 10$
$A = 314$ in.² The area is about 314 square inches (in.²). This is because we used 3.14 as an approximation of pi instead of its actual value, which does not end.

Example 6 Use a Calculator to Find the Area of a Circle

Find the area of a circle with a radius of 8 meters.

Press 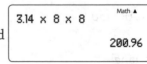.

The display will read

3.14 × 8 × 8   Math ▲

200.96

The area of the circle is about 200.96 square meters (m²).

THINK ABOUT MATH

Directions: Find the area of each figure.

1.

13 cm   13 cm

12 cm

10 cm

2.

11 ft.

14 ft.

3.

7 in.

4.

8 mm

Area of Complex Shapes

Complex shapes are composed of two or more shapes. To find the area of a complex shape, divide the shape into two or more shapes—such as rectangles, squares, triangles, or circles—find the area of each shape, and then add the areas.

Example 7 Area of a Shape Composed of a Rectangle and Triangle

Find the area of the polygon.

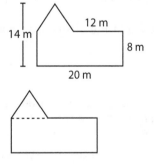

14 m   12 m   8 m

20 m

Step 1 Divide the whole into smaller shapes. Divide it into a rectangle and a triangle.

Step 2 Identify the dimensions of each shape.

rectangle: 20 m length and 8 m width
triangle: 8 m base (20 m − 12 m) and 6 m height (14 m − 8 m)

Step 3 Find the area of each shape.

rectangle: $A = 20 \times 8 = 160$ m^2
triangle: $A = \frac{1}{2}(8 \times 6) = \frac{1}{2}(48) = 24$ m^2

Step 4 Find the total area of the polygon. Add together the areas of the rectangle and the triangle.

160 m^2 + 24 m^2 = 184 m^2
The area of the polygon is 184 square meters.

Example 8 Area of a Shape Composed
of a Rectangle and Semicircle

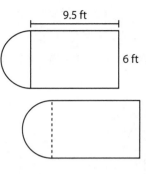

Find the area of the shape.

Step 1 Divide the whole into smaller shapes.
Divide it into a rectangle and semicircle.

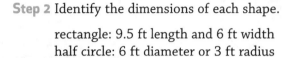

Step 2 Identify the dimensions of each shape.

rectangle: 9.5 ft length and 6 ft width
half circle: 6 ft diameter or 3 ft radius

Step 3 Find the areas of each shape. Multiply length and width for the rectangle. Multiply π times radius times radius for the circle.
Since it is only a half circle, divide that area by 2.

rectangle: $A = 9.5 \times 6 = 57$ ft^2
half circle: $A = (3.14 \times 3 \times 3) \div 2 = 28.26 \div 2 = 14.13$ ft^2

Step 4 Find the area of the complex shape. Add together the areas of the rectangle and the semicircle.

57 ft$^2 + 14.13$ ft$^2 = 71.13$ ft^2

The area of the shape is about 71.13 square feet.

Vocabulary Review

Directions: Fill in the blanks with one of the words or phrases below.

area   base   height   length   width

1. The _____ of a triangle is the length of the segment perpendicular to the base and extending to the top of the triangle.

2. _____ is the measure of the longer dimension in a rectangle.

3. The amount of surface covered by a figure is its _____, and it is measured in square units.

4. _____ is the measure of the shorter dimension in a rectangle.

5. The _____ of a triangle is the side that it rests on.

Skill Review

Directions: Find the area of each shape.

1.
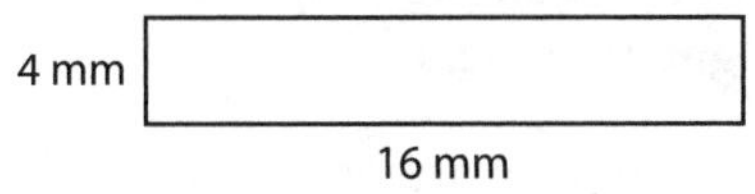
4 mm
16 mm

2.
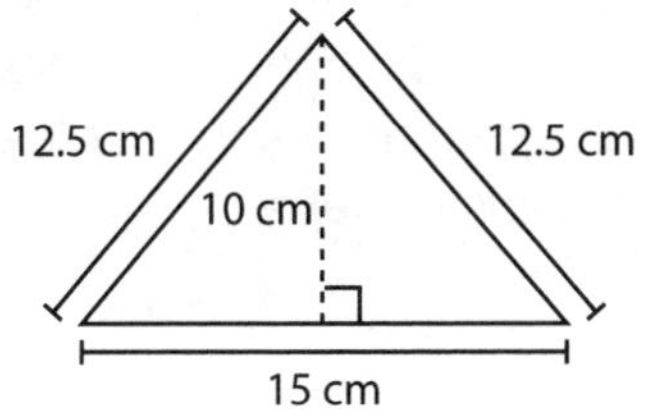
12.5 cm
12.5 cm
10 cm
15 cm

3.
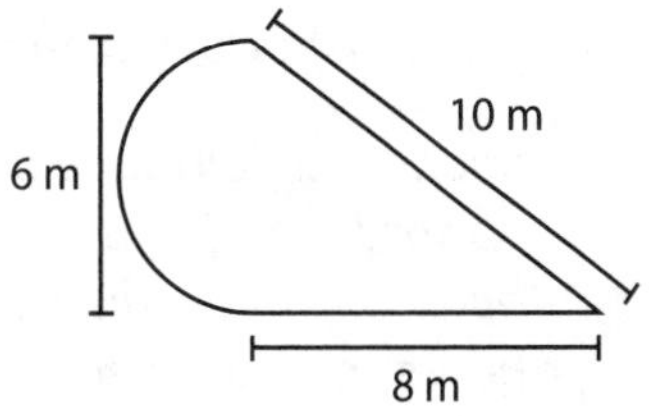
10 m
6 m
8 m

Skill Practice

Directions: Choose the best answer for each question.

1. Which of these is the area of the triangle in square meters?

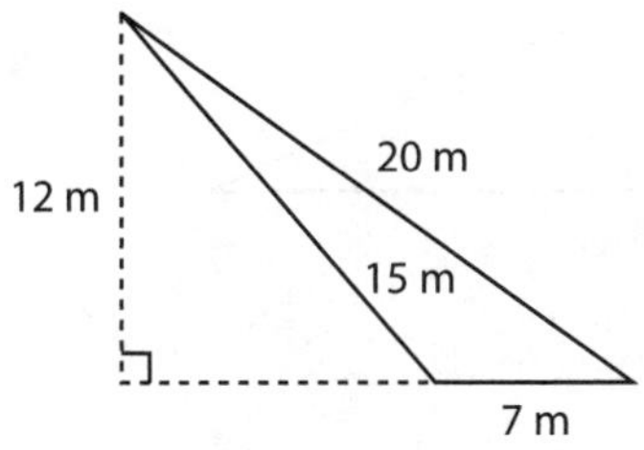
20 m
12 m
15 m
7 m

A. 42
B. 52.5
C. 70
D. 96

2. Cho Hee plans to use wallpaper on the four rectangular walls of her bedroom. Two walls measure 12 ft by 8 ft. The other two walls measure 9 ft by 8 ft. How many square feet of wallpaper does she need?

A. 168
B. 336
C. 384
D. 672

3. The diagram shows the shape of Daichi's yard. How many square feet does he have to mow?

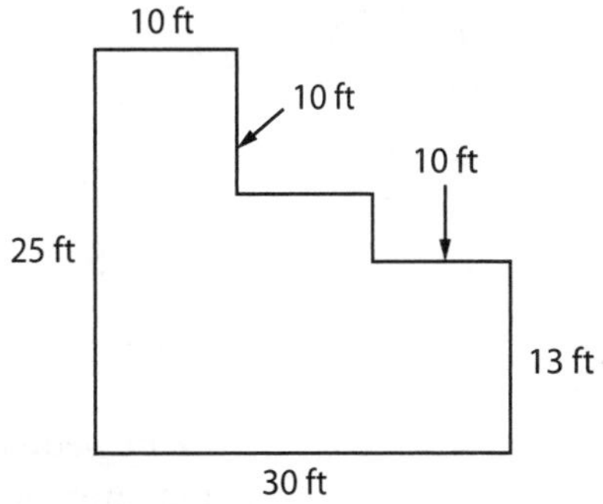
10 ft
10 ft
10 ft
25 ft
13 ft
30 ft

A. 690
B. 590
C. 560
D. 530

4. The area of a triangle is 30 square centimeters. Its base is 6 centimeters. What is its height in centimeters?

A. 5
B. 7.5
C. 10
D. 15

Pythagorean Theorem

Lesson Objectives

You will be able to

- Explain the Pythagorean theorem

- Apply the Pythagorean theorem to solve problems

Skills

- **Core Skill:** Analyze Events and Ideas

- **Core Skill:** Solve Quadratic Equations

Vocabulary

congruent
hypotenuse
leg
proof
Pythagorean theorem
quadratic equation
theorem

KEY CONCEPT: The Pythagorean theorem shows a special relationship between the sides of a right triangle.

Imagine an airplane beginning to leave a runway. Its nose points upward, and for a brief moment, the airplane's body forms an acute angle with the runway. An acute angle, like all other angles, is defined by its measurement.

The acute angle formed by the airplane in that moment measured less than 90°. An angle that measures exactly 90° is called a right angle. Next in size comes an obtuse angle, which measures greater than 90° but less than 180°. Finally, an even larger angle, which measures exactly 180°, is a straight angle.

Right Triangles

Triangles are classified, or grouped, by the measures of their angles. Every triangle has three sides, and every triangle has three angles. It is also always true that the sum of any triangle's angles is 180°.

If one of the angles in a triangle measures 90°, the triangle is called a right triangle. The longest side of a right triangle is opposite the right angle and is called the **hypotenuse**. The other two sides are called the triangle's **legs**.

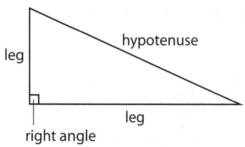

Pythagorean Theorem

Ancient peoples knew about right angles and how to make them long before anyone had written about them. They used ropes with evenly spaced knots in their building projects. Builders observed that if they used the 3-length, 4-length, and 5-length sides to make a triangle, the shorter sides always formed the same kind of angle, a right angle. That angle named the right triangle. The right triangle was considered a special triangle because it helped mathematicians understand mathematical relationships.

Early mathematicians learned that it was possible to use the lengths of the shorter sides, or legs, of a right triangle to calculate the length of the longest side, or hypotenuse. For example, consider the 3, 4, and 5 length sides of the knotted rope again. Square the shortest side 3 to get 9. Square the next shortest side 4 to get 16. Then add. The sum equals the hypotenuse squared, or 5^2.

$$3^2 + 4^2 = ?^2$$
$$9 + 16 = 25$$
$$\sqrt{25} = 5$$
$$3^2 + 4^2 = 5^2$$

The hypotenuse is 5 lengths long.

The characteristics of a special triangle were generally accepted, but until Pythagoras, no one had ever devised a **theorem**, or mathematical statement, describing the relationship. The **Pythagorean theorem** says that the sum of the squares of the legs equals the square of the hypotenuse: $a^2 + b^2 = c^2$. In other words, if you square the lengths of the shorter sides and add them, the sum equals the length of the longest side squared.

Not only did Pythagoras provide a theorem for the special triangle, he went a step further and provided a **proof**, or logical progression of true statements, that the relationship that existed in a special triangle exists in every right triangle.

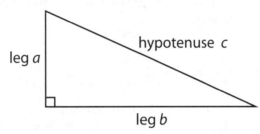

The Proof of the Pythagorean Theorem

Look closely at the image shown. You will see four congruent, right triangles that have been rotated and placed so that their hypotenuses form a square in the middle. *Congruent* means "identical in form." The two figures are congruent because you can flip, turn, or slide one exactly on the other.

The following statements are true about the shape:

- Each of the sides of the outside square has a length of $a + b$.

- The area of the outside square is $(a + b)(a + b)$.

- The area of each triangle is $\frac{1}{2}(ab)$.

- There are four triangles, so the total area of the four triangles is $4(\frac{1}{2})(ab)$, or $2ab$.

- The area of the inside square is c^2.

- The area of the four triangles ($2ab$) plus the area of the inside square, (c^2) is equal to the area of the large square $(a + b)(a + b)$.

The last statement gives the equation:

$$2ab + c^2 = (a + b)(a + b)$$

Core Skill
Analyze Events and Ideas

Pythagoras was a Greek philosopher who lived from about 570 BC to 490 BC. He studied astronomy, music, politics, religion, and mathematics.

Pythagoras traveled from his home in Greece to study in Egypt. He later settled in southern Italy and founded a society called the Order of Pythagoreans.

The Pythagoreans kept their work a secret, and for many years they did not write down any of their theories and discoveries. As a result, no one is certain how many of the mathematical discoveries attributed to Pythagoras actually belonged to him. However, as leader of the society, Pythagoras has received credit for their work, including the discovery of the first proof of the relationship of right angles.

Research Pythagoras and his society. Analyze how his theories have affected the way people look at and measure things in their surroundings.

A **quadratic equation** contains a variable to the power of 2 but no higher powers. The variable x^2 is an example. Knowing how to solve quadratic equations is an important part of applying the Pythagorean theorem.

Try solving the following example: $x^2 = 3^2 + 4^2$. The first step toward finding the value of x is to simplify terms. Look to the right of the equation. Simplify the squared values.

$3^2 = 9$ and $4^2 = 16$, so you can rewrite the right side of the equation as:

$9 + 16 = 25$

Now you can rewrite the equation:

$x^2 = 25$

To find the value of x, take the square root of each side of the equation.

$\sqrt{x^2} = \sqrt{25}$

$x = 5$ and -5

In the equation, x represents the length of one side of a triangle. A length can only be positive, so what is the value of x?

When you simplified 3^2 and 4^2, you restructured the problem to make it easier to solve. When you completed the solution process, you successfully solved a quadratic equation.

Several steps are listed below. Notice that the first letter of the first word in each step is bold. Together, the letters spell FOIL. Use the word to help you remember the order of steps you should follow to find the solution to an equation.

First: Multiply the first term in each set of parentheses.

Outer: Multiply the outer term in each set of parentheses.

Inner: Multiply the inner terms in each set of parentheses.

Last: Multiply the last term in each set of parentheses.

Use the FOIL method to expand the right side of the equation to get:

$$2ab + c^2 = a^2 + 2ab + b^2$$

Subtract $2ab$ from each side of the equation to simplify the equation:

$$c^2 = a^2 + b^2$$

This logical progression of true statements leads to the Pythagorean theorem.

Identifying Right Triangles

The proof of the Pythagorean theorem shows that if a and b are legs of a right triangle, and c is the hypotenuse, then $a^2 + b^2 = c^2$.

It is also true that if the sides of a triangle are a, b, and c, and $a^2 + b^2 = c^2$ is true, then triangle abc, or $\triangle ABC$, must be a right triangle.

Say that a triangle has sides that measure 10, 12, and 15 inches. Is the triangle a right triangle? Apply the Pythagorean theorem to find the answer.

Step 1 Write the Pythagorean theorem.

$a^2 + b^2 = c^2$

Step 2 Substitute the length of the sides.

The hypotenuse is always the longest side, so 15 must be substituted for c, and 10 and 12 must be substituted for a and b.

$10^2 + 12^2 = 15^2$

Step 3 Solve the equation.

$100 + 144 = 225$
$244 \neq 225$

Conclusion: The result of your calculations is a false statement—244 does not equal 225. So the triangle formed by side lengths 10, 12, and 15 does not form a right triangle.

Recognizing Pythagorean Triples

A Pythagorean triple is a set of three positive integers, a, b, and c, where $a^2 + b^2 = c^2$. That is, they are lengths of the sides of a right triangle. The Pythagorean triple with the smallest set of whole numbers is 3, 4, 5. They are the lengths of sides of a right triangle: $3^2 + 4^2 = 5^2$.

If you find a multiple of each number in a Pythagorean triple, you have another Pythagorean triple. For example, multiply 3, 4, and 5 by 2 to get 6, 8, and 10. This is also a Pythagorean triple, since $36 + 64 = 100$. If you multiply each number by 3, you get the Pythagorean triple 9, 12, and 15.

Missing Side Lengths

When you know two sides of a right triangle, you can use the Pythagorean theorem to find the length of the third side.

Example: Find the missing side length of the triangle.

In this example, the lengths of the legs are given, but the length of the hypotenuse is unknown. Although you may recognize that this is a 5, 12, 13 Pythagorean triple and solve by inspection, the algebraic solution follows.

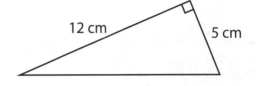

12 cm   5 cm

Step 1 Write the Pythagorean theorem.

$a^2 + b^2 = c^2$

Step 2 Substitute the known information.

Since the leg lengths are 5 and 12, substitute the values into a and b. It does not matter which number is substituted for each letter.

$5^2 + 12^2 = c^2$

Step 3 Solve the equation for the positive root.

$5^2 + 12^2 = c^2$
$25 + 144 = c^2$
$169 = c^2$
$13 = c$

The length of the hypotenuse is 13 cm.

Example: A 10-foot ladder is placed 3 feet from a wall. How high on the wall does the ladder reach? Round your answer to the nearest tenth.

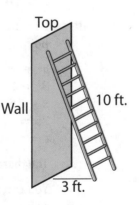

Top

Wall   10 ft.

3 ft.

Step 1 Draw a picture.

Step 2 Write the Pythagorean theorem.

$a^2 + b^2 = c^2$

Step 3 Substitute the known information.

The hypotenuse is 10 and a leg is 3. Substitute these values into c and a. Remember, 3 can be substituted for either a or b.

$3^2 + b^2 = 10^2$

Step 4 Solve the equation for the positive root.

$3^2 + b^2 = 10^2$
$9 + b^2 = 100$
$b^2 = 91$
$b \approx 9.5$ ft

The top of the ladder reaches 9.5 feet above the wall.

THINK ABOUT MATH

The sides of a triangle measure 3, 7, and 8. Use the Pythagorean theorem to determine if they form a right triangle.

The Distance Between Two Points on a Coordinate Graph

There are several ways to find the distance between two points on a graph. One strategy is to use the Pythagorean theorem.

Example: Find the distance between the points (3, –1) and (–1, 2).

Step 1 Plot each point on a coordinate axis.

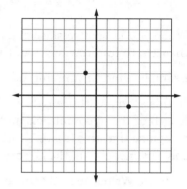

Step 2 Draw a right triangle so that the hypotenuse connects the points, one leg is vertical, and the other leg is horizontal.

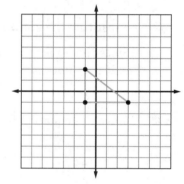

Step 3 Count the length of the horizontal and vertical legs.

3 and 4

Step 4 Use the Pythagorean theorem to find the distance between the points, or the hypotenuse of the triangle.

$a^2 + b^2 = c^2$
$3^2 + 4^2 = c^2$
$9 + 16 = c^2$
$25 = c^2$
$c = 5$

The distance between the points is 5.

You can use this method to find the distance between any two points on a coordinate graph.

THINK ABOUT MATH

Directions: Use the Pythagorean theorem to answer the following question.

What is the distance between the points (2, 2) and (6, 5) rounded to the nearest tenth?

Vocabulary Review

Directions: Match each term to its definition.

1. _____ congruent
2. _____ hypotenuse
3. _____ leg
4. _____ proof
5. _____ Pythagorean theorem
6. _____ quadratic equation
7. _____ theorem

A. one of the two shorter sides of a right triangle

B. the longest side of a right triangle

C. a mathematical statement

D. a logical progression of true statements that show something new is true

E. a mathematical statement that says in a right triangle with legs a and b and hypotenuse c, $a^2 + b^2 = c^2$

F. an equation containing a variable to the power of 2 but no higher power

G. equal or identical when one figure is placed over the other

Skill Review

Directions: Solve the quadratic equations.

1. $x^2 = 36$

2. $121 = c^2$

3. $h^2 + 4 = 13$

Directions: Examine each set of numbers. Label each set as "a Pythagorean triple" or "not a triple."

4. _____ 4, 5, 6

5. _____ 30, 40, 50

6. _____ 7, 10, 149

7. Use the Pythagorean theorem to find the length of the missing side.

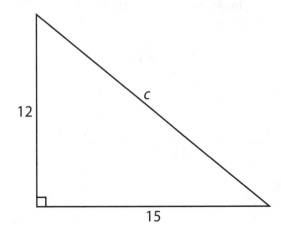

8. A designer sketches the floor plan for a bedroom. The bedroom forms a rectangle. The distance from one wall to the opposite wall is 12 feet. The distance between the remaining walls is 10 feet. The designer wants to know the length of a diagonal across the room. Draw and label a picture of the floor plan. Then use the Pythagorean theorem to find the length of the diagonal.

9. _____ Use a coordinate graph to find the distance between the points (3, 1) and (0, −3).

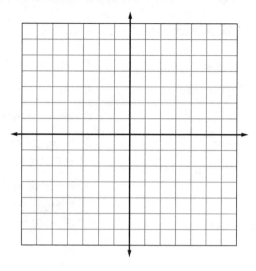

Skill Practice

Directions: Solve the quadratic equations.

1. $7^2 + b^2 = 12^2$

2. $a^2 + 6^2 = 9^2$

3. $7^2 + x^2 = 25^2$

Directions: Examine each set of numbers. Label each set as "a Pythagorean triple" or "not a triple."

4. _____ 20, 48, 52   5. _____ 4, 4, 32   6. _____ 12, 16, 20

7. Use the Pythagorean theorem to find the length of the missing side.

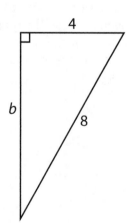

8. Use the Pythagorean theorem to find the length of the missing side.

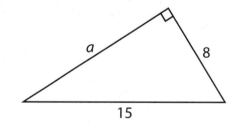

Skill Practice (continued)

9. _____ Use a coordinate graph to find the distance between the points (3, 7) and (5, 1).

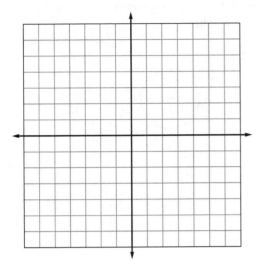

10. _____ Use a coordinate graph to find the distance between the points (−4, 6) and (2, −2).

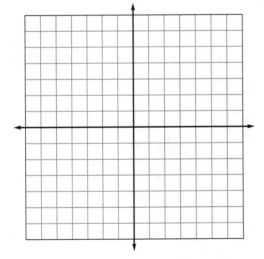

11. _____ Three friends are playing catch. Josué is standing 20 feet south of Vanessa, and Sam is standing 12 feet west of Vanessa. What is the distance between Sam and Josué? Draw a picture to help you find the solution.

12. _____ The length of a slide is 6 feet. The bottom of the slide is 4 feet away from the bottom of the ladder. What is the height of the ladder?

Geometric Solids and Volume

KEY CONCEPT: Extend understanding of geometric figures to include solids and the concept of volume.

Identify each shape.

1.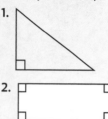

2.

3.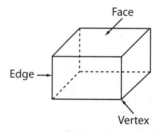

Solids

A solid is a **three-dimensional figure** that has length, width, and height. It is a figure that occupies space.

The two-dimensional surfaces of solids are **faces**. A surface has only length and width, but no depth, or height. The line segment where two faces **intersect** (meet) is an **edge**. The point where edges come together is a **vertex** (*vertices* is the plural). The figure below is called a **rectangular solid** or **rectangular prism**, because all of its faces are rectangles. We commonly call it a box.

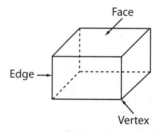

Volume of Rectangular Solids

Volume is a measure of the space inside a three-dimensional figure. It is similar to capacity, the amount a container can hold. Volume is measured in cubic units.

To find the volume of a rectangular solid, multiply the length, width, and height. The formula is

$$V = l \times w \times h$$

Example 1 Find the Volume of a Rectangular Solid

Find the volume of the rectangular prism shown.

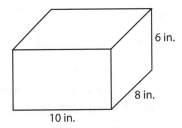

Step 1 Identify the length, width, and height of the rectangular solid.
length = 10 in.; width = 8 in.; height = 6 in.

Step 2 Use the volume formula.
Volume = length × width × height
$V = 10 \times 8 \times 6$
$V = 480$
The volume of the rectangular prism is 480 cubic inches (in.^{3}).

Core Skill
Solve Real-World Problems

When you solve mathematical problems in the real world, you will have to make sure you understand what the problem is asking and what is given.

You can use a variety of appropriate sources to solve volume problems. Possible sources include knowing which formula to use to determine volume (or missing side lengths) and having the correct formula for the appropriate shape. You will also need references for conversions of units if the problem requires converting between units.

Consider the following problem. Suppose you are given the task of creating a solid concrete cube statue to stand in the town square. You are given a specified amount of concrete (in pounds) and a specified size of cube that is to be created (in cubic feet).

Write down the information you need to determine if you were provided enough concrete for the job.

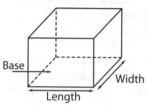
Example 2 Find Volume in a Real-World Problem

A board foot is a unit used for lumber. It measures 12 inches by 12 inches by 1 inch. What is the volume, in cubic inches, of one board foot?

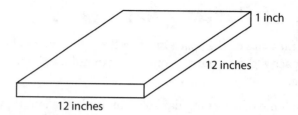

1 inch

12 inches

12 inches

Step 1 Find the length, width, and height of the board foot.
$l = 12$, $w = 12$, $h = 1$

Step 2 Use the formula for volume. $V = l \times w \times h$
$12 \times 12 \times 1 = 144$

Step 3 Write the volume using cubic units.
The volume of a board foot is 144 cubic inches (in.³).

Example 3 Find One of the Dimensions of a Rectangular Solid

The volume of a rectangular solid is 364 cubic centimeters. Its length is 13 centimeters, and its height is 7 centimeters. What is the width of the rectangular solid?

Step 1 Identify the information you know about the rectangular solid.
$V = 364$ cm³, $l = 13$ cm, $h = 7$ cm

Step 2 Use the volume formula. By replacing the values you know to find the width.
$364 = 13 \times 7 \times w$

Step 3 Solve the equation for w.
$364 = 13 \times 7 \times w$
$364 = 91 \times w$
$\frac{364}{91} = \frac{91 \times w}{91}$
$4 = w$

Step 4 Check the solution.
$13 \times 7 \times 4 = 364$ is true.
4 centimeters (cm) is the width of the rectangular solid.

Volume of Cubes

A **cube** is a rectangular solid in which each face is a square. So, if the side, or edge, of the cube is s, the volume of the cube is s^3. The formula for the volume of a cube is $V = s^3$.

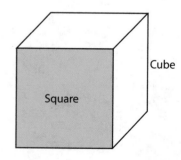

Cube

Square

Example 4 Find the Volume of a Cube

Find the volume of the cube shown.

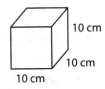

Step 1 Identify the length of one edge or side of the cube.
The side length is equal to 10 cm.

Step 2 Find the volume of the cube.
$V = 10^3 = 10 \times 10 \times 10$
$V = 1,000$
The volume of the cube is 1,000 cubic centimeters (cm³).

Example 5 Find the Edge of a Cube

The volume of a cube is 343 cubic feet. What is the measure of an edge or side of the cube?

Step 1 Use the formula for the volume of a cube.
$V = s^3$
$343 = s^3$

Step 2 Use a calculator to find the cube root of 343.

$\sqrt[3]{343} = s$

Press 3   enter

Step 3 The cube root of 343 is 7.
An edge of the cube is 7 feet.

Core Skill
Calculate Volume

The volume of a three-dimensional solid is the amount of space that would fill the shape if it were hollow, just as area is the amount of "space" that fills a two-dimensional shape.

Consider the following problem. When metal is recycled, it is crushed into cubes of size 1 yd × 1 yd × 1 yd for transport. Suppose a new law has passed that states that these cubes must now be 1 ft × 1 ft × 1 ft. In a notebook, determine how many of the new cubes will be made from the old cubes because of the new law.

THINK ABOUT **MATH**

Directions: Find the volume of each figure.

1.

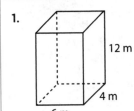

2.

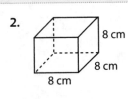

Vocabulary Review

Directions: Match each word to one of the phrases below.

1. _____ cube
2. _____ edge
3. _____ face
4. _____ rectangular prism
5. _____ rectangular solid
6. _____ three-dimensional figure
7. _____ vertex
8. _____ volume

A. a measure of the space inside a three-dimensional figure, measured in cubic units

B. the intersection of two faces of a solid

C. a rectangular prism whose faces are squares in which the length, width, and height are equal

D. a three-dimensional figure whose faces are longer than they are wide

E. a two-dimensional surface of a solid

F. another name for rectangular prism

G. a figure with length, width, and height

H. the point where edges of a three-dimensional figure come together

Skill Review

Directions: Compare and contrast.

1. A square and a cube.

_____

_____

2. A cube and a rectangular solid.

_____

_____

Skill Practice

Directions: Choose the best answer for each question.

1. The volume of a rectangular prism is 126 cubic feet. If the length is 7 feet and the width is 3 feet, what is the height in feet?

   A. 18
   B. 12.6
   C. 6.3
   D. 6

2. What is the volume, in cubic inches, of a cube with a side of 14 inches?

   A. 2,744
   B. 196
   C. 42
   D. 28

3. A planter box with the shape of a cube has a volume of 1,728 cubic inches. What is the length of a side in inches?

   A. 144
   B. 24
   C. 12
   D. 6

4. Milo is sending a package that has a length of 18 inches, a width of 15 inches, and a height of 8 inches. What is the volume, in cubic inches, of the box?

   A. 41
   B. 120
   C. 270
   D. 2,160

Volume of Cones, Cylinders, and Spheres

Lesson Objectives

You will be able to
- Calculate the volumes of cones, cylinders, and spheres
- Calculate the volumes of complex 3-D objects

Skills

- **Core Practice:** Model with Mathematics
- **Core Practice:** Make Use of a Structure

Vocabulary

apex
base
cone
cylinder
frustum
sphere

KEY CONCEPT: The volume of a cone, cylinder, or sphere is the amount of measurable space inside the object. These objects are three-dimensional, meaning they have length, width, and height. So, their units of measurement are cubed, such as in.³, ft³, and m³.

Recall that the area of a circle depends on the circle's radius (r). The constant π, which is approximately equal to 3.14, relates the area of a circle to its radius. The formula for the area, A, of a circle is

$$A = \pi r^2$$

Area is measured in units of length squared. For example, if a circle's radius is measured in inches, the units of its area are inches squared, or in.².

Imagine that a circular ice-skating rink has a diameter of 20 meters. The area of the rink, or the total amount of space on its surface, is A = πr². Remember that the radius is one-half the diameter, or $\frac{d}{2}$.

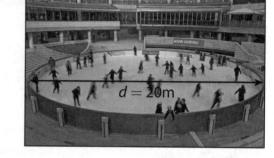

$d = 20m$

A = 3.14 × 10 × 10

A = 314 m²

The ice-skating rink has an area of approximately 314 m².

Volume and 3-D Shapes

A flat object, like the surface of an ice-skating rink has two dimensions—length and width. But not all objects are flat. Some objects, like computers, books, school buses, and desks, have a third dimension—height. These objects are three-dimensional, or 3-D.

Rectangular prisms, cubes, and pyramids are 3-D shapes. They have square, rectangular, or triangular faces that intersect at edges. Cylinders, cones, and spheres also are 3-D objects, but their surfaces are curved.

All 3-D shapes take up space, and the amount of space they occupy, their volume, is measurable. When you measure an object's length, width, and height, you have three units of measure. So, the total volume is expressed in units cubed.

The units may be customary units such as inches, feet, and yards. They may also be metric units, such as centimeters and meters.

Volume of a Cylinder

A **cylinder** is a solid with two circular ends and one curved side. The formula for finding the volume of a cylinder is given by the following equation:

$$V = \pi r^2 h$$

The length r is the horizontal distance from the center of the cylinder to its curved edge. The height, h, is the vertical, or perpendicular, distance between the circular ends that form the cylinder's top and bottom surfaces.

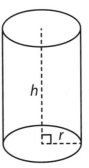

Suppose your family buys a new kitchen trash can. The new can has a diameter of 20 inches and a height of 34 inches. You can use the formula $V = \pi r^2 h$ to find the can's volume. That is, you can find how much space the trash can will occupy in your kitchen.

$$V = \pi r^2 h$$
$$r = \frac{d}{2}$$
$$\frac{r}{2} = \frac{20}{10}$$
$$\pi \times 10^2 \times 34 \text{ in.} =$$
$$3.14 \times (10 \times 10) \times 34 = 10,676 \text{ in.}^3$$

20 in.

34 in.

An Example of Calculating the Volume of a Cylinder

A bakery owner sells a variety of birthday candles. His large candles measure 0.5 inch wide and 2.5 inches tall. What is the volume of each candle?

$$V = \pi r^2 h$$
$$r = \frac{d}{2}, \text{ or } \frac{0.5}{2} = 0.25$$
$$V = 3.14 \times (0.25 \times 0.25) \times 2.5$$
$$V \approx 0.49 \text{ in.}^3$$

THINK ABOUT MATH

Directions: Read the problem. Then answer the question.

A food manufacturer packages cereal in cardboard cylinders. The cylinder measures 6 inches wide and 12 inches tall. What is the cylinder's volume?

You can add or subtract volumes of 3-D shapes to determine the volume of more complex shapes. For example, consider a cone with height h_1 and radius r_1. If you cut off the top of this cone, the part you cut off is also a cone. It has a height h_2 and radius r_2. The shape you are left with is called a **frustum**.

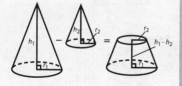

Say the original cone's height is $h_1 = 12$ in. and its radius is $r_1 = 8$ in. The part of the original cone that was cut off has height of $h_2 = 4$ in. and radius of $r_2 = 3$ in.

In a notebook, use the formula for the volume of a cone twice to determine the volume of the frustum described above.

Volume of a Cone

A **cone** has a circular base and an apex. The **base** is the flat surface on which a 3-D object sits. The **apex** is the point at which all of the straight lines that form the cone meet.

The formula for finding the volume of a cone is given by the following equation:

$$V = \tfrac{1}{3}\pi r^2 h$$

In the formula, the length r represents the radius of the base of the cone, and h represents the cone's height. The height is the vertical distance between the base and the apex.

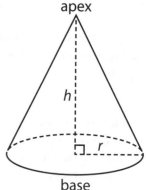

apex

base

Imagine there is a water cooler in the school's gym. The cooler comes with cone-shaped drinking cups. The cups are 10 cm high and have a maximum diameter of 7.5 cm. What is the volume of one cup?

$$V = \tfrac{1}{3}\pi r^2 h$$
$$r = \tfrac{d}{2}$$
$$r = \tfrac{7.5}{2} = 3.75$$
$$\tfrac{1}{3} \times 3.14 \times (3.75)^2 \times 10$$
$$\approx 147 \text{ cm}^3$$

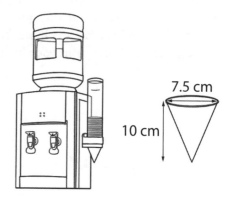

7.5 cm

10 cm

An Example of Calculating the Volume of a Cone

The same bakery owner that sells birthday candles also sells conical party hats for children and adults. Adult party hats have a diameter of 7 inches and a height of 9 inches. What is the volume of an adult party hat?

$$V = \tfrac{1}{3}\pi r^2 h$$
$$r = \tfrac{d}{2}$$
$$r = 3.5 \text{ in.}$$
$$V = \tfrac{1}{3}(3.14) \times (3.5)^2 \times 9$$
$$\approx 115 \text{ in.}^3$$

THINK ABOUT MATH

Directions: Read the problem. Then answer the question.

A gym teacher orders a set of solid disc cones to create lanes on an outdoor field. Each cone stands 2.25 inches tall and its base has a diameter of 7.5 inches. What is the volume of one disc cone?

_____

Volume of a Sphere

A **sphere** is a round object like a ball or globe. Every point on a sphere's surface is the same distance to its center. The formula for finding the volume of a sphere is given by the following equation:

$$V = \tfrac{4}{3}\pi r^3$$

In the formula, the sphere's radius is represented by the letter r.

You may have seen large spherical tanks standing on legs. Some are used to store water. Others are used to store natural gas that has been changed from a gaseous to a liquid state.

If a spherical natural gas tank has a radius of 8 meters, what is its volume?

$$r = 8 \text{ meters}$$
$$V = \tfrac{4}{3}\pi r^3$$
$$= \tfrac{4}{3}\pi 8^3$$
$$= \tfrac{4}{3} \times 3.14 \times (8 \times 8 \times 8)$$
$$\approx 2{,}144 \text{ m}^3$$

An Example of Calculating the Volume of a Sphere

Eartha is a 3-D scale model of Earth, located in Yarmouth, Maine. It took engineers two years to construct the model, which measures 41.5 feet in diameter. What is Eartha's volume?

$$d = 41.5 \text{ feet}$$
$$r = \tfrac{d}{2} = \tfrac{41.5}{2} = 20.75 \text{ feet}$$
$$V = \tfrac{4}{3}\pi r^3$$
$$= \tfrac{4}{3}\pi \times 20.75^3$$
$$= \tfrac{4}{3} \times 3.14 \times (20.75 \times 20.75 \times 20.75)$$
$$\approx 37{,}404 \text{ ft.}^3$$

MATH LINK

In the same way that you can cut off the top of a cone to make a frustum, you can subtract the volume of one cylinder from another to find the volume of a tube, or hollow cylinder.

For example, say you have a tube 4 inches high, with a diameter of 3 inches. The tube's walls are one-half inch thick.

One way to calculate the tube's volume is to calculate the volume of two cylinders. Both cylinders are 4 inches tall, like the tube. But one has a diameter of 3 inches and the other has a diameter of 2 inches. Subtract the lesser volume from the greater volume to find the volume of the tube.

$$= 3.14 \times (1.5^2) \times 4$$
$$\approx 28 \text{ in.}^3$$
$$= 3.14 \times (1^2) \times 4$$
$$\approx 13 \text{ in.}^3$$

Tube: $28 \text{ in.}^3 - 13 \text{ in.}^3 = 15 \text{ in.}^3$

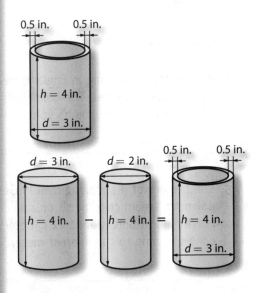

Core Practice
Make Use of a Structure

Look again at the formulas for finding the volume of a cylinder, a cone, and a sphere.

cylinder: $V = \pi r^2 h$

cone: $V = \tfrac{1}{3}\pi r^2 h$

sphere: $V = \tfrac{4}{3}\pi r^3$

If you look closely, you see that the term πr^2 is a factor that is common to each formula. Recall that πr^2 is the area of a circle with a radius of r.

This tells you that the volumes of these 3-D shapes are associated with circles: A circle is the base of a cone. A circle is the surface at each end of a cylinder. A circle shares its center with the center of a sphere.

Draw an example of each 3-D figure. Identify the circle in each shape.

Directions: Use the following terms to complete each sentence.

apex   base   cone   cylinder   frustum   sphere

1. A 3-D shape with one end that ends in a point is a _____.

2. A _____ has two equally sized circular ends.

3. The _____ of a cone is a circle.

4. A _____ is a cone with its top cut off.

5. The point of a cone is the _____.

6. All of the points on the outside surface of a _____ are the same distance from the shape's center.

Skill Review

Directions: Use the following terms to complete each sentence.

1. Which of the following values represents the volume of a 3-D shape?

   A. 7.65 in.   C. 12.2 m³
   B. 8.1 ft¹ᐟ²   D. 65.9 mm²

2. Calculate the volume of a can of vegetables that has a height of 12 cm and a diameter of 8 cm.

   _____

3. A billiard ball has a radius of $1\frac{1}{8}$ inch. What is its volume to the nearest cubic inch?

   _____

4. An apartment building in England has 25 stories and is shaped like an enormous cylinder. The building is 81 meters tall and 32 meters wide. What is the building's volume to the nearest meter?

   _____

5. The owners of a shopping mall in Germany hired artists to design an enormous, upside-down ice cream cone to sit on one corner of the mall's roof. The cone, made of steel, plastic, and wood, stands 12.1 meters tall and has a diameter of 5.8 meters. What is the cone's volume to the nearest meter?

Skill Practice

Directions: Read the problem. Then follow the directions.

1. A helium tank used to fill balloons consists of a half-sphere atop a cylinder. The height of the cylinder is 0.5 meters, and its radius is 0.1 meters. The radius of the half-sphere is also 0.1 meters. What is the volume of the helium tank?

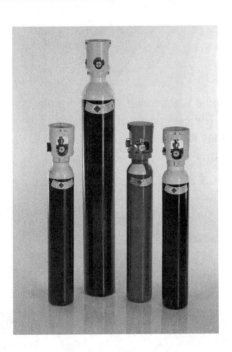

2. A prescription medicine capsule consists of a cylinder with two half-spheres at each end. The cylinder is 9 millimeters long and has a radius of 4 millimeters. Each half-sphere also has a radius of 4 millimeters. What volume of medicine can one capsule hold to the nearest millimeter?

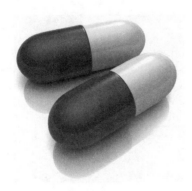

3. A soft-drink cup is a common example of a frustum. Suppose a cone has a base with a radius of 2 inches and a height of 25 inches. You cut off part of the cone to make a drinking cup. The cup's base has a radius of 1.5 inches and its height is 6.25 inches. What is the volume of the soft-drink cup to the nearest inch?

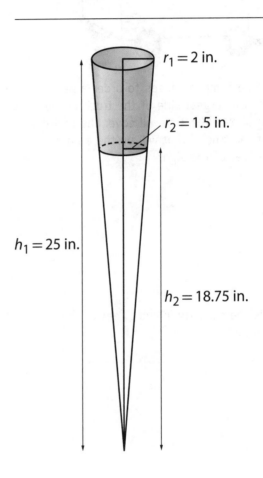

$r_1 = 2$ in.

$r_2 = 1.5$ in.

$h_1 = 25$ in.

$h_2 = 18.75$ in.

Directions: Choose the best answer for each question.

Use the information in the following picture to answer Questions 1 and 2.

14 ft

13 ft

6 ft

1. A trucking company needs to order a new sign for the largest side of the truck in the picture above. The sign will cover the entire side of the truck. What will be the area in square feet of the sign?

   A. 182
   B. 91
   C. 54
   D. 27

2. What is the capacity in cubic feet of the truck above?

   _____

3. Which statement is true?

   A. All rectangles are squares.
   B. All right triangles have three right angles.
   C. All squares are rectangles.
   D. All triangles have three sides of equal length.

4. The width of a rectangular window frame is 2.5 feet. The perimeter of the frame is 15 feet. What is the length, in feet, of the window frame?

   _____

5.

   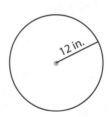

   12 in.

   What is the area, in square inches, of the circle above?

   A. 37.68
   B. 75.36
   C. 113.04
   D. 452.16

Review

Use the following figure to answer Questions 6 and 7.

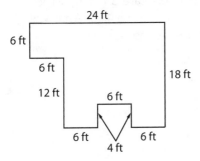

6. Julian is planning to put crown molding around the ceiling of a room. He drew the above diagram of the room. How many feet of crown molding does Julian need?

A. 432   C. 84
B. 92   D. 80

7. Julian also plans to paint the ceiling. How many square feet will he paint?

A. 84
B. 252
C. 336
D. 432

Use the following figure to answer Questions 8 and 9.

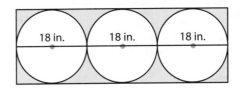

8. If the diameter of each circle is 18 inches, what is the length in inches of the rectangle?

A. 27
B. 54
C. 72
D. 81

9. What is the area in square inches of the rectangle?

A. 972
B. 818
C. 324
D. 144

10. The aspect ratio (length-to-width ratio) of older televisions is 4 to 3. If the diagonal of a television is 35 inches long, what is the width?

    A. 21 inches
    B. 28 inches
    C. 49 inches
    D. 7 inches

11. What is the volume of a sphere with radius $r = 21$ mm?

    A. 29,106 mm^3
    B. 38,808 mm^3
    C. 4,851 mm^3
    D. 3,638.25 mm^3

12. Which of the following is a Pythagorean triple?

    A. 11, 60, 61
    B. 4, 6, 8
    C. 8, 15, 17
    D. 7, 23, 24

13. A stamp is to be made using the Mona Lisa. The dimensions (height to width) of the actual Mona Lisa are 30 in. by 21 in. The stamp height is to be 1 in. What will the width of the stamp be?

    A. 1.43 inches
    B. 2.3 inches
    C. .7 inches
    D. .43 inches

14. What will happen to the volume of a cylinder if the radius is doubled?

    A. The volume will stay the same.
    B. The volume will double.
    C. The volume will quadruple (be multiplied by 4).
    D. The volume will octuple (be multiplied by 8).

Review

Check Your Understanding

On the following chart, circle the number of any item you answered incorrectly. Near each lesson title, you will see the pages you can review to learn the content covered in the question. Pay particular attention to reviewing those lessons in which you missed half or more of the questions.

Chapter 12: Geometry	**Procedural**	**Conceptual**	**Application/ Modeling/ Problem Solving**
Geometric Figures pp. 326–331		3	
Perimeter and Circumference pp. 332–337			4, 6, 8
Scale Drawings and Measurement pp. 338–345	5		1, 7, 9
Area pp. 346–351			2
Pythagorean Theorem pp. 352–359	6		1, 4
Geometric Solids and Volume pp. 360–365	3		1
Volume of Cones, Cylinders, and Spheres pp. 366–371	2	5	

Mathematics

Directions: The Mathematics Posttest consists of **50** problems. It will help you to determine how well you have learned the material in this book. You should take the posttest only after you have completed all of the chapters. Answer each question as carefully as possible. If a question seems to be too difficult, do not spend too much time on it. Work ahead and come back to it later when you can think it through carefully.

When you have completed the test, check your work with the answers and explanations on pages **386–388**. Use the Evaluation Chart on page **389** to determine which areas you need to review.

Mathematics

Directions: Choose the best answer to each question.

1. Eva lives in Wyoming, which has a general state sales tax on purchases. If Eva buys a car for $18,000, how much money will she pay in sales tax?

   A. $180
   B. $900
   C. $1,800
   D. Not enough information is given.

The following table shows the cost to purchase books for a marketing class. Use the table for Questions 4 and 5.

Number of Books	Cost ($)
1	95
2	190
3	285
4	380
5	475

2. The stem-and-leaf plot shows the number of hours Jamal worked in the last 10 weeks.

Stem	Leaf
0	8 8
1	5 5
2	0 0 0 4 4
3	0

   $2 \mid 0 = 20$ hours

   What is the median number of hours Jamal worked last month?

   A. 8
   B. 18.4
   C. 20
   D. 22

4. Which statement describes the pattern for finding the cost of the books?

   A. Add the number of books to 95.
   B. Multiply the cost of the books by 95.
   C. Multiply the number of books by 95.
   D. Multiply the number of books by 10 and add 85.

5. If the pattern for the cost of the books continues, what will be the cost of 8 books?

   A. $570   C. $760
   B. $665   D. $855

3. Sonia is building a fence for a rectangular area of her yard. She wants to use all 80 feet of fencing that she bought. She will place a post every 5 feet, including one at each corner. Which dimensions will work for the rectangular area she will fence?

   A. 2 feet by 40 feet
   B. 5 feet by 35 feet
   C. 8 feet by 32 feet
   D. 12 feet by 28 feet

6. Which pair of equations has an infinite number of solutions?

   A. $12x + 4y = 28$,
      $y = 3x + 7$
   B. $5x + 5y = 30$,
      $y = -x + 10$
   C. $4x + 2y = 6$,
      $y = -2x + 3$
   D. $7x + y = 8$,
      $y = 7x + 8$

7. One quilt block is shown. The shaded areas represent the dark fabric. Karina will sew together 12 of these quilt blocks to make the quilt. What fraction of the quilt is made of the dark fabric?

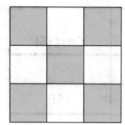

- **A.** $\frac{5}{12}$
- **B.** $\frac{5}{9}$
- **C.** $\frac{2}{3}$
- **D.** Not enough information is given.

8. How can $\frac{a}{7} + 4 = -12$ be solved for a?

- **A.** Multiply both sides of the equation by 7, and then add 4 to both sides.
- **B.** Subtract 4 from both sides of the equation, and then divide both sides by 7.
- **C.** Multiply both sides of the equation by 7, and then subtract 4 from both sides.
- **D.** Subtract 4 from both sides of the equation, and then multiply both sides by 7.

9. In a survey asking whether people listen to AM radio, 45% of the people responded "yes," while the rest responded "no." How many people responded "no"?

- **A.** 45
- **B.** 55
- **C.** at least 55
- **D.** Not enough information is given.

10. The ratio of black keys to white keys on a standard piano is 9:13. There are 88 keys altogether. How many keys on a standard piano are black?

- **A.** 9   **C.** 52
- **B.** 36   **D.** 75

11. Which answer choice lists the lengths from least to greatest?

A 840 centimeters
B 6 meters
C 2,843 millimeters
D 0.005 kilometers

- **A.** C, A, B, D
- **B.** C, D, B, A
- **C.** D, B, A, C
- **D.** A, B, D, C

12. Jasmine needs $3\frac{3}{4}$ yards of red velvet and $\frac{5}{6}$ yard of blue velvet. How many more yards of red velvet than blue velvet does she need?

- **A.** $2\frac{11}{12}$ **C.** $4\frac{1}{2}$
- **B.** $3\frac{1}{8}$ **D.** $4\frac{7}{12}$

13. Convert 3.53942×10^{-4} to standard notation.

- **A.** 353.942
- **B.** 0.0353942
- **C.** 35394.2
- **D.** 0.000353942

Mathematics

Directions: Refer to the line graphs below for Questions 14 and 15.

ABBY'S HEIGHT CHART

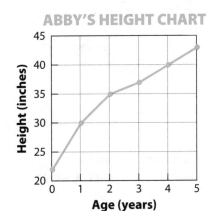

ABBY'S WEIGHT CHART

14. How tall was Abby when she was two years old?

- A. about 24 inches
- B. about 28 inches
- C. about 30 inches
- D. about 35 inches

15. How much did Abby weigh when she was about 40 inches tall?

- A. about 35 pounds
- B. about 40 pounds
- C. about 44 pounds
- D. Not enough information is given.

16. Round 3141.5926 to the nearest hundredth.

- A. 3100
- B. 3141.59.
- C. 3141.6
- D. 3140

17. Which line below passes through the points (1, 7) and (6, −3)?

- A. $y = -2x + 7$
- B. $y = -2x + 9$
- C. $y = 2x + 5$
- D. $y = 2x - 3$

18

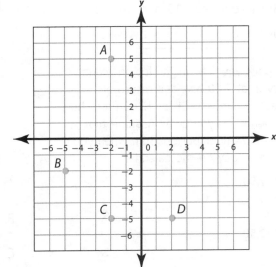

Which point is located at (−2, −5)?

- A. *A*   C. *C*
- B. *B*   D. *D*

19. Thiu runs 3 miles in 25 minutes. Kai runs an 8-minute mile. Lin runs $7\frac{1}{2}$ miles per hour. Which is true?

- A. Thiu runs the fastest.
- B. Thiu runs faster than Kai.
- C. Kai and Lin run at the same rate.
- D. Lin runs faster than Kai.

Mathematics

20. The admission to a carnival for a family of three children and two adults is $45. Each ride ticket costs $3. What is the maximum number of ride tickets that can be bought if the family has $155 to spend, including admission, at the carnival?

A. 51   C. 36
B. 37   D. 15

21. The following sign appears in front of an airport parking lot.

Parking Lot Rates
$1.50 per hour
Not to exceed $15 per day
Not to exceed $75 per week

If you park your car in the lot on Monday, January 2, at 7:00 A.M. and return on Wednesday, January 11, at 12:00 noon, how much money do you owe? Assume that 1 day is a 24-hour period from when you enter, not 1 day on the calendar.

A. $112.50
B. $114
C. $120
D. $142.50

22. Which of the following scenarios would be linearly related?

A. In a 100-mile trip, the time traveled, t, to the speed s.
B. The area, A, of a square to its side s.
C. The circumference, C, of a circle to its diameter d.
D. The volume, V, of a sphere to its radius r.

23. Which result is least likely to happen?

A. flipping a coin and getting "heads"
B. drawing an ace from a standard deck of 52 cards
C. rolling a number cube and getting a "4"
D. rolling two number cubes and getting a sum of 5

24. The line plot below shows the number of hours volunteers worked at an animal shelter last month.

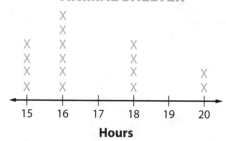

VOLUNTEER HOURS AT ANIMAL SHELTER

Which statement is true?

A. The range of the data is greater than the mean.
B. The mean of the data is less than the median.
C. The two modes of the data are 15 hours and 18 hours.
D. The median of the data is the same as the mode.

25. A bag contains 15 yellow and blue marbles. The probability of randomly drawing a blue marble is $\frac{1}{3}$. How many yellow marbles are in the bag?

A. 12   C. 5
B. 10   D. 3

Mathematics

26. Rami says that the bar graph below shows that twice as many people at his workplace drive to work than take the bus. Justine says the graph is misleading and does not support Rami's claim.

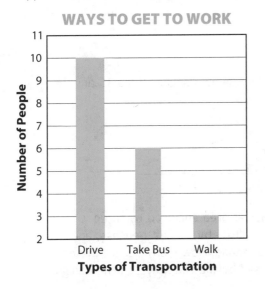

WAYS TO GET TO WORK

Which statement *best* explains the results shown in the bar graph?

A. Rami's claim is correct because the bar for people driving to work is twice as tall as the bar for people who take the bus.
B. The graph is misleading because there are too few bars for alternative ways to get to work, such as riding a bicycle or taking the subway.
C. The graph is misleading because the intervals should be 5 instead of 1.
D. The graph is misleading because the vertical axis does not start at 0. It appears that twice as many people drive as take the bus, when 6 take the bus and 10 drive.

Refer to the following table for Question 27.

Human Activity	Quarts of Blood Pumped per Minute	Number of Heart Beats per Minute
Rest	5–6	60–80
Mild exercise	7–8	100–120
Strenuous exercise	30	Up to 200

27. How many times does an average person's heart beat while sleeping for 8 hours?

A. 3,600
B. 4,200
C. 33,600
D. 57,600

28. Choose the phrase that represents the largest amount.

A. 300 tenths 5 hundreths
B. 4 and 70 hundredths
C. 100 tenths 91 thousandths
D. 2345 hundredths

29. A function is an equation that has exactly one _____ for every _____.

A. output; input
B. input; output
C. variable; input
D. output; variable

Mathematics

30. What expression below equals the expression $10^3 + 9^3$?

    A. $(10 + 9)^3$
    B. $12^3 + 1^3$
    C. $(12 + 1)^3$
    D. $10 \times 3 + 9 \times 3$

31. The median age of a set of 11 employees is 36. The employee who figured out the median age will not reveal his or her own age. The ages of the other 10 employees are as follows: 32, 35, 24, 50, 29, 45, 36, 38, 25, and 36. Which of these is a possible age for the eleventh employee?

    A. 25   C. 33
    B. 29   D. 36

32. Which three statements are true?

    Opposite sides of a rectangle

    A are perpendicular
    B are parallel
    C are the same length
    D are usually not the same length
    E intersect
    F never intersect

    A. A, C, and F
    B. A, D, and E
    C. B, C, and F
    D. B, C, and E

33. What is the solution for x in the equation?

    $-6x + 12 = 54$

    A. $x = -21$
    B. $x = -11$
    C. $x = -7$
    D. $x = 7$

34. What is the value of the expression $-3x + 2y$ when $x = -4$ and $y = 8$?

    A. -32
    B. -28
    C. 4
    D. 28

Use the diagram below for Questions 35 and 36.

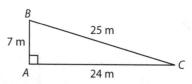

35. The triangular area enclosed by three walkways will feature trees, benches, and a concession stand. How many square meters are enclosed by the walkways on the design?

    A. 84
    B. 87.5
    C. 168
    D. 175

36. Alejandro is helping with the design of a zoo. A water fountain is located at A, a bird exhibit at B, and an elephant exhibit at C. Walkways will connect each of the points. How many meters of walkways connect the three points on the design? Write your answer on the line provided.

_____

Mathematics

37. Maria asked 50 people, "What's your favorite color: blue, red, or yellow?" Results are shown in the table below.

Color	Number of People
Blue	22
Red	15
Yellow	13

Maria draws a circle graph to display the data. What percent of the circle will represent red as a favorite? Write your answer on the line provided.

38. Lena spun the needle on the spinner below. Then she rolled a six-sided number cube numbered 1–6. What is the probability that the dial of the spinner landed on a number greater than 2, and she rolled an odd number? Write your answer on the line provided. Express any fractions in lowest terms.

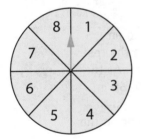
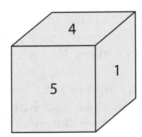

39. Dekentra wants to hire a disc jockey for her wedding reception. The disc jockey charges two-thirds the cost of Best Sounds, a local company that provides music. If the disc jockey charges $540 for three hours, what does Best Sounds charge for three hours, in dollars? Write your answer on the line provided.

40. Cassie needs to buy mulch to place around her flowers. She can only spend up to $33.06 after taxes. The bags of mulch cost $3.48 per cubic foot after tax. How many cubic feet of mulch can Cassie buy? Write your answer on the line provided.

Mathematics

41. Mt. Aconcagua in Argentina is 6,960 meters high. Valdes Peninsula in Argentina is 40 meters below sea level. What is the difference, in meters, in the elevation between these two places? Write your answer on the line provided.

_____

42. What is the area, in square meters, of the figure shown below?

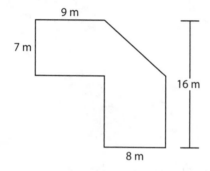

_____

43. A paperweight in the shape of a cube has a side length of 4.1 centimeters. What is the volume of the cube, rounded to the nearest whole number, in cubic centimeters? Write your answer on the line provided.

_____

44. A square has the same area as a rectangle that measures 4 feet by 16 feet. What is the length of a side of the square, in feet? Write your answer on the line provided.

_____

45. Simplify the following decimal division.
0.032 ÷ 1.28

_____

46 Tianran has $10,000 to invest in two accounts. He invests part of it into an account promising 5% return, and the rest in an account promising 8% return. At the end of the year, he has made $702.50 in interest. How much did Tianran invest in the 8% account? Write your answer on the line provided.

_____

Mathematics

47. Jin plotted the vertices of a square on a coordinate plane. The coordinates of three of the vertices are $A(-3, 2)$, $B(3, 2)$, and $C(3, -4)$. What is the location of D, the fourth vertex? Write your answer on the line provided.

_____

49. Jarod goes to the store to buy a few items. He buys a gallon of milk for $3.10, a small picture frame for $9.87, and 3 cartons of eggs for $2.34 after tax. How much was Jarod's entire bill? Write your answer on the line provided.

_____

48. What is the probability that a family with 3 children has 3 girls? Assume that a boy and a girl are equally likely. Write your answer on the line provided.

_____

50. The area of a square is 361 square feet. What is the length of one side? Write your answer on the line provided.

_____

Answer Key

1. **A, B, C.** assumed sales tax rate when rate is not given
   D. KEY: You need to know the sales tax rate to solve the problem.

2. **A.** chose minimum instead of median
   B. chose mean instead of median
   C. KEY: the two middle values are 20
   D. chose range instead of median

3. **A.** $2 + 40 + 2 + 40 = 84$; she does not have enough fencing.
   B. KEY: $5 + 35 + 5 + 35 = 80$; because both 5 and 35 are multiples of 5, a post can be placed in each corner and every 5 feet.
   C, D. found correct perimeter but side lengths are not multiples of 5

4. **A.** Money and books cannot be added together to get the cost of books.
   B. Since $95 is the cost of one book, multiplying $95 by 95 would give the cost of 95 books, not one book.
   C. KEY: Multiply the number of books by 95.
   D. This only works for 1 book. For 2 books it would be $2(10) + 85 = 105$, but according to the table, the cost of 2 books is $190.

5. **A.** found the cost for 6 books instead of 8
   B. found the cost for 7 books instead of 8
   C. KEY: multiply the number of books, 8, by the cost of one book, $95, to get $760
   D. found the cost of 9 books instead of 8

6. **A.** These lines intersect once.
   B. These lines are parallel.
   C. KEY: these lines are identical. Multiply the second by 2 and add the x term over to see it. Two lines that have an infinite number of solutions are identical.
   D. These lines intersect once.

7. **A.** compared number of shaded squares in one quilt block to the number of quilt blocks
   B. KEY: 5 of 9 squares are shaded in each quilt block, so $\frac{5}{9}$ of the quilt is dark fabric.
   C. used row 1, column 1, row 3, or column 3 to find 2 out of 3 squares are shaded
   D. Enough information is given.

8. **A.** switched order in which operations are performed and undid addition by addition
   B. chose to undo division by division
   C. switched order of operations
   D. KEY: Subtract 4 from both sides:
   $\frac{a}{7} + 4 - 4 = -12 - 4; \frac{a}{7} = -16;$
   then multiply both sides by 7: $7(\frac{a}{7}) = 7(-16); a = -112; -\frac{112}{7} + 4 \stackrel{?}{=} -12;$
   $-16 + 4 \stackrel{?}{=} -12; -12 = -12$

9. **A.** assumed 100 people responded and chose number who said "yes" instead of "no"
   B. assumed 100 people responded
   C. assumed at least 100 people responded
   D. KEY: You need to know how many people responded in order to solve the problem.

10. **A.** chose 9 since 9:13 is the ratio of black keys to white keys
    B. KEY: $\frac{\text{black}}{\text{total}} = \frac{\text{black}}{\text{black} + \text{white}} = \frac{9}{9 + 13} = \frac{9}{22}$;
    $\frac{9}{22} = \frac{b}{88}; b = \frac{88 \cdot 9}{22} = 36$
    C. found the number of white keys
    D. subtracted 13 from 88

11. **A.** listed smallest to largest units
    B. KEY: A: 840 cm = 8.4 m; B: 6 m; C: 2,842 mm = 2.842 m; D: 0.005 km = 5 m; Order: C, D, B, A
    C. listed least to greatest not looking at units
    D. chose the order from greatest to least

12. **A.** KEY: $3\frac{9}{12} - \frac{10}{12} = 2\frac{21}{12} - \frac{10}{12} = 2\frac{11}{12}$
    B, C, D. choose wrong operation

13. **A, C.** wrong direction
    B. right direction, not enough places
    D. KEY: 10^{-4} means moving the decimal point to the left 4 places, which gives 0.000353942

14. **A.** found approximate weight on bottom graph at 1 year
    B. found approximate weight on bottom graph at 2 years
    C. found approximate height on top graph at 1 year
    D. KEY: On top graph, find 2 years of age, then find the height, about 35 inches.

15. **A.** found her approximate weight at about 3 years instead of 4
    B. KEY: On top graph, find 40 inches and her age, about 4 years. On bottom graph, find 4 years and her weight, about 40 pounds.
    C. found approximate weight at 5
    D. Enough information is given.

Answer Key

16. **A.** rounded to hundreds
    B. KEY: hundredths is the second digit to the right of the decimal point so 3141.59
    C. rounded to tenths
    D. rounded to tens

17. **A.** correct slope, incorrect intercept
    B. KEY: the slope is $\frac{(-3-7)}{(6-1)} = \frac{-10}{5} = -2$. Using slope-intercept form, you get $7 = -2 \times 1 + b$, so $b = 9$.
    C. incorrect slope and intercept
    D. incorrect slope and intercept

18. **A, B, D.** inverted numbers or used negatives incorrectly
    C. KEY: identified point at $(-2, -5)$

19. **A.** Thiu: $\frac{3\text{ miles}}{25\text{ minutes}} \times \frac{60\text{ minutes}}{\text{hour}} = \frac{7.2\text{ miles}}{\text{hour}}$, which is slower than Lin's rate at $\frac{7.5\text{ miles}}{\text{hour}}$.
    B. Kai: $\frac{1\text{ mile}}{8\text{ minutes}} \times \frac{60\text{ minutes}}{\text{hour}} = \frac{7.5\text{ miles}}{\text{hour}}$, Thiu is slower than Kai
    C. KEY: Kai and Lin run at $\frac{7.5\text{ miles}}{\text{hour}}$.
    D. Kai and Lin both run at $\frac{7.5\text{ miles}}{\text{hour}}$, so Lin does not run faster than Kai.

20. **A.** found the solution to $155 \geq 3x$
    B. found $36.67 \geq x$ but incorrectly rounded up to 37
    C. KEY: $155 \geq 3x + 45$, $110 \geq 3x$, $36.6 \geq x$ so the greatest number of tickets is 36.
    D. found $45 \geq 3x$ as $15 \geq x$

21. **A.** KEY: Jan. 2, 7:00 A.M. to Jan. 9, 7:00 A.M. is 1 week = \$75; from Jan. 9, 7:00 A.M. to Jan. 11, 7:00 A.M. is 2 days, and $2 \times \$15 = \$30$; Jan. 11 from 7:00 A.M. to 12:00 noon is 5 hours, and $5 \times \$1.50 = \$7.50$; $\$75 + \$30 + \$7.50 = \$112.50$
    B, C, D. did not calculate correctly

22. **A.** represents $100 = ts$, or $t = \frac{100}{s}$, which is not linear.
    B. represents $A = s^2$, not linear
    C. KEY: represents $C = \pi \times d$, which is linear
    D. represents $V = \left(\frac{4}{3}\right) \pi \times r^3$

23. **A, C, D.** identified probabilities more likely to happen than choice B
    B. KEY: probability $= \frac{4}{52}$ or about 8% and is the least likely to happen of the choices

24. **A.** thought range was 20 and greater than mean of 16.75
    B. switched mean of 16.75 and median of 16
    C. misunderstood mode to mean the same number of hours
    D. KEY: Median and mode are both 16 hours.

25. **A, D.** thought $\frac{1}{3}$ of 15 was 3
    B. KEY: $\frac{1}{3} = \frac{5}{15}$, so there are 5 blue marbles; 15 total $-$ 5 blue $= 10$ yellow
    C. found the number of blue marbles

26. **A, B, C.** doesn't understand misleading graphs
    D. KEY: Vertical axis starts at 2 not 0, which makes bar length for "Drive" twice as long as bar length for "Take the Bus." 6 take the bus and 10 drive, and $6 \times 2 \neq 10$.

27. **A.** used 60 beats per minute and 1 hour sleep
    B. used 70 beats per minute and 1 hour of sleep
    C. KEY: $\frac{70\text{ beats}}{\text{minute}} \times \frac{60\text{ minutes}}{\text{hour}} \times 8\text{ hours} = 33{,}600$ beats
    D. used 120 beats per minute instead of 70

28. **A.** KEY: 300 tenths 4 hundredths equals 30.05, which is the largest of the rest.
    B. this equals 4.7
    C. this equals 10.091
    D. this equals 23.45

29. **A.** KEY: A function has exactly one output for every input.
    B. reversed the order of the blanks
    C. incorrect term; correct term and position
    D. two incorrect terms and place

30. **A.** confused with the distributive property
    B. KEY: Both equal 1729
    C. confused with the distributive property using the equivalent expression
    D. multiplied powers instead of exponentiation

31. **A, B, C.** chose an age that results in a median less than 36
    D. KEY: 36 must be the 6th value in the list of 11 values. The missing age must be at least 36.

32. **A.** Opposite sides of a rectangle are not perpendicular.
    B. Opposite sides of a rectangle are not perpendicular, are the same length, and do not intersect.
    C. KEY: Opposite sides of a rectangle are parallel, the same length, and never intersect.
    D. Opposite sides of a rectangle do not intersect.

Answer Key

33. **A, B, D** did not use signs correctly
    C. KEY: $-6x + 12 = 54$; $-6x = 42$; $x = -7$
34. **A.** switched the values for x and y
    B. added -12 and -16 instead of 12 and 16
    C. added -12 instead of 12 to 16
    D. KEY: 28; $-3(-4) + 2(8) = 12 + 16 = 28$
35. **A.** KEY: Area of triangle $= \frac{1}{2}bh$; base $= 24$ m, height $= 7$ m; $\frac{1}{2} \times 24 \times 7 = 84$ m²
    B. used 25 m instead of 24 m for the base of the triangle
    C. did not multiply the base and height of the triangle by $\frac{1}{2}$
    D. used 25 m instead of 24 m for the base of the triangle and did not multiply by $\frac{1}{2}$
36. 56; 7 m $+ 25$ m $+ 24$ m $= 56$ m
37. 30; 15 out of 50 chose red so $\frac{15}{50} = \frac{x}{100}$, $50x = 1,500$, $x = 30$
38. $\frac{3}{8}$; Probability of spinning a number greater than 2: $\frac{6}{8} = \frac{3}{4}$. Probability of rolling an odd number: $\frac{3}{6} = \frac{1}{2}$. Probability of both: $\frac{3}{4} \times \frac{1}{2} = \frac{3}{8}$.
39. 810; $\frac{2}{3}x = 540$; $2x = 1,620$; $x = 810$
40. 9.5; $33.06 \div 3.48 = 9.5$ bags
41. 7,000: $6,960$ m $- (-40$ m$) = 6,960$ m $+ 40$ m $= 7,000$ m
42. 163; divide into 9×7 rectangle, 8×9 rectangle, and base 8 and height 7 triangle. $(9 \times 7) + (8 \times 9) + \frac{1}{2}(8 \times 7) = 63 + 72 + 28 = 163$ m².
43. 69; $4.1 \times 4.1 \times 4.1 = 68.921$; round to 69 cm³.
44. 8; area of rectangle $= 4 \times 16 = 64$; area of square $= 64 = 8 \times 8$; each side of square $= 8$ ft.
45. 0.025; $0.032 \div 1.28 = 0.025$
46. \$6,750; Solve equations $x + y = 10000$ and $.05x + .08y = 702.5$. Substitute $x = 10000 - y$ into the other to get $.05(10000 - y) + .08y = 702.5$, which is $500 + .03y = 702.5$, or $.03y = 202.5$. So $y = 6,750$.
47. $(-3, -4)$; the fourth vertex will have the same x-coordinate as A; the same y-coordinate as C.
48. $\frac{1}{8}$; 8 possible outcomes: BBB, BBG, BGB, GBB, GGB, GBG, BGG, GGG. One favorable outcome: GGG; Probability $= \frac{\text{favorable outcomes}}{\text{possible outcomes}} = \frac{1}{8}$
49. \$15.31; $\$3.10 + \$9.87 + \$2.34 = \$15.31$
50. 19; $s^2 = 361$, so $s = \sqrt{361} = 19$ feet

Evaluation Chart

Check Your Understanding

On the following chart, circle the number of any problem you got wrong. After each problem, you will see the name of the section where you can find the skills you need to solve the problem.

Problem	Unit: Section	Starting Page
3, 21, 39 16, 28 49 40, 45 12	**Number Sense and Operations** Problem Solving Introduction to Decimals Add and Subtract Decimals Divide Decimals Add and Subtract Fractions	40 50 54 64 82
41 18, 47 34 8, 33 20 4 6, 17, 22 46 29	**Basic Algebra** Subtract Integers The Coordinate Grid Expressions Solve Two-Step Equations Solve One- and Two-Step Inequalities Identify Patterns Linear Equations Pairs of Linear Equations Functions	114 124 134 144 150 156 166 184 200
5, 7, 10, 19 1, 9, 37 30 50 13	**More Number Sense and Operations** Ratios and Rates Use Percents in the Real World Exponents Roots Scientific Notation	212 242 250 254 260
24, 31 14, 15 2, 26 23, 25, 48 38	**Data Analysis and Probability** Measures of Central Tendency and Range Graphs and Line Plots Plots and Misleading Graphs Introduction to Probability Compound Events	270 274 282 296 302
27 11 32 36 35, 42, 44 43	**Measurement and Geometry** Customary Units Metric Units Geometric Figures Perimeter and Circumference Area Geometric Solids and Volume	312 316 326 332 346 360

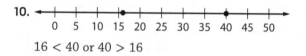

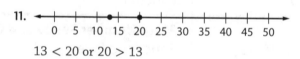

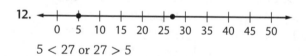

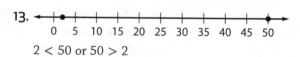

CHAPTER 1 Whole Numbers

Lesson 1.1

Key Concept, page 12
1. 9
2. 72
3. 70
4. 35

Think About Math, page 14 (top)
1. 7 tens or 70
2. 8 ten thousands or 80,000
3. 6 ones or 6
4. 3 hundred millions or 300,000,000
5. 8 billions or 8,000,000,000

Think About Math, page 14 (bottom)
1. E.
2. D.
3. A.
4. C.
5. B.

Think About Math, page 16
1. Compare the digits of the numbers from left to right until the digits in the same column are different. The digits in the thousands place, 3 and 4, are not the same. Compare those digits. 204,210 > 203,478.
2. 698,321; 698,432; 701,286

Vocabulary Review, page 16
1. value
2. whole numbers
3. periods
4. digits
5. approximate
6. number line

Skill Review, pages 16–17
1. 2 millions or 2,000,000
2. 4 hundred thousands or 400,000
3. 7 ten thousands or 70,000
4. 3 thousands or 3,000
5. 0 hundreds
6. 1 ten or 10
7. 5 ones or 5
8.

Thousands	Hundreds	Tens	Ones
6	7	2	9

9.

Millions	Hundred thousands	Ten thousands	Thousands	Hundreds	Tens	Ones
1	0	0	0	0	3	5

10. 16 < 40 or 40 > 16

11. 13 < 20 or 20 > 13

12. 5 < 27 or 27 > 5

13. 2 < 50 or 50 > 2

14. There are several ways to compare numbers correctly.
15. Digits represent numbers.

Answer Key

(Lesson 1.1 cont.)

Skills Practice, page 17
1. **C.** Answers (A) and (D) are in order from least to greatest. Answer (B) is not in any order.
2. **A.** Answer (B) written as a standard number is 240, Answer (C) is 214, and Answer (D) is 21.
3. **D.** 22,201 < 22,345 < 23,456 < 23,712 so Car D at $23,712 is the most expensive.
4. **B.** 0 is in the ten thousands place.

Lesson 1.2

Key Concept, page 18
1. thirty-seven
2. one thousand, eight
3. one hundred fifty-two
4. thirty-two thousand
5. <
6. >
7. >
8. <

Think About Math, page 19
1. 79
2. 85
3. 553

Think About Math, page 20
1. 25
2. 1,932
3. 629

Vocabulary Review, page 20
1. difference
2. calculate
3. sum
4. operations

Skill Review, page 21
1. increased by; 3,642
2. combined; 6,063
3. deductions; $387
4. depreciated; $6,850
5. withdraws; $466
6. total; 806
7. greater than; $5,334

Skill Practice, page 21
1. **D.** add 456, 482, 449, 479, and 468 to find the total miles for the week
2. **A.** subtract 937,642 from 1,000,000 to find how many more
3. **A.** subtract 23,470 from 31,067 because *how much more* means to find a difference
4. **B.** add 380, 407, 298, and 321 because *how many in total* means to find a sum

Lesson 1.3

Key Concept, page 22
1. 12
2. 70
3. 231
4. 1,020
5. 33
6. 637
7. 46
8. 12

Think About Math, page 23
1. 68
2. 414
3. 1,560
4. 19,593
5. 11,426
6. 684
7. 3,552
8. 9,936
9. 48,200
10. 51,345

Think About Math, page 24
1. 13 R2
2. 31
3. 32
4. 202 R14
5. 200
6. 34 R1
7. 20
8. 15 R8
9. 109 R3
10. 16 R1

Vocabulary Review, pages 24–25
1. divisor
2. factor
3. product
4. dividend
5. multiplication
6. quotient
7. division

Skill Review, page 25
1. same amount, each month; $40
2. how much money did he collect; $540
3. equally divided; 811 stamps
4. monthly; after two years; $1,800
5. per ticket; $1,748,250
6. tickets are $3 each; 144 tickets
7. each table; 510 petals

(Lesson 1.3 cont.)

Skill Practice, page 25

1. **C.** Multiply the number of employees by their pay.
$5 \times 589 = 2{,}945$

2. **D.** 1 year is equal to 12 months. Multiply 12 by the rent per month, $525, to get $6,300.

3. **B.** Divide the number of tables by the number of people at the table. $320 \div 16 = 20$

4. **A.** Divide the amount in ticket sales by the amount the band earns per ticket. $1{,}315 \div 5 = 263$

Lesson 1.4

Think About Math, page 29

The factors of 63 are: 3, 7, 9, and 21.

The factors of 28 are: 2, 4, 7, 14, and 28.

The only common factor is 7.

So, the greatest common factor is 7.

$63 - 28 = (7 \times 9) - (7 \times 4)$

$(7 \times 9) - (7 \times 4) = 7 \times (9 - 4)$

Finally, you can rewrite the original expression as:

$63 - 28 = 7 \times (9 - 4)$

Math Link, page 30

$2 \times 10 = 20$

$4 \times 5 = 20$

$5 \times 4 = 20$

$10 \times 2 = 20$

The factors of 20 are 2, 4, 5, and 10.

The Commutative Property of Multiplication states that the product remains the same regardless of the order of the numbers that are multiplied. During factoring, the Commutative Property becomes evident in the list of equations. Its appearance tells you that you have found all of the factors for a number and need not go any further.

Vocabulary Review, page 30

1. C.
2. A.
3. E.
4. B.
5. F.
6. G.
7. D.
8. H.

Skill Review, page 30

1. 2, 3, 4, 6
2. 2, 4, 8, 16
3. 3, 5, 9, 15
4. 2, 4, 11, 22
5. 2, 4, 8, 11, 22, 44
6. 4
7. 6
8. 7
9. 22
10. 50
11. $12 \times (2 + 3)$
12. $9 \times (5 - 3)$
13. $4 \times (5 + 16)$
14. $24 \times (2 + 3)$
15. $22 \times (3 - 2)$

Skill Practice, page 31

1. List the equations with the number as a product.

↓

Apply the Commutative Property of Multiplication to cross out equivalent equations.

↓

Cross out equations containing the factors 1 and the number itself.

↓

Order the remaining factors.

2. B.
3. B.
4. C.
5. B.
6. D.
7. E.

Answer Key

Lesson 1.5

Math Review, page 32
1. 898
2. 2,300
3. 560
4. 7
5. 11
6. 16,416
7. 376
8. 45

Think About Math, page 32
1. 60
2. 90
3. 130
4. 1,350

Think About Math, page 34
Sample answers:
1. 700
2. 5
3. 1,100
4. 110

Vocabulary Review, page 35
1. D.
2. B.
3. A.
4. C.

Skill Review, page 35
1. Since Jamie is overestimating the number of people as 60 for each bus and underestimating the number of people going to the picnic, 1,200, he will get 20 buses. The actual answer of 24 buses means that Jamie will not have enough buses. Jamie should choose another method of estimating.

2. Mai will use the numbers 400 and 200 for her estimation and subtract. The estimate will be $200, which would mean she could not write the check. The actual amount is $274, so she could write the check. The conclusion is that Mai should not use front-end estimation.

Skill Practice, page 35
1. **C.** Round 2,067 to 2,000 and 478 to 500. Then subtract to get 1,500.
2. **C.** Divide with compatible numbers 5,400 ÷ 90 = 60
3. **A.** Round the numbers up so he knows he can cover the cost of the items he buys.
4. **D.** Round 365 to 400 and multiply by 6 to get 2,400.

Lesson 1.6

Key Concept, page 36
1. 97
2. 29,887
3. 2,580
4. 7,344
5. 1,013
6. 200 R13
7. 42
8. 2,304

Think About Math, page 37
1. 2
2. 16
3. 7
4. 44

Think About Math, page 39
1. 336
2. 4,500

Vocabulary Review, page 39
1. compensation
2. strategy
3. order of operations
4. mental math

Skill Review, page 39
1. 1,115; a week is 7 days
2. $712; a car has 4 tires
3. $53; there are 12 months in a year

(Lesson 1.6 cont.)

Skill Practice, Page 39

1. **D.** Use the order of operations: $(8 + 3) = 11$, $11 \times 4 = 44$; $44 - 1 = 43$.

2. **10** Use the order of operations, $2 \times 14 = 28$, $3 + 5 = 8$, $44 - 28 = 16$, $16 \div 8 = 2$; $2 + 8 = 10$

3. **C.** $100 \times 36 = 3{,}600$

Lesson 1.7

Key Concept, page 40

1. 28,033
2. 323,850
3. 945
4. 19
5. 21,967
6. 5,473,269
7. 24
8. 4,683

Think About Math, page 42
addition; $7

Think About Math, page 43

1. guess and check; 4
2. draw a picture; 25 times

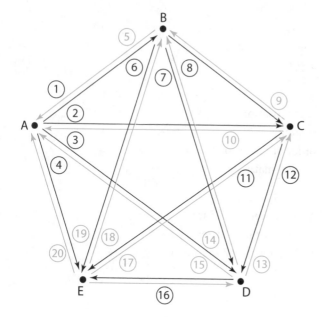

Vocabulary Review, page 44

1. solution
2. reasonable
3. irrelevant

Skill Review, page 45

1. Manny spent $3.15; 12 pieces
2. Mr. Martinez worked 32 hours; 13 hours
3. He handed the clerk $12.43; 10 pounds
4. He also sold his DVD player for $20; $36

Skill Practice, page 45

1. **B.** $88 + 77 = 165$.
   $87 + 65 = 152$.
   $86 + 75 = 161$.
   $85 + 67 = 152$

2. **445** The current rent $415 plus the increase $30 is $415 + 30 = 445$.

3. **B.** *How many more* indicates subtraction.

4. There are 12 months in one year. Divide the cost of garbage service for one year by the number of months in one year, 12.
   156 divided by 12 is equal to 13.

Answer Key

Chapter 1 Review, pages 46–47

1. **C.** 3 is in the ten thousands place, so its value is 30,000.

2. **C.** Factors of 36 are 2, 3, 4, 6, 9, 12, 18. Factors of 48 are 2, 3, 4, 6, 8, 12,16, and 24. Factors of 60 are 2, 3, 4, 5, 6, 10, 12, 15, 20, and 30.

3. **B.** In the order of operations, parentheses come first.

4. **A.** $124 - 14 - 12 = 98$

5. **C.** 2 is in the thousands place in 4,572,013. The digit to the right is 0, so 2 stays the same. The number rounds to 4,572,000

6. **D.** Round $2.79 to $3.00. Subtract $3.00 from $5.00. $5.00 - $3.00 = $2.00.

7. **B.** Factors of 14 are 2 and 7. Factors of 21 are 3and 7. Factors of 42 are 2, 3, 6, 7, 14, and 21. The greatest factor is 21.

8. **C.** 11,260 rounds to 11,000; $5 \times 11,000 = 55,000$.

9. **B.** $215 + 3 \times 65 = 410$

10. **C.** Multiplication undoes division because they are inverse operations.

11. **D.** Scores in order of greatest to least are 248, 187, 114. The people who match these scores are Uppinder, Marietta, James.

12. **A.** $8 \div 4 = 2$, $7 - 2 = 5$, $26 - 6 = 20$, $20 \div 5 = 4$, $15 + 4 = 19$

13. **D.** $842 \div 27 = 31$ R5

14. **A.** $2,000 - 582 - 491 - 361 - 500 = 66$

15. **B.** $12,398 - 762 = 11,636$

CHAPTER 2 Decimals

Lesson 2.1

Key Concept, page 50

1. 7 hundreds or 700
2. 0 thousands or 0
3. 8 ones or 8
4. 2 millions or 2,000,000
5. 60
6. 190
7. 300
8. 6,380

Think About Math, page 51

1. >
2. >
3. =
4. <
5. =

Think About Math, page 52

1. 6.1
2. 3.0
3. 16
4. 5.01
5. 4.24
6. 12.37

Vocabulary Review, page 53

1. decimals
2. tenth
3. decimal point
4. hundredths
5. cent

Skill Review, page 53

1. 2 thousands or 2,000
2. 4 hundreds or 400
3. 6 tens or 60
4. 1 one or 1
5. 8 tenths or 0.8
6. 5 hundredths or 0.05

7.

Ones	.	Tenths	Hundredths
1	.	4	5

8.

Tens	Ones	.	Tenths	Hundredths	Thousandths
3	2	.	0	9	1

9.

Tens	Ones	.	Tenths	Hundredths
2	4	.	3	1

10.

Hundreds	Tens	Ones	.	Tenths	Hundredths
1	0	0	.	0	2

Skill Practice, page 53

1. C. The digit in the thousandths place is 6. The digit to the right of it is 7. Since 7 > 5, round the digit in the thousandths place up to 7. So 5412.8367 rounds to 5412.837.

2. C. 2.99 > 2.45 > 2.39 > 1.89. Mocha at $2.99 costs the most.

Answer Key

Lesson 2.2

Key Concept, page 54

1. 70
2. 176
3. 6,941
4. 6,066

5. $>$
6. $<$
7. $>$
8. $=$

Think About Math, page 56

1. $0.60
2. 7.9
3. 2.005

Think About Math, page 57

1. 3.3
2. 8.53
3. 1.985

Vocabulary Review, page 57

1. place value
2. vertically
3. align
4. organize
5. annexed

Skill Review, pages 58

1. $$\begin{array}{r} 17.3\,5\,0 \\ +50.9\,2\,7 \\ \hline \end{array}$$

2. $$\begin{array}{r} 42.0\,0\,0 \\ -36.4\,9\,8 \\ \hline \end{array}$$

3. $$\begin{array}{r} 3.8\,9\,0 \\ -1.4\,2\,6 \\ \hline \end{array}$$

4. $$\begin{array}{r} 0.1\,8\,0 \\ 8.9\,2\,1 \\ +39.6\,0\,0 \\ \hline \end{array}$$

5. $$\begin{array}{r} 1.563 \\ +8.030 \\ \hline \end{array}$$

6. $$\begin{array}{r} 29.00 \\ -\ 0.25 \\ \hline \end{array}$$

7. $$\begin{array}{r} 7.500 \\ -1.004 \\ \hline \end{array}$$

8. $$\begin{array}{r} 0.230 \\ 1.006 \\ +80.000 \\ \hline \end{array}$$

9. Marco is incorrect. The difference he got, 4.31, is greater than number he subtracted from, 4.28.

10. Lucy is incorrect. She did not regroup when adding the digits in the tenths place and ones place.

Skill Practice, page 59

1. D. $1.5 + 1.8 + 2.75 + 2.9 = 8.95$

2. C. $$\begin{array}{r} {}^{3}\,{}^{1} \\ 2.\cancel{4}\,0 \\ -0.3\,5 \\ \hline 2.0\,5 \end{array}$$

3. $13.1 - 2.4 = 10.7$

4. $12 + 15.25 = 27.25$

Lesson 2.3

Key Concept, page 60

1. 36
2. 160
3. 456
4. 199,076
5. 6
6. 12.6

Think About Math, page 61

1. $2.30
2. $10
3. $3.75
4. $3.50
5. 0.15
6. 20.4
7. 0.0325
8. 2.294

Vocabulary Review, page 62

1. factor
2. product
3. multiplication

Skill Review, page 63

1. 3
2. 6
3. 3
4. 2
5. 180; 1.8
6. 984; 0.984
7. 63; 6.3
8. 1,457; 0.01457

Skill Practice, page 63

1. **C.** The product, 0.036, has three decimal places. To place the decimal point three places to the left of 6 in the product, a zero needs to be inserted.
2. **A.** She should place the decimal point two places from the far right of the product because the sum of the decimal places in the product is two.
3. **B.** $2.5 \times 2.38 = 5.95$

Lesson 2.4

Key Concept page 64

1. 167
2. 17
3. 3,709
4. 400
5. 5
6. 50
7. 500
8. 5,000

Think About Math, page 67

1. 0.09
2. 30
3. 60
4. 0.9
5. 4.8
6. $6
7. 23
8. $0.80

Vocabulary Review, page 68

1. dividend
2. quotient
3. divisor
4. evaluate
5. reasoning

Skill Review, page 68

1. Correct
2. Correct
3. Error, the incorrect answer is based on the dividend, not the divisor; 1 place to the right.
4. Correct
5. Error; the incorrect answer is based on the dividend, not the divisor; the dividend is already a whole number, so the decimal does not need to move.
6. Error, the incorrect answer is based on the dividend, not the divisor; 1 place to the right.
7. Correct
8. Error, the answer is based on incorrect counting; 2 places to the right.
9. When dividing a decimal by a whole number, first place the decimal point in the quotient directly above the decimal point in the dividend. Then divide as you would with whole numbers.
10. To check a division problem, multiply the quotient you got times the original divisor. If you did the division correctly, the product you get should equal the original dividend.

Answer Key

(Lesson 2.4 cont.)

Skill Practice, page 69
1. **B.** $5.44 \div 8 = 0.68$
2. **A.**
$$9\overline{)165.6}$$ with quotient 18.4
3. **A.** $2.34 \div 6 = 0.39$
4. **B.** $227.25 \div 4.5 = 50.5$

Chapter 2 Review, pages 70–71
1. **C.** 0.7 is 100 times greater than 0.007.
2. **B.** $5.43 \div 1.2 = 4.525$
3. **C.** $3 \times 0.69 + 1.2 \times 3.95 + 2.5 \times 4.50 = 18.06$
4. **C.** $0.315 - 0.206 = 0.109$
5. **B.** $2.5 + (0.1 \times 56) \div (3 + 5) = 2.5 + 5.6 \div 8 =$
   $2.5 + 0.7 = 3.2$
6. **D.** $1.3 + 12.502 + 0.045 = 13.847$
7. **B.** to make it easier to align addition and subtraction
8. **D.** 10.5
9. **D.** 7 is in the hundredths place so its value is 0.07.
10. **C.** The answer should be 4.294.
11. **B.** $\$64.54 < \$65.97 < \$71.90 < \$90.15$; May is the month with the smallest amount.
12. **D.** In 67.142, 4 is the digit to the right of the tenths place. So, the digit in the tenths place stays the same.
13. **B.** Subtract the amount in April from the amount in July; $\$90.15 - \$65.97 = \$24.18$.
14. **B.** $7.25 \times 22 = 159.50$

CHAPTER 3 Fractions

Lesson 3.1

Key Concept, page 74

1. 0.3
2. 0.23
3. <
4. >
5. =

Think About Math, page 76

1. $\frac{5}{8}$
2. $\frac{1}{8}$
3. $\frac{3}{10}$
4. $\frac{11}{12}$

Think About Math, page 78

1. C.
2. A.
3. E.
4. D.
5. B.

Think About Math, page 79

1. Draw a number line from 0 to 1 and divide it into fourths and eighths. Locate each fraction on the line. The fraction farther to the right is the greater fraction.

2. After finding a common multiple to rewrite the fractions with a common denominator, compare the numerators. The fraction with the smaller numerator is less than the fraction with the greater numerator.

3. $\frac{5}{6}, \frac{7}{9}, \frac{2}{3}$

Vocabulary Review, page 80

1. fraction
2. denominator
3. lowest terms
4. numerator
5. common multiple
6. equivalent fractions

Skill Review, page 80

1. 1: One way to find an equivalent fraction is to multiply a fraction by a form of 1; 2: A form of 1 is any fraction in which the numerator and denominator are the same, such as $\frac{5}{5}$; 3: Another way to find an equivalent fraction is to divide by a form of 1.

   The main idea is that equivalent fractions have the same value. Sentences 1 and 3 explain how to find equivalent fractions; 2 is a detail, but it does not support the main idea.

2. 1: One way is by finding a common denominator; 2: List the multiples of each denominator; 3: The first common multiple is the least common denominator of the two fractions; 4: Rewrite the fractions with the common denominator. 5: Then compare the numerators of the fractions to determine the lesser or greater fraction.

   The main idea is that there are several ways to compare fractions. Sentences 1, 2, 3, 4, and 5 are all details that describe one way to compare fractions.

3.

   6 out of 7 sections in the diagram are shaded. The shaded sections represent the numerator of the fraction, and the total sections represent the denominator.

4.

   5 out of 6 sections in the diagram are shaded. The shaded sections represent the numerator of the fraction, and the total sections represent the denominator.

5.

   3 out of 8 sections in the diagram are shaded. The shaded sections represent the numerator of the fraction, and the total sections represent the denominator.

6.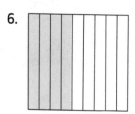

   4 out of 9 sections in the diagram are shaded. The shaded sections represent the numerator of the fraction, and the total sections represent the denominator.

Answer Key

(Lesson 3.1 cont.)

7.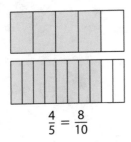

$$\frac{9}{12} = \frac{3}{4}$$

8.

$$\frac{4}{5} = \frac{8}{10}$$

Skill Practice, page 81

1. **A.** $\frac{2}{5} = \frac{16}{40}$, $\frac{1}{2} = \frac{20}{40}$, $\frac{5}{8} = \frac{25}{40}$; $16 < 20 < 25$, so $\frac{2}{5} < \frac{1}{2} < \frac{5}{8}$

2. **C.** $\frac{5}{6}$: 5 out of 6 parts are shaded.

3. **B.** $\frac{4}{7} \times 42 = 24$

4. **D.** $\frac{14}{25}$

Lesson 3.2

Key Concept, page 82

1. **C.**   4. $\frac{4}{5}$

2. **B.**   5. $\frac{3}{7}$

3. **A.**   6. $\frac{13}{25}$

Think About Math, page 83

1. $\frac{2}{3}$

2. $\frac{13}{18}$

3. $\frac{1}{3}$

4. $\frac{9}{10}$

Think About Math, page 85

1. 8; 3; 24; $\frac{17}{24}$

2. 6; 3; 6; $\frac{1}{2}$

Vocabulary Review, page 86

1. **C.**

2. **D.**

3. **A.**

4. **B.**

Skill Review, page 86

1. 1. Check the denominators. The denominators are different, so find a common denominator.
   2. Rewrite the fractions using the common denominator.
   3. Add the new numerators.
   4. Simplify, if necessary. The fraction is already in lowest terms.
   $\frac{7}{10}$

2. 1. Check the denominators. They are the same.
   2. Subtract the numerators.
   3. Simplify. Divide by $\frac{3}{3}$
   $\frac{3}{5}$

3. 1. Press the on button.
   2. Press the fraction button, then 7, the down button, then 8
   3. Press the right button, then press the subtraction button.
   4. Press the fraction button, then 3, the down button, then 10
   5. Press the equal button.
   6. Read the fraction in the lower right corner of the screen. $\frac{23}{40}$

Skill Practice, page 87

1. **A.** $\frac{7}{12} - \frac{1}{6} = \frac{5}{12}$

2. **B.** $\frac{3}{4} - \frac{1}{4} = \frac{1}{2}$

3. $\frac{1}{2} - \frac{3}{12} = \frac{6}{12} - \frac{3}{12} = \frac{1}{4}$

4. $\frac{3}{4} + \frac{1}{8} = \frac{6}{8} + \frac{1}{8} = \frac{7}{8}$

Lesson 3.3

Key Concept, page 88

1. $\frac{4}{7}$

2. $\frac{9}{16}$

3. $\frac{11}{12}$

4. $\frac{3}{4}$

Think about Math, page 89

1. C.
2. D.
3. A.
4. B.

Think About Math, page 90

1. $9 \div \frac{1}{3}$; $9 \times \frac{3}{1} = 27$; There are 27 thirds in 9.
2. 18

Vocabulary Review, page 90

1. multiplicative inverse
2. reciprocal
3. invert

Skill Review, pages 90–91

1. Enter the numerator and denominator of the dividend. Select the division sign. Then enter the numerator and denominator of the divisor. Select the equals sign. Then the quotient will appear on screen.

2. Divide the Celsius temperature by $\frac{5}{9}$, and then add 32 to find the temperature in Fahrenheit since multiplication and division, as well as addition and subtraction, are inverse operations.

   $104 - 32 = 72$, and $72 \times \frac{5}{9} = 40°C$

   $40 \div \frac{5}{9} = 40 \times \frac{9}{5} = 72$, and $72 + 32 = 104°F$

Skill Practice, page 91

1. B. $\frac{2}{3} \div \frac{4}{5} = \frac{2}{3} \times \frac{5}{4} = \frac{10}{12} = \frac{5}{6}$

2. D. $10 \div \frac{1}{2} = 10 \times 2 = 20$

3. A. $15 \div \frac{3}{5} = 15 \times \frac{5}{3} = 25$

4. B. $\frac{1}{4} \div \frac{5}{8} = \frac{1}{4} \times \frac{8}{5} = \frac{8}{20} = \frac{2}{5}$

Lesson 3.4

Key Concept, page 92

1. $\frac{9}{10}$

2. $\frac{13}{24}$

3. $\frac{5}{21}$

4. $\frac{4}{5}$

Think About Math, page 95

1. division; 22,000 pound

Vocabulary Review, Page 96

1. improper
2. mixed numbers
3. rename
4. proper fraction
5. reduce

Skill Review, pages 96

Sample answers:

1. When dividing fractions or mixed numbers, change mixed numbers to improper fractions. Then, invert the divisor and multiply. Simplify the result and rewrite improper fractions as mixed numbers or whole numbers. The summary gives the steps to follow when dividing fractions or mixed numbers, so details about specific fractions and definitions are not necessary.

2. Adding fractions and mixed numbers are nearly the same. Check whether the denominators in the fractions are the same or different. If different, rewrite the fractions with common denominators. Then add the numerators, and for mixed numbers, the whole numbers. Finally, simplify the results by writing the fraction in lowest terms and/or rewriting the improper fraction as a mixed number or whole number.

3. I already knew that multiplying fractions consists of multiplying the numerators and multiplying the denominators, and then simplifying the results if necessary. All I had to learn for multiplying mixed numbers is to first change the mixed number to an improper fraction. Once I do that, the steps are exactly the same. So, making the connections made it easier to learn how to multiply mixed numbers.

(Lesson 3.4 cont.)

Skill Practice, page 97

1. **C.** $2\frac{2}{3} \times 1\frac{1}{2} = \frac{8}{3} \times \frac{3}{2} = \frac{24}{6} = 4$

2. **A.** $10\frac{3}{4} = 10\frac{15}{20} = 9\frac{35}{20}$

   $-\frac{64}{5} = 6\frac{16}{20} = 6\frac{16}{20}$

   $ 3\frac{19}{20}$

3. **D.** Change the mixed numbers to improper fractions.

4. **B.** $6\frac{1}{2} \div \frac{3}{4} = \frac{13}{2} \times \frac{4}{3} = \frac{52}{6} = 8\frac{4}{6} = 8\frac{2}{3}$ This means 8 full recipes can be made from the secret sauce.

Chapter 3 Review, pages 98–99

1. **C.** 4 and 15 do not have any common factors

2. **D.** Fractions must have a common denominator to be added. Find a common denominator.

3. **A.** $4\frac{2}{3} + 2\frac{7}{8} = 4\frac{16}{24} + 2\frac{21}{24} = 6\frac{37}{24} = 7\frac{13}{24}$

4. **D.** Multiply $2\frac{3}{8}$ and $7\frac{1}{2}$. $2\frac{3}{8} \times 7\frac{1}{2} = \frac{19}{8} \times \frac{15}{2} = \frac{285}{16} = 17\frac{13}{16}$

5. **A.** a fraction whose value is less than either factor

6. **B.** Write mixed numbers as improper fractions.

7. **B.** $2\frac{2}{3} \times 2 = \frac{8}{3} \times \frac{2}{1} = \frac{16}{3} = 5\frac{1}{3}$

8. **B.** Rename $2\frac{1}{5}$ as $1\frac{6}{5}$ so the fractional part of the mixed number can be subtracted.

9. **D.** $6 \times 3\frac{1}{3} = \frac{6}{1} \times \frac{10}{3} = \frac{60}{3} = 20$

10. **B.** $7\frac{3}{5} - 5\frac{2}{3} = 7\frac{9}{15} - 5\frac{10}{15} = 6\frac{24}{15} - 5\frac{10}{15} = 1\frac{14}{15}$

11. **C.** She did not rename the fractions correctly.

    $10\frac{5}{7} - 8\frac{2}{9} = 10\frac{45}{63} - 8\frac{16}{63} = 2\frac{31}{63}$

12. **B.** $4\frac{2}{7} = 7 \times 4 + \frac{2}{7} = \frac{30}{7}$

13. **A.** $\frac{4}{5} = \frac{144}{180}; \frac{7}{9} = \frac{140}{180}; \frac{3}{4} = \frac{135}{180}; \frac{4}{6} = \frac{120}{180}$

14. **C.** Change any mixed numbers to improper fractions.

CHAPTER 4 Integers

Lesson 4.1

Key Concept, page 104

1-4.
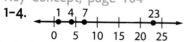

5. <
6. >
7. >
8. =

Think About Math, page 106

1. −4
2. +19
3. not an integer
4. −3
5. not an integer
6. >
7. <
8. >
9. <

Think About Math, page 106

1. 9
2. 12
3. 13
4. 25

Vocabulary Review, page 106

1. opposite
2. integer
3. infinite
4. absolute value

Skill Review, page 107

1. less than
2. greater than
3. greater than
4. less than
5. −6°F, −2°F, 0°F, 4°F, 15°F, 18°F, 20°F

6. +194 < +600 or +600 > +194

Skill Practice, page 107

1. **C.** The opposite of −3 is −(−3) = +3.
2. **D.** −32, −10, 0, +24, +316; They order from left to right on a number line with value increasing from left to right.
3. **B.** Ariana was 89 feet *below* the surface, which is represented by −89.
4. **C.** The absolute value of an integer is the distance that integer is from 0.

Lesson 4.2

Key Concept, page 108

1. 9   5. <
2. 10   6. >
3. 2   7. >
4. 6   8. >

Think About Math, page 110

1. D.   4. B.
2. A.   5. C.
3. E.   6. F.

Think About Math, page 111

1. E.   4. C.
2. A.   5. D.
3. F.   6. B.

Vocabulary Review, page 112

1. negative
2. positive
3. addends
4. sign

Skill Review, page 112

1.
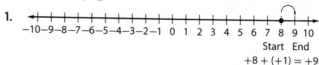
+8 + (+1) = +9

2.
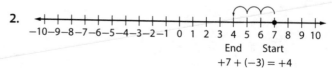
+7 + (−3) = +4

3.
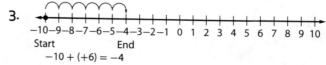
−10 + (+6) = −4

Answer Key

(Lesson 4.2 cont.)

4.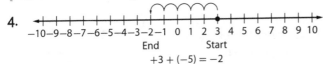
End   Start
$+3 + (-5) = -2$

5.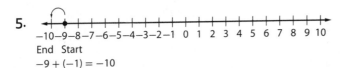
End Start
$-9 + (-1) = -10$

6.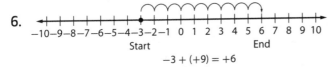
Start   End
$-3 + (+9) = +6$

7. First, determine that the addends have the same sign. Then add the absolute values of +8 and +12 to get 20. Finally, use the sign of the addends, so the result is +20,

8. First, determine that the addends have different signs. Then subtract the absolute values of +14 and −9 to get 5. Finally, use the sign of +14 since it has the greater absolute value, so the result is +5.

9. First, determine that the addends have different signs. Then subtract the absolute values of −15 and +7 to get 8. Finally, use the sign of −15 since it has the greater absolute value, so the result is −8.

10. (1) First; (2) Then; (3) After that; (4) finally

11. (3) had wanted; (2) earlier; (1) would have; (5) had subtracted; (4) After all

Skill Practice, page 113

1. **D.** +16 has a greater absolute value than −9, and its sign is positive.

2. **C.** $-5 + (+6) = +1$

3. **B.** $-8 + (+4)$; −8 has a greater absolute value than +4. The sum is negative.

4. **B.** $-35 + 10 = -25$

Lesson 4.3

Key Concept, page 114

1. −6
2. −2
3. −8
4. −6
5. −5
6. +6
7. +2
8. −10

Think About Math, page 116

1. E.
2. F.
3. C.
4. D.
5. A.
6. B.

Vocabulary Review, page 116

1. tic mark(s)
2. point(s)
3. solve

Skill Review, page 116

1.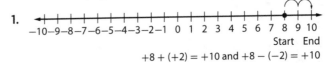
Start End
$+8 + (+2) = +10$ and $+8 - (-2) = +10$

2.
End   Start
$+4 + (-5) = -1$ and $+4 - (+5) = -1$

3. $+12 + (+11) = +23$ and $+12 - (-11) = +12 + (+11) = +23$; Compare: they both are solved as addition problems and have the same answer; Contrast: the first is a positive plus a positive, and the second is a positive minus a negative.

4. $+7 + (-3) = +4$ and $+7 - (+3) = +7 + (-3) = +4$; Compare: they both are solved as addition problems and have the same answer; Contrast: the first is a positive plus a negative, and the second is a positive minus a positive.

5. Since a positive minus a negative is the same as adding a positive and a positive, and since a positive plus a positive is always a positive, I know that $+10 - (-3)$ equals a positive number.

Skill Practice, page 117

1. **C.** $3 - 8 = -5$
2. **B.** Locate +7 on a number line, then move right 5 units.
3. **D.** $-40 - 8 = -48$
4. **A.** $245 - 302 = -57$
5. **B.** $231 - (-218) = 231 + 218 = 449$
6. **A.** $-92 - 114 = -206$
7. **D.** $-23 - 24 = -47$
8. **C.** A negative number minus a positive number is the same as adding two negative numbers.

Lesson 4.4

Key Concept, page 118
1. −8
2. −16
3. +18
4. +8
5. −6
6. −5
7. +7
8. +10

Think About Math, page 120
1. B.
2. A.
3. D.
4. C.

Think About Math, page 121
1. D.
2. C.
3. B.
4. A.

Vocabulary Review, page 121
1. title
2. inverse
3. repeated
4. columns
5. rows
6. table

Skill Review, page 122
1. +63
2. +16
3. +9
4. +2
5. −6
6. The second column tells you the population of Smithville in 1990 was 4,200.
7. The second row tells the population count for each of the years 1990, 2000, and 2010.
8. 1,300 people
9. 8,100 people

Skill Practice, page 123
1. D. −3 × 2 is equal to −3 + (−3). The sum of −3 + (−3) is a negative number.
2. B. −36 ÷ −2
3. A. A negative integer times a negative integer is a positive integer, and 12 times 7 is 84.
4. A. Divide 120 by 20 to find how many times the diver goes down 10 feet. Then multiply this number, 6, by −10 to get −60.

Lesson 4.5

Key Concept, page 124
1. to the right of 0
2. to the left of 0

Think About Math, page 127
1. J
2. H
3. M
4. B
5. (2, 1)
6. (−8, 3)
7. (−4, −9)
8. (5, −8)

Vocabulary Review, page 128
1. A.
2. H.
3. F.
4. G.
5. C.
6. D.
7. B.
8. E.

Skill Review, page 128
1. A.
2. Start at the origin, go right 9 units and then move up 2 units.
3. A.

Skill Practice, page 129
1. D. The x-coordinates are the same, so the line that contains both points is vertical.
2. C. Start at the origin. Move 3 units to the right and 7 units down to get to the point (3, −7).
3. Point A is 6 units to the left of the origin and 5 units up from there.
4. The location of point B is 4 units to the right of the origin and 3 units down from there.

Answer Key

Chapter 4 Review, pages 130–131

1. **D.** From the origin, move 3 to the left. Then from there, move 4 down to point *B*.

2. **A.** Move down from point *E* to find the *x*-coordinate, –3. Then move to the right from point *E* to find the *y*-coordinate, 4. So, the coordinates of point *E* are (–3, 4).

3. **D.** All four of the integers are negative. Example: –2 × (–3) × (–4) × (–5) = 6 × (–4) × (–5) = –24 × (–5) = 120

4. **B.** 102 + (–24) + 89 + (–225) = –58

5. **D.** To find (–8, 9), from the origin, move 8 units to the left, then from there, move 9 units up.

6. **D.** –300 × 10 = –3,000

7. **A.** 5,280 – (–10) = 5,280 + 10 = 5,290

8. **C.** –120 ÷ (–20)

9. **A.** 178 + (+250) + (–60) + (–2) + (–187) = 179; The starting balance and deposit are represented by positive integers. The withdrawal, fee, and check are represented by negative integers.

10. **B** –35, –2, 0, 14, 31

11. **C.** 7 and 5

    D. 0 and –2

CHAPTER 5 Expressions and Equations

Lesson 5.1

Key Concept, page 134
1. 9.54
2. −9
3. −15
4. 4.2
5. 14
6. 18

Think About Math, page 137
Note: Any variable is acceptable.
1. $p + 12$
2. $2s − 250$

Sample answers:
3. four times a number t divided by two
4. a number c minus nine

Think About Math, page 138
1. 10.7
2. −6
3. −3
4. 4

Vocabulary Review, page 138
1. D.
2. E.
3. A.
4. G.
5. F.
6. B.
7. C.

Skill Review, page 139
1. 1. algebraic expression, verbal expression, key words
   2. unknown, variable
   3. numbers, operations
2. 1. evaluating, expression, substitute, given values, variables
   2. substituted, correct values, variable
   3. operation, order of operations
3. *Sample answer:* I would apply the definitions of *evaluate* and *expression* and the knowledge of how to multiply and add integers. The value of the expression is −2.

Skill Practice, page 139
1. B. Substitute for s and t to get $3 + 5(−2)$. Then multiply 5 times −2 first and add 3 to get −7.
2. C. Translate each verbal expression to an algebraic expression.
3. D. The expression shows the cost of the ladder ($48.75) plus the cost of the rototiller per day ($18) times the number of days it is needed (d).
4. C. Translate −17 less than the product of −12 and some number. *The product of −12 and some number is* $−12x$. *−17 less than* is $− (−17)$. So the entire expression is $−12x − (−17)$.

Lesson 5.2

Key Concept, page 140
1. $n + 4$
2. $3n − 1$
3. $n ÷ 8$ or $\frac{n}{8}$
4. −4
5. 5
6. 6

Think About Math, page 141
1. $n + 2 = 3$
2. $a − 5 = 12$
3. 7(8) is not equal to 42, so $c = 8$ is not a solution for the equation.
4. $−4 + 7 = 3$, so $y = −4$ is a solution for the equation.

Think About Math, page 142
1. 7
2. 11
3. 6
4. 1

Vocabulary Review, page 143
1. D.
2. C.
3. A.
4. E.
5. B.

Answer Key

(Lesson 5.2 cont.)

Skill Review, page 143

Sample answers:

1. To solve equations you need to understand that you use inverse operations to solve the equation and that you perform the same operation on each side of the equation. For example, if the equation is $x + 2 = 3$, subtract 2 from both sides of the equation to solve it.

2. inverse operations

Skill Practice, page 143

1. **A.** To find the total number of lunches, 80, Maemi ordered, multiply the cost of each lunch, 16, times the number of lunches ordered. $16t = 80$

2. **C.** $2 + b = 14$
$\underline{-2 \qquad -2}$
$\qquad b = 12$

3. **D.** To check the solution, $n = 36$, Nizioni should substitute *36 for n* in the equation. $\frac{36}{6} = 6$.

4. **A.** Solve the equation $7x = 84$ by dividing both sides by 7 to get $x = 12$.

Lesson 5.3

Key Concept, page 144

1. **B.**   3. **A.**
2. **D.**   4. **C.**

Think About Math, page 148

1. $x = 8$
2. $x = 12$
3. $x = -64$
4. $x = 1$

Vocabulary Review, page 148

1. two–step equation
2. isolate
3. affect

Skill Review, page 148

1. Sample answer: Understanding sequence in solving equations helps me to determine which inverse operation to do first and which to do second. Knowing this allows me to correctly solve two–step equations.

2. Sample answer: Since the operations in the equation are multiplication and addition, I would first subtract 3 from both sides of the equation and then divide both sides of the equation by 6.

3. Sample answer: To solve any two–step equation, I would follow this sequence:
   1. Identify the two operations in the equation.
   2. Identify the inverse operations.
   3. Do the inverse operations in the reverse order, usually addition or subtraction first and division or multiplication second.

Skill Practice, page 149

1. **C.** *A number divided by eight plus three is fifty-one is* $\frac{n}{8} + 3 = 51$.

2. **A.** $2m - 7 = 49$

3. $\frac{n}{7} + 12 = 58$
$\frac{n}{7} + 12 - 12 = 58 - 12$
$(7)\frac{n}{7} = (7)46$
$n = 322$

4. $150t + 2{,}000 = 3{,}650$
$150t + 2{,}000 - 2{,}000 = 3{,}650 - 2{,}000$
$\frac{150t}{150} = \frac{1{,}650}{150}$
$t = 11$

Lesson 5.4

Key Concept, page 150

1. $=$ 4. $>$
2. $<$ 5. $<$
3. $>$ 6. $>$

Think About Math, page 151

1. $n - 8 > 12$
2. $c + 3 \geq 10$
3. $2n + 4 < 25$
4. $3h + 25 \geq 310$

(Lesson 5.4 cont.)

Think About Math, page 153
1. $t < -5$
2. $x \leq 2$
3. $a \leq -6$
4. $c < 7$

Think About Math, page 154
1. $x < -1$
2. $y < 45$
3. $m \leq -28$
4. $b < -3$

Vocabulary Review, page 155
1. **B.**
2. **A.**
3. **C.**

Skill Review, page 155
1. By connecting what I know about translating a verbal statement into an inequality, I know that x is a variable that stands for an unknown number, the symbol $-$ means subtraction, and the symbol $>$ means greater than. I can use these connections to write *a number minus four is greater than six.*
2. By connecting the ideas for solving equations and inequalities, I learned that I can solve both for the variable by using the inverse operations of addition and subtraction and multiplication and division to isolate the variable on one side of the equation or inequality. The only difference I had to learn is that you reverse the direction of the inequality symbol if you multiply or divide by a negative number.

Skill Practice, page 155
1. **B.** $\leq$ means at most.
2. **C.** $4s + 35 \geq 180$
   $4s + 35 - 35 \geq 180 - 35$
   $\frac{4s}{4} \geq \frac{145}{4}$
   $s \geq 36.25$
3. **D.** Remember that when dividing by a negative number, the inequality sign must be reversed.
   $-5x > 5$
   $\frac{-5x}{-5} < \frac{5}{-5}$
   $x < -1$
4. **A.** To represent *no more than*, use $\leq$.
   $100f + 15 \leq 65$

Lesson 5.5

Key Concept, page 156
1. $2n = 10$
2. $3n - 5$
3. 48
4. 30

Think About Math, page 158
1.

Position of Term, n	1	2	3	4	5	6
Number in Sequence	4	8	12	16	20	24

$4n$

2.

Position of Term, n	1	2	3	4	5	6
Number in Sequence	5	8	11	14	17	20

$3n + 2$

Vocabulary Review, page 160
1. generalize
2. term
3. output variable; input variable
4. common difference
5. numerical pattern

Skill Review, page 160
1. To make it easier to find the pattern, I would make a table that relates the position of each term and the number in the sequence. Then I would look at the common differences in both rows and use that to find a rule. Then I would test the rule on several numbers in the sequence. The rule is multiply the position number by 3 and subtract 2 to get the next term; $3n - 2$.
2. If you know the rule, then you can substitute any term in the expression to find its value in the sequence. In an equation, you can substitute any value for the input variable to get the output variable.

Skill Practice, page 161
1. **D.** Test the all of the numbers in the table with each equation.
2. **B.** $4n + 1$ is the rule for the table.
3. **C.** The rule is to multiply the number of each term by 38 and add 1.
   $6 \times 38 + 1 = 228 + 1 = 229$
4. **A.** Test each set of data with the equation.

Answer Key

1. **B.** $-8x + 11 = 35$

   $-8x + 11 - 11 = 35 - 11$

   $\frac{-8x}{-8} = \frac{24}{-8}$

   $x = -3$

2. **C.** $5(7) - 3(-8) = 35 - (-24) = 59$

3. **B.** Let x stand for Atian's son's age. The equation to solve is

   $5x - 7 = 43$

   $5x - 7 + 7 = 43 + 7$

   $\frac{5x}{5} = \frac{50}{5}$

   $x = 10$

4. **A.** $t = 13d + 45$

5. **D.** To undo division by -12, you must multiply by -12 on both sides of the equation.

6. **D.** $y = 5x + 2$. Replace x and y with values in table.

7. **C.** $5(10) + 2 = 52$. You can also count by 5's starting with 7. The tenth number will be 52.

8. **C.** The rule is to multiply the term by 3 and then add 1.

9. **C.** $-5(-11) - 23 = 55 - 23 = 32$

10. **D.** One value for one set of brakes that fits the solution of the inequality, $2b + 85 \leq 265$ is $90.

11. **A.** Solve each inequality to find the one that has the solution $x \geq -8$.

12. **B.** Solve $465 = 2w + 15$ to get $225.

13. **D.** The one equation that fits is $c = 45t$.

14. **A.** Undo addition, then undo division by subtracting 17 and multiplying by 6 on both sides of the equation.

CHAPTER 6 Linear Equations and Functions

Lesson 6.1

Complete a Data Table, p.169

Number of Extra Text Messages	Text-Message Charge	Coordinate Pair
0	5	(0, 5)
5	6	(5, 6)
10	7	(10, 7)
15	8	(15, 8)
20	9	(20, 9)
25	10	(25, 10)
30	11	(30, 11)

Think About Math, p. 171
The slope of the line is 25.

Vocabulary Review, p. 173
1. E.
2. B.
3. D.
4. A.
5. G.
6. F.
7. C.

Skill Review, p. 173
1. linear because a straight line connects the points
2. linear because a straight line connects the points
3. nonlinear because a straight line cannot connect the points

4. The independent variable is degrees Celsius.
5. The dependent variable is degrees Fahrenheit.
6. The slope is 1.8.
7. The *y*-intercept is 32.

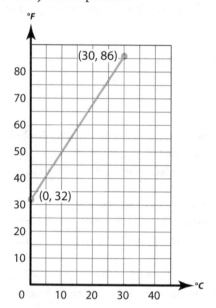

Skill Practice p. 175

1.

1. Select two points on the line. → 2. Determine the horizontal distance between the points to identify the run.

3. Determine the vertical distance between the points to identify the rise. → 4. Calculate the slope. Slope = $\frac{\text{rise}}{\text{run}}$

2. C
3. B
4. D

Answer Key

Lesson 6.2

Think About Math, p. 178

The rise = $45 - 35 = 10$.

The run = $10 - 5 = 5$.

The slope = $\frac{10}{5} = 2$.

Vocabulary Review, p. 181

1. point-slope form
2. intersects
3. slope-intercept form
4. two-point form
5. subscript

Skill Review, p. 181–182

1. A.
2. C.
3. B.
4.

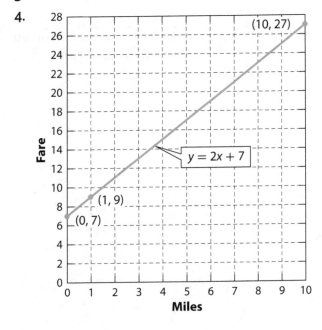

Skill Practice, p. 183

1. The statement is incorrect. If you know only the slope of the line, there are many different lines that you can draw, so you cannot plot a line with so little information. Besides the slope, you also need to know a point that the line goes through.

2. It will take five months to pay off the credit card. At the end of month 4, the balance is $100, so the next month (the fifth month), you pay the balance.

3. The slope of the line is -150. As each month passes (run = 1), the balance (rise) decreases by 150. slope = $\frac{\text{rise}}{\text{run}} = \frac{-150}{1} = -150$

4. The y-intercept is 700. This represents the starting balance on the credit card (at month 0), which is $700 – the cost of the mountain bike.

Lesson 6.3

Think About Math, p. 186

Option 1: The total monthly cost would be $105.00. Your half of the bill would be $52.50.
Option 2: The total monthly cost would be $45.00. Your half of the bill would be $22.50.

Vocabulary Review, p. 188

1. B.
2. A.
3. E.
4. C.
5. F.
6. D.

Skill Review, p. 188–189

1. 1 solution
2. 0 solutions
3. infinite number of solutions
4. $x + y = 100,000$
5. $0.03x + 0.01y = 1,800$
6. $40,000
7. $60,000

Skill Practice, p. 189

1. D.
2. B.
3. D.
4. $20,000 = x + y$; $0.06x + 0.04y = 1,000$

Lesson 6.4

Think About Math, p. 195

In general, as height increases, so does weight. There is a positive linear relationship between the variables.

Vocabulary Review, p. 196

1. If the value of one variable increases while the value of the second variable decreases, there is a <u>negative correlation</u> between the variables.

2. If the value of one variable increases as the value of the other variable increases, there is a <u>positive correlation</u> between the variables.

3. A <u>scatter plot</u> is a visual display of the relationship between two variables.

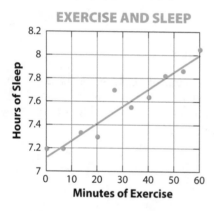

EXERCISE AND SLEEP

4. If two variables follow a clearly recognizable pattern, then there is a <u>correlation</u> between the two variables.

5. If points in a scatter plot increase or decrease proportionally, then there is a <u>linear correlation</u> between the variables that they represent.

6. An <u>outlier</u> is located further away from the trend line than the other points in a scatter plot.

7. A <u>cluster</u> is a grouping together of points on a scatter plot.

8. If the trend line on a scatter plot is exponential or quadratic, then there is a <u>nonlinear correlation</u> between the variables.

9. The line or curve around which the points in a scatter plot appear is called a <u>trend line</u>.

1   There are two outliers in the scatter plot. They are outliers because they are farther away from the trend line than the other points.

2. There is a positive linear correlation between the variables of length and width.

3.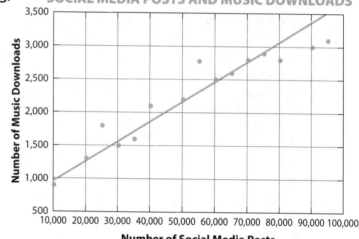

SOCIAL MEDIA POSTS AND MUSIC DOWNLOADS

Answer Key

(Lesson 6.4 cont.)

Skill Practice, p. 198

1. Scatter Plots

3. A. strong negative correlation

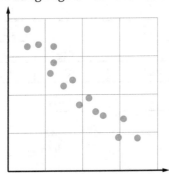

1. B. no correlation

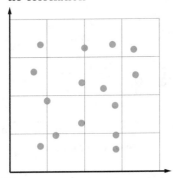

2. C. weak positive correlation

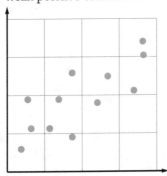

2.

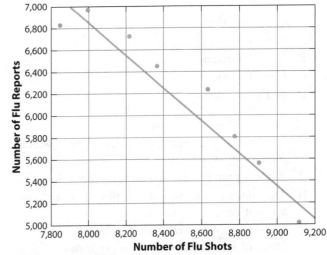

There is a negative linear correlation between the number of reported cases of flu and the number of flu shots.

3.

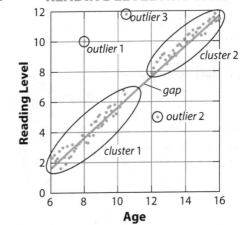

Outlier #1 represents an 8-year old student who has a higher than average reading level for her age.

Outlier #2 represents an 11-year old student who has a higher than average reading level for his age.

Outlier #3 represents a $12\frac{1}{2}$ year old student who has a lower than average reading level for her age.

Answers will vary. Sample answer: Possible explanations for the outliers include there was an error in the age or score of the student representing each outlier. Other influencing factors could include the possibility of poor physical health on testing day, students were absent for part of the testing period, or those students experience test anxiety.

Lesson 6.5

Think About Math, p. 203

A. Function

B. Not a Function

C. Not a Function

D. Not a Function

Vocabulary Review, p. 204

1. An equation is a <u>function</u> if there is only one <u>output</u> for each <u>input</u>.
2. A <u>linear function</u> has the form $y = mx + b$.
3. The points of a <u>nonlinear function</u> are not all on a straight line.
4. A <u>vertical line test</u> can help determine whether the graph of an equation is a <u>function</u> or not.

Skill Review, p. 204

1. B.
2. B.
3. Square the input, x

   Multiply the input, x, by 2

   Add the two numbers

   Add 1

   The result is the output, y.

Skill Practice, p. 205

1. **A.** Nonlinear Function

   B. Nonlinear Function

   C. Linear Function

2.

Mass (in kg)	g	Height (in m)	PE (in kJ)
90	9.8	0	0
90	9.8	10	8.82
90	9.8	20	17.64
90	9.8	30	26.46
90	9.8	40	35.28
90	9.8	50	44.1

3.

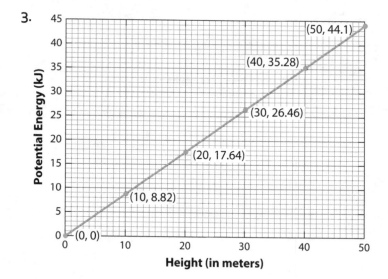

4. This graph represents a function because it passes the vertical line test. The function is linear because all the points lie on a straight line.

Answer Key

1. **D.** the term $\frac{1}{x}$ makes this equation nonlinear.

2. **D.** the slope, m, is the coefficient of D in the equation $C = 0.1\,D + 5$, thus $m = 0.1$. The y-intercept, b, is the constant term in the equation, thus $b = 5$.

3. **B.** the time and temperature data both increase proportionally, that is, the temperature increases 4 degrees each hour. Thus this data set follows a linear pattern.

4. **D.** a line with a positive slope means that the ratio of the rise over the run is positive. Thus as the independent variable increases (or decreases), the dependent variable increases (or decreases).

5. **C.** use the formula for the slope of a line: $m = \frac{\text{rise}}{\text{run}} = (y_2 - y_1)/(x_2 - x_1)$, where $(x_1, y_1) = (5, 9)$ and $(x_2, y_2) = (-1, -3)$; $m = \frac{(-3 - 9)}{(-1 - 5)} = \frac{-12}{-6} = 2$.

6. $(3, 4)$: to check: $2(3) - 2(4) = -2$; $6 - 8 = -2$
$4(3) + 4 = 16$; $12 + 4 = 16$

7. **B.** select two points on the graph to calculate the slope: $(x_1, y_1) = (0, 70)$ and $(x_2, y_2) = (30, 130)$. Use the formula for slope: $m = (y_2 - y_1)/(x_2 - x_1)$ $= \frac{(130 - 70)}{(30 - 0)} = 60/30 = 2$.

8. **B.** is the only graph that passes the vertical line test.

9. **C.** is the only data set whose values do not increase proportionally. Thus it is nonlinear.

10. **C.** is the only equation that has a linear term, r. All others have their independent variables raised to a power (and are thus nonlinear).

11. **B.** since the trend line is linear and the slope is negative, the data are linearly and negatively correlated.

12. **C.** is the trend line, since it minimizes the average distance between the points and the trend line.

CHAPTER 7 Ratios, Proportions, and Percents

Lesson 7.1

Key Concept, page 212
1. $\frac{3}{5}$
2. $\frac{1}{2}$
3. $\frac{2}{3}$
4. $\frac{3}{5}$
5.

Think About Math, page 214
1. 6:7, $\frac{6}{7}$
2. 1 to 50, $\frac{1}{50}$
3. 10 to 19, 10:19
4. $\frac{1 \text{ can concentrate}}{3 \text{ cans water}}$, $\frac{1}{3}$, 1:3, or 1 to 3
5. $31 - 23 = 8$, so the ratio is $\frac{23 \text{ snow days}}{8 \text{ no-snow days}}$, $\frac{23}{8}$, 23:8, or 23 to 8.

Think About Math, page 215
1. 22.5 miles per gallon
2. $6.50 per hour
3. $0.30
4. $0.09

Vocabulary Review, page 216
1. rate
2. unit price
3. unit rate
4. ratio

Skill Review, page 216
1. B.
2. D.
3. E.
4. A.
5. C.

Skill Practice, page 217
1. A. Divide 2.79 by 15 to get unit price.
2. C. The total number of people is 3. So, 2 out of 3 are female.
3. C. $\frac{366}{582} = \frac{366 \div 6}{582 \div 6} = \frac{61}{97}$; So, the ratio of Emilio's votes to Marshall's votes is 61 to 97.
4. B. Divide 50 by 6 to get miles per minute.

Lesson 7.2

Think About Math, page 219
The unit rate is $3 \div \frac{1}{2} = 6$ miles/hour

Think About Math, page 221
To calculate the time it will take to fill the pool:
$x = \frac{y}{k} = \frac{12{,}000 \text{ gallons}}{1{,}200 \text{ gallons/hour}} = 10 \text{ hours}$

Vocabulary Review, page 222
1. A *proportional relationship* exists between two variables if the ratio between them is always the same.
2. The *constant of proportionality* is the value of the ratio between two variables that are proportionally related.

Skill Review, page 222
1. D.
2. A.
3. C.
4. B.

Skill Practice, page 223
1. a. unit rate = scale of map = $\frac{200 \text{ miles}}{2.5 \text{ inches}} = \frac{80 \text{ miles}}{\text{inch}}$
   b. $y = 80x$
   c. $y = \frac{80 \text{ miles}}{\text{inch}} \times 8.75 \text{ inches} = 700 \text{ miles}$
2. The slope of a line that represents a proportional relationship is equal to the unit rate, which is equal to the constant of proportionality. Therefore line A has a greater constant of proportionality, since it has a greater slope.
3. Model A
   $\frac{30 \text{ miles}}{0.7 \text{ gallons}} = \frac{42.9 \text{ miles}}{\text{gallon}}$

   Model B
   slope = $\frac{\text{rise}}{\text{run}} = \frac{20 \text{ miles}}{1.5 \text{ gallons}} = \frac{13.3 \text{ miles}}{\text{gallon}}$

Answer Key

Lesson 7.3

Key Concept, page 224

1. 3:5, 3 to 5

2. 9 to 4, $\frac{9}{4}$

3. 5:9, $\frac{5}{9}$

4. $\frac{3}{5}$

5. $\frac{2}{5}$

6. $\frac{3}{13}$

Think About Math, page 226

1. yes, $\frac{50}{20} = \frac{10}{4}$

2. yes, $\frac{8}{3} = \frac{24}{9}$

3. no

Sample Answers:

4. $\frac{12}{13} = \frac{24}{26}$

5. $\frac{6}{3} = \frac{2}{1}$

6. $\frac{25}{20} = \frac{5}{4}$

Think About Math, page 227

1. $a = \frac{1}{30}$

2. $b = 54$

3. $c = 36$

4. $d = 21$

Vocabulary Review, page 228

1. cross multiplication

2. equivalent

3. value

4. proportion

Skill Review, page 228

1. $\frac{5 \text{ circles}}{8 \text{ triangles}} = \frac{10 \text{ circles}}{16 \text{ triangles}}$,

   $\frac{8 \text{ triangles}}{16 \text{ triangles}} = \frac{5 \text{ circles}}{10 \text{ circles}}$,

   $\frac{8 \text{ triangles}}{5 \text{ circles}} = \frac{16 \text{ triangles}}{10 \text{ circles}}$

   $\frac{5 \text{ circles}}{10 \text{ circles}} = \frac{8 \text{ triangles}}{16 \text{ triangles}}$

2. $\frac{\$2}{7 \text{ miles}} = \frac{\$6}{21 \text{ miles}}$, $\frac{\$2}{\$6} = \frac{7 \text{ miles}}{21 \text{ miles}}$,

   $\frac{7 \text{ miles}}{\$2} = \frac{21 \text{ miles}}{\$6}$

   $\frac{\$6}{\$2} = \frac{21 \text{ miles}}{7 \text{ miles}}$

3. $\frac{80 \text{ seeds}}{15 \text{ feet}} = \frac{32 \text{ seeds}}{6 \text{ feet}}$, $\frac{80 \text{ seeds}}{32 \text{ seeds}} = \frac{15 \text{ feet}}{6 \text{ feet}}$,

   $\frac{6 \text{ feet}}{15 \text{ feet}} = \frac{32 \text{ seeds}}{80 \text{ seeds}}$

   $\frac{15 \text{ feet}}{80 \text{ seeds}} = \frac{6 \text{ feet}}{32 \text{ seeds}}$

4. $\frac{7 \text{ cups flour}}{5 \text{ T. sugar}} = \frac{10.5 \text{ cups flour}}{7.5 \text{ T. sugar}}$,

   $\frac{7 \text{ cups flour}}{10.5 \text{ cups flour}} = \frac{5 \text{ T. sugar}}{7.5 \text{ T. sugar}}$,

   $\frac{5 \text{ T. sugar}}{7 \text{ cups flour}} = \frac{7.5 \text{ T. sugar}}{10.5 \text{ cups flour}}$

   $\frac{10.5 \text{ cups flour}}{7 \text{ cups flour}} = \frac{7.5 \text{ T. sugar}}{5 \text{ T. sugar}}$

5. $\frac{72 \text{ chairs}}{8 \text{ tables}} = \frac{126 \text{ chairs}}{14 \text{ tables}}$,

   $\frac{14 \text{ tables}}{8 \text{ tables}} = \frac{126 \text{ chairs}}{72 \text{ chairs}}$,

   $\frac{72 \text{ chairs}}{126 \text{ chairs}} = \frac{8 \text{ tables}}{14 \text{ tables}}$

   $\frac{8 \text{ tables}}{72 \text{ chairs}} = \frac{14 \text{ tables}}{126 \text{ chairs}}$

6. no; $\frac{3}{5} \neq \frac{4}{6}$ because $3 \times 6 \neq 4 \times 5$

7. yes; $\frac{3 \text{ blue}}{1 \text{ yellow}} = \frac{6 \text{ blue}}{2 \text{ yellow}}$ because

   $3 \times 2 = 6 \times 1$

8. yes; $\frac{650 \text{ words}}{10 \text{ minutes}} = \frac{780 \text{ words}}{12 \text{ minutes}}$ because

   $650 \times 12 = 780 \times 10$

Skill Practice, page 229

1. **D.**

   $\frac{1 \text{ in.}}{50 \text{ mi.}} = \frac{2\frac{1}{2} \text{ in.}}{x \text{ mi.}}$

   $2\frac{1}{2} \times 50 = 1(x)$

   $125 = x$

2. **D.** $\frac{100}{8} = \frac{d}{10}$

3. **B.** This is the only true statement because a proportion is formed by equivalent fractions. The statements (A), "2:3 and 15:10 are equal ratios," and (C), "2 × 10 and 3 × 15 are equal," are both false because 2 × 20 is not equal to 3 × 15. Because $\frac{2}{3} = \frac{10}{15}$ is a proportion, (D) is false.

4. **C.** Use the proportion $\frac{3}{8} = \frac{x}{400}$ to find the number of rock songs. So,

   $\frac{3}{8} = \frac{x}{400}$

   $3 \times 400 = 8 \cdot x$

   $1{,}200 = 8x$

   $150 = x$

Answer Key

Lesson 7.4

Key Concept, page 230
1. 46
2. 4.6
3. 0.46
4. 0.046

Think About Math, page 232
1. 0.67
2. $3\frac{4}{6}$
3. 0.03
4. $\frac{4}{10}$

Think About Math, page 234
1. $0.25, \frac{3}{12} = \frac{1}{4}$
2. $66\frac{2}{3}\%, 0.\overline{6}, \frac{2}{3}$
3. $6.8\%, 0.068, \frac{68}{1,000} = \frac{17}{250}$
4. $240\%, 2.4, 2\frac{4}{10} = 2\frac{2}{5}$
5. $37\%, 0.37, \frac{37}{100}$
6. $0.2\%, 0.002, \frac{2}{1,000} = \frac{1}{500}$

Vocabulary Review, page 234
1. percent
2. similarity
3. repeating decimal

Skill Review, page 235
1–6

Percent	Decimal	Fraction
45%	0.45	$\frac{45}{100} = \frac{9}{20}$
80%	0.8	$\frac{8}{10} = \frac{4}{5}$
35%	0.35	$\frac{7}{20}$
206%	2.06	$2\frac{6}{100} = 2\frac{3}{50}$
24.1%	0.241	$\frac{241}{1,000}$
137.5%	1.375	$1\frac{3}{8}$

45% and 24.1% were similar because they changed to decimals and fractions. 0.8 and 2.06 were similar because they changed to fractions and percents. $\frac{7}{20}$ and $1\frac{3}{8}$ were similar because they changed to decimals and percents.

7. These are both the same number. The first is written as a decimal, and the second is a fraction. Both are equal to 75%.
8. These are both the same number. The first is written as a mixed number, and the second is a repeating decimal. Both are equal to $783\frac{1}{3}\%$.
9. These are both the same number. The first is written as a fraction, and the second is a percent. Both are equal to $0.\overline{6}$.
10. These are both the same number. The first is written as a fraction, and the second is a repeating decimal. Both are equal to $11\frac{1}{9}\%$.

Skill Practice, page 235
1. B. $2\% = \frac{2}{100} = \frac{1}{50}$
2. C. $\frac{3}{10} = \frac{30}{100} = 30\%$
3. B. $\frac{4}{5} = \frac{4 \times 20}{5 \times 20} = \frac{80}{100} = 80\%$
4. D. $\frac{12}{30} = \frac{4}{10} = 0.4 = 40\%$

Lesson 7.5

Key Concept, page 236
1. 81.7
2. 2.8
3. 7
4. 120
5. 21
6. 2
7. 6
8. 240

Think About Math, page 237
1. 87
2. 470
3. 12
4. $25

Think About Math, page 239
1. 9
2. 62.5%
3. 35

Vocabulary Review, page 240
1. portion
2. means
3. extremes

Answer Key

(Lesson 7.5 cont.)

Skill Review, page 240

1. (What number) is 25% of 80?; $\frac{\square}{80} = \frac{25}{100}$; 20

2. 10% of what number is (8)?; $\frac{8}{\square} = \frac{10}{100}$; 80

3. What percent of 44 is (11)?; $\frac{11}{44} = \frac{\square}{100}$; 25%

4. (9) is what percent of 100?; $\frac{9}{100} = \frac{\square}{100}$; 9%

5. 16% of what number is (200)?; $\frac{200}{\square} = \frac{16}{100}$; 1,250

6. 3% of 500 is (what number?); $\frac{\square}{500} = \frac{3}{100}$; 15

7. (7) is 1% of what number?; $\frac{7}{\square} = \frac{1}{100}$; 700

8. What percent of 16 is (12)?; $\frac{12}{16} = \frac{\square}{100}$; 75%

9. Annabelle drank 340 cups of coffee in one year. She drank (73 cups) during January alone. What percent of cups of coffee did Annabelle drink during January?
$\frac{73}{340} = \frac{\square}{100}$; about 21.47%

10. Lucio received 52% of the votes to win an election. There were 215,400 voters. (How many people) voted for Lucio?
$\frac{\square}{215,400} = \frac{52}{100}$; 112,008

11. Panya bought some mittens on sale for ($12). She paid only 80% of the original price. What was the original price of the mittens? How much money did she save by buying the mittens on sale? $\frac{12}{\square} = \frac{80}{100}$; $15. She saved $3.

Skill Practice, page 241

1. **C.**
$$\frac{x}{12,500} = \frac{6}{100}$$
$$12,500 \times 6 = 100x$$
$$75,000 = 100x$$
$$750 = x$$

2. **B.** $25\% = \frac{1}{4}$, find $\frac{1}{4}$ of 80 or $\frac{1}{4} \times 80$

3. **D.**
$$\frac{70}{100} = \frac{x}{30}$$
$$x = 70 \times \frac{30}{100}$$
$$x = 21$$

4. **C.**
$$x/100 = \frac{16}{24}$$
$$x = 16 \times \frac{100}{24}$$
$$x = 66.6\overline{6} \text{ or about } 67\%$$

Lesson 7.6

Key Concept, page 242

1. $\frac{28}{52} = \frac{7}{13}$

2. $\frac{55}{365} = \frac{11}{73}$

3. $\frac{9}{12} = \frac{3}{4}$

4. $27.50

5. 0.04

6. 2.5

Think about Math, page 243

1. $4,000

2. $50

3. $324,000

4. $9.04

Vocabulary Review, page 244

1. principal

2. time

3. formula

4. convert

5. rate

6. interest

Skill Review, page 244

1. $200

2. about $150.68

3. $7,492.50

4. $1,770

Skill Practice, page 245

1. **B.** $4,000 \times 0.04 \times 3 = 480$

2. **A.** $75,000 \times 0.08 \times 30 = 180,000$

3. **A.** Loan A: time $\frac{26}{52} = \frac{1}{2}$
$0.05 \times \frac{1}{2} \times 12,500 = 312.50$
Loan B: time $\frac{18}{52} = \frac{9}{26}$
$0.065 \times \frac{9}{26} \times 12,500 = 281.25$
$312.50 - 281.25 = 31.25.$

4. **B.** $100 \times 0.24 \times \frac{12}{73} = 3.95$

Answer Key

Chapter 7 Review, pages 246–247

1. **D.** 12 black cars:20 total cars,
   $12 \div 4{:}20 \div 4 = 3{:}5$

2. **C.** To find x, multiply 6 times 32, then divide by 24. To solve a proportion, cross multiply opposite numerator and denominator, then divide by the number that is cross multiplied by the variable.

3. **B.** $15\% = 0.15$
   $0.15 \times 40 = 6$

4. **B.** $\frac{12}{14} = \frac{x}{100}$
   $12 \times 100 = 14x$
   $\frac{1,200}{14} = \frac{14x}{14}$
   $85.71 = x$
   85.71 compared to 100 is about equal to 86%.

5. **B.** $\frac{2}{47} = \frac{130}{x}$
   $47 \times 130 = 2x$
   $\frac{6,110}{2} = \frac{2x}{2}$
   $3,055 = x$

6. **A.** 6 years, 3 months is equal to 6.25 years.
   $3\% = 0.03$ so $I = 6,700 \times 0.03 \times 6.25$

7. **C.** $\frac{8}{15,000} = \frac{7}{x} = \frac{\text{hours}}{\text{boxes of candy}}$

8. **D.** $0.73 \times \$12,455 = \$9,092.15$

9. **D.** 8:3, 24:9 because $3 \times 24 = 8 \times 9$

10. **A.** $12,5000 \times 0.065 = 812.5$ or $\$812.50$

11. **B.** 133 miles/minutes

12. **B.** 19.6 gallons

13. $0.68 \times 50 = 34$

14. **C.** $0.15 \times 19 + 19 = 2.85 + 19 = 21.85$

15. **D.** 275 miles at 55 miles per hour is equal to 5 hours driving time. One hour of stops makes the trip 6 hours. 6 hours before 4:00 P.M. is 10:00 A.M.

16. 56
    $$\frac{8}{y} = \frac{3}{21}$$
    $21 \times 8 = 3y$
    $$\frac{168}{3} = \frac{3y}{3}$$
    $56 = y$

Answer Key

CHAPTER 8 Exponents and Roots

Lesson 8.1

Key Concept, page 250
1. 1
2. 3
3. −2
4. 28
5. −4
6. −22

Think About Math, page 251
1. 64
2. 32
3. 25
4. 27
5. 2,097,152
6. 7,962,624
7. 3,418,801
8. 2,985,984

Think About Math, page 252
1. 36
2. 21
3. 5
4. 14
5. 4
6. 18

Vocabulary Review, page 252
1. B.
2. C.
3. A.

Skill Review, page 253
1. By understanding that there is a sequence, or order of operations, for finding the value of an expression, I can apply that order to find the correct value of the expression. If I did not understand the sequence, then I would probably perform the operations from left to right and get an incorrect value.
2. For the expression $4^2 + 3^3 \div 9$, I would first find the value of the exponents, from left to right. Next, I would divide by 9. Then I would add. The value of the expression is 19: $16 + 27 \div 9$; $16 + 3$; 19.
3. For the expression $2 \times (14 - 7^0) + 28 \div 2^2$, I would first evaluate the exponent in the parentheses; second, do the operations in the parentheses; third, evaluate the other exponent; and last, multiply, divide, and add.
   The value of the expression is 33: $2 \times (14 - 1) + 28 \div 2^2$; $2 \times 13 + 28 \div 2^2$; $2 \times 13 + 28 \div 4$; $26 + 28 \div 4$; $26 + 7$; 33.

Skill Practice, page 253
1. B. $4^3 = 4 \times 4 \times 4 = 64$; $8^2 = 64$ so $4^3 = 8^2$
2. D. Addition in parentheses should be first.
3. D. $2s \times 2s \times 2s = 8s^3$
4. C. $2^5 = 32$, $3^3 = 27$, $32 - 27 = 5$

Lesson 8.2

Key Concept, page 254
1. 49
2. 32
3. 81
4. 216
5. 15,625
6. 16,777,216
7. 20,736
8. 115,856,200

Think About Math, page 257
1. 2
2. 10
3. 3
4. 15
5. 6
6. 25
7. 4
8. 18

Vocabulary Review, page 258
1. radical sign
2. square root
3. perfect square
4. squared
5. cube root
6. perfect cube

(Lesson 8.2 cont.)

Skill Review, page 258

1. The labels in the columns show that columns 1 and 3 are whole numbers that are perfect cubes, and columns 2 and 4 are the cube roots of the whole numbers. The numbers in the rows show that the first two columns are perfect cubes and cube roots, respectively, from 1–10; and the third and fourth columns are perfect cubes and cube roots, respectively, from 11–20.

2. One pattern in the table is that the cubes increase more rapidly than the cube roots. The cube roots are consecutive numbers while the cubes are not. Another pattern is that if the ones digit in the cube is 1, 4, 5, 6, 9, or 0, the ones digit in the cube root is also 1, 4, 5, 6, 9, or 0. If the ones digit in the cube is 8, 7, 3, or 2, then the ones digit in the cube root is 2, 3, 7, or 8.

3. If you know a number is a perfect cube and the cube root is a number from 1–20, you could use the data in the table to find the cube root. You can use the data to approximate cube roots, and you might be able to use the patterns in the table to find cube roots of numbers not in the table.

4. You could look in column 1 to find the perfect cubes between which 326 lies. The number 326 falls between 216 and 343. Then you could look in the second column to find that the cube roots of 216 and 343 are 6 and 7. This means that the cube root of 326 is between 6 and 7.

Skill Practice, page 259

1. **C.** The number 33 lies between the perfect squares 25 and 36. So the square root of 33 lies between the square roots of 25 and 36, which are 5 and 6.

2. **B.** $\sqrt{289 - 225} = \sqrt{64} = 8$

3. $\sqrt{6,400} = 80$

4. $\sqrt[3]{2,744} = 14$

Lesson 8.3

Key Concept, page 260

1. 20,736
2. 100
3. 64
4. 10
5. 64
6. 243

Think About Math, page 261

1. 1.84×10^4
2. 4.5326×10^8
3. 2×10^7
4. 8.7×10^5
5. 1.265×10^{10}
6. 9.348×10^6

Think About Math, page 262

1. 310,000
2. 7,000,000,000,000
3. 4,060,000
4. 291,300,000
5. 664,100,000
6. 10,020,000
7. 5,900,000,000
8. 82,200

Vocabulary Review, page 262

1. scientific notation
2. powers of ten
3. standard notation
4. annex zeros

Skill Review, page 262

1. The numbers in scientific notation are 3.786×10^9 and 9.2433×10^4.

2. The numbers in standard notation are 4,000,000,000,000 and 19,236,000.

Skill Practice, page 263

1. **B.**
2. **A**
3. **C.**
4. **D.**
5. **C.**

Answer Key

Chapter 8 Review, pages 264–265

1. **D.** $2^{10} = 1,024$, $5^4 = 625$, $1^{200} = 1$, $200^1 = 200$

2. **D.** $3,245 \times 10^2$ and 324.5×10^3 are not in scientific notation because 3,245 and 324.5 are not a numbers greater than or equal to 1 and less than 10. 3.245×10^3 is in scientific notation, but is not equal to 32,450.

3. **B.** Find the two perfect squares closest to 90. The number $\sqrt{90}$ is between the square roots of those perfect squares.

4. **A.** $\sqrt[3]{64} = 4$ because $4 \times 4 \times 4 = 64$.

5. **A.** $3^2 + 4^2 = 9 + 16 = 25$ and $5^2 = 25$

6. **D.** $1.2 \times 10^5 = 120,000$; $8.3 \times 10^4 = 83,000$; $6.7 \times 10^3 = 6,700$; $4.3 \times 10^3 = 4,300$

7. **D.** $9.3 \times 10^7 = 93,000,000$

8. **B.** Move the decimal point five places to the right and annex 3 zeros to the right of the 4. $6.04 \times 10^5 = 604,000$

9. **B.** $25^2 = 625$

10. **D.** 6 is the base, and 3 is the number of times the base is multiplied: $6 \times 6 \times 6$.

11. $3^2 + 6 \times 2 - 15 = 9 + 12 - 15 = 6$

12. **D.** $5^3 = 125$; $6^3 = 216$; $125 < 145 < 216$; so $5 < \sqrt[3]{145} < 6$

13. **A.** $\sqrt[3]{3,375} = 15$

14. **D.** $12 \times 12 = 144$ square inches

CHAPTER 9 Data

Lesson 9.1

Key Concept, page 270
1. 104
2. 37
3. 45
4. 70
5. 18
6. 9.2 or $9\frac{1}{5}$

Think About Math, page 272
1. mean: 3, median: 3, mode: 3 and 4
2. mean: 37.6, median: 34, mode: 35
3. $525
4. 21

Vocabulary Review, page 272
1. mean
2. data
3. mode
4. median
5. measures of central tendency
6. range

Skill Review, page 273
1. median: 65; modes: 67 and 92
2. median: 60.5; modes: 37 and 95
3. median: 42; no mode
4. The mean, median, and mode of a data set can be equal when all of the values in a data set are the same; for example, 24, 24, 24.

Skill Practice, page 273
1. **B.** List the house selling prices in order from least to greatest. Since there are an odd number of data, 7, find the middle value. 85,000, 95,500, 99,900, 105,000, 108,000, 120,000, 124,000. The middle value is 105,000.
2. **C.** Add the grades, then divide by 8.
$\frac{75 + 72 + 88 + 90 + 85 + 100 + 77 + 86}{8} = \frac{673}{8} =$ 84.125 to the nearest percent is 84%.
3. **A.** Add the age range to the youngest age to find the greatest age. $10 + 15 = 25$.
4. **D.** Swimming occurs 2 times, skiing occurs 3 times, and scuba diving, fishing, rafting, and sailing each occur once. Since the sport *skiing* occurs more often than any other sport, skiing is the mode of the data set.

Lesson 9.2

Key Concept, page 274
1. mean: $17, median: $16, mode: $14, range: $8
2. mean: 58, median: 58.5, mode: 35, 63, range: 58

Think About Math, page 279
1. nitrogen and carbon dioxide
2. 78%

Vocabulary Review, page 280
1. circle graph
2. line graph
3. horizontal axis
4. vertical axis
5. bar graph
6. line plot
7. graph
8. trend

Skill Review, page 281
1. The amount of money will be exactly $46,609.57. The graph supports this, so any number in the range of $45,00–$48,000 would be an acceptable answer.
2. The amount of money at 30 years will be exactly $100,626.57. An estimate between $100,000–$105,000 is appropriate due to the labeling of the graph. Therefore, the number of times greater than before any interest is calculated will be $100,626.57/10,000 = 10.062657$. Using the estimate range stated before can give a range of 10–10.5.

Skill Practice, page 281
1. **D.**
$\frac{x}{12,383} = \frac{13}{100}$
$100x = 12,383 \times 13$
$x = \frac{160,979}{100}$
$x = 1,609.79$ or about 1,610
2. **D.** The 2005 stock price was closest $20.
3. **B.** $5 - 2 = 3$
4. **C.** Look at the item with the greatest number of X's above it. The mode is the item that appears most often.

Answer Key

Lesson 9.3

Key Concept, page 282
1. mean: 36.7; median: 34; mode: 28; range: 33
2. 50
3. 24

Think About Math, page 286
1. A circle graph could be misleading if the percents do not add up to 100%.
2. The scale does not increase evenly.
3. 16.9

Vocabulary Review, page 286
1. stem-and-leaf plot
2. stem
3. leaf
4. outlier
5. key
6. mislead

Skill Review, page 286
1. Ian may be trying to show that Dora, Elly, and Ian all walked nearly the same number of miles. He used a scale that had large gaps, thereby making the data appear to be all pretty much in the same range.
2. Dora may be trying to show that she walked twice as many miles as Ian. She did not start her scale at 0.

Skill Practice, page 287
1. **D.** The scale is numbered in reverse order. This makes it look as if the boys had the greater number of summer jobs.
2. **B.** *Key: 131|4 means 1,314.* There should be only one leaf for each piece of data.
3. **A.** The percents do not add up to 100%.
   30% + 18% + 10% + 35% = 93%
4. **B.** The scale does not begin at 0. It looks as if there are no senior citizens in Millvale in 2010.

Chapter 9 Review, pages 288–289
1. **B.** The temperature went up from 10° to 19°, a difference of 9°.
2. **D.** The temperature either dropped or stayed the same between 1 A.M. and sunrise.
3. **D.** Both $150 and $300 have 4 Xs.
4. **C.** There are 9 Xs for amounts paid that are less than $300.
5. **B.** Multiply each money amount by the number of X's and add, then divide by 16 to get $253.13 (rounded to the nearest penny).
6. **D.** The outlier is 18; the median is 35 whether the outlier is one of the data or not.
7. **D.** Office has 6 employees, press has 6 employees, and the Editorial department has 22 employees; 6 + 6 + 22 = 34.
8. **C.** Subtract the department with the fewest number of employees, 6, in the Press or Office departments, from the department with the most employees, 23, in the Mail department, to get the range.
9. **D.** The graph is misleading because the scale does not start at 0.

Answer Key

CHAPTER 10 Probability

Lesson 10.1

Key Concept, page 292
1. 112
2. 180
3. 168
4. 120

Think About Math, page 294
1. 36
2. 125

Vocabulary Review, page 294
1. E.
2. F.
3. A.
4. C.
5. D.
6. B.

Skill Review, page 295
1. A tree diagram allows you to visually see every combination possible and allows you determine the outcome by a visual display.
2. The tree diagram confirms the statement in the text that there are 8 possible outcomes. The tree diagram expands upon the text by showing all of the specific outcomes.
3.

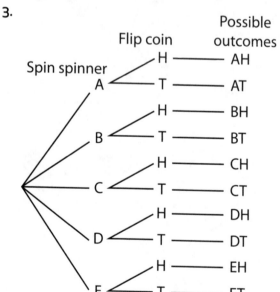

10 possible outcomes

Skill Practice, page 295
1. **B.** There should be 3×3 or 9 outcomes in the sample space: 1, 1; 1, 2; 1, 3; 2, 1; 2, 2; 2, 3; 3, 1; 3, 2; 3, 3.
2. **C.** There are two possible outcomes (boy or girl) used as a factor three times (three children): $2 \times 2 \times 2 = 8$.
3. **A.** There are 6 outcomes for the cube and 4 outcomes for the spinner: $6 \times 4 = 24$.
4. **C.** There are 10 possible outcomes (digits 0–9) used as a factor 4 times (4-digit PIN number): $10 \times 10 \times 10 \times 10 = 10{,}000$.

Lesson 10.2

Key Concept, page 296
1. 36
2. 27
3. 8
4. 12

Think About Math, page 299
1. 0
2. $\frac{1}{3}$, about 0.33, or about 33%
3. $\frac{1}{28}$, about 0.036, or about 3.6%

Vocabulary Review, page 300
1. trials
2. impossible event
3. permutation
4. probability
5. combination
6. certain event
7. theoretical probability
8. experimental probability

Answer Key

(Lesson 10.2 cont.)

Skill Review, page 300

1. Yes; the probability of drawing a red marble out of the bag is $\frac{4}{9}$ or about 44%, and 44% of 150 is about 67.

2. The probability is $\frac{1}{120}$. The brochures are placed in an arrangement in which order is important. This means that the number of possible outcomes is a permutation. So you use permutations to find the number of possible outcomes. There are 5 choices for the first brochure, 4 for the second, 3 for the third, 2 for the fourth, and 1 for the fifth. If you multiply those numbers together, you get 120 as the number of possible outcomes. Since there is only 1 favorable outcome, the probability is $\frac{1}{120}$.

3. To predict an outcome, you first find the probability of the particular outcome, and then you use the probability to make a prediction. For theoretical probability, you use the ratio of favorable outcomes to total possible outcomes to find the probability. For experimental probability, you use the ratio of the number of successes to the number of trials to find the probability. Theoretical probability and experimental probability will differ because theoretical probability is based on equally likely results, while experimental probability is based on actual results.

Skill Practice, page 301

1. **A.** If she buys 5 out of the 100 tickets sold, she has a 1 in 20, or 0.05, or 5% chance of winning the car.

2. **D.** This is permutation since the order matters. There are 10×9 or 90 possible outcomes. There is 1 favorable outcome, so the probability is $\frac{1}{90}$.

3. $\frac{8 \text{ successes}}{20 \text{ trials}} = 8 \div \frac{4}{20} \div 4 = \frac{2}{5} = 0.40$ or 40%

4. $\frac{42 \text{ favorable}}{150 \text{ possible}} = \frac{7}{25} = 0.28$
   $0.28 \times 50 = 14$

Lesson 10.3

Key Concept, page 302

1. $1\frac{1}{10}$

2. $\frac{1}{4}$

3. $\frac{5}{14}$

4. $\frac{1}{24}$

5. $\frac{6}{8} = \frac{3}{4}$, 0.75, or 75%

6. $\frac{4}{6} = \frac{2}{3}$, about 0.67, or about 67%

Think About Math, page 304

1. overlapping; $\frac{7}{12} + \frac{6}{12} - \frac{3}{12} = \frac{10}{12} = \frac{5}{6}$

2. mutually exclusive; $\frac{1}{12} + \frac{6}{12} = \frac{7}{12}$

3. mutually exclusive; $\frac{3}{12} + \frac{1}{12} = \frac{4}{12} = \frac{1}{3}$

4. overlapping; $\frac{6}{12} + \frac{5}{12} - \frac{2}{12} = \frac{9}{12} = \frac{3}{4}$

Vocabulary Review, page 305

1. E.
2. C.
3. A.
4. D.
5. B.

Skill Review, page 306

Sample answers:

1. For color tiles that are not the same color as the tile selected in the first pick, the likelihood goes up. This is because every time you remove a tile, the number that goes in the denominator is reduced by one, but the number in the numerator remains the same, so the percentage increases. For color tiles that are the same color as the tile selected in the first pick, the likelihood goes down. This is because the numbers in both the numerator and denominator are reduced by one, leading to a lower percentage. For example, if the probability in the first pick was $\frac{2}{3}$, or about 67 percent, the probability in the second pick would be $\frac{1}{2}$, or 50 percent.

Answer Key

(Lesson 10.3 cont.)

2. The first step in solving this problem is to use definitions to find out the type of compound event. The definitions of mutually exclusive and overlapping do not apply to the situation, so the compound event is either independent or dependent. Each digit in the password can repeat, so one event does not affect the other event. This is the definition of independent events. The probability is $\frac{1}{9}$ for the first digit and $\frac{1}{9}$ for the second, so the probability of the first two numbers being 2 is $\frac{1}{81}$.

Skill Practice, page 306

1. **A.** $\frac{6}{18} \times \frac{5}{17} = \frac{30}{306} = \frac{5}{51}$

2. **C.** $\frac{3}{14} \times \frac{7}{14} = \frac{21}{196} = \frac{3}{28}$

3. **D.** Since events are overlapping, add the probabilities of either event happening and subtract the probability of both events happening.

4. **B.** The probability of the number being less than 4 is $\frac{3}{6}$, and the probability that the number is equal to 6 is $\frac{1}{6}$. Since these are mutually exclusive events, add the probabilities together: $\frac{3}{6} + \frac{1}{6} = \frac{4}{6} = \frac{2}{3}$

Chapter 10 Review, pages 307–308

1. **B.** $12 + 15 + 12 + 18 = 57$ total coupons. There are 15 clothing store coupons. $\frac{15 \div 3}{57} \div 3 = \frac{5}{19}$

2. **D.** There are 12 restaurant coupons and 12 home improvement coupons. Since there are the same number of each type in the packet, they are equally likely to be pulled at random.

3. **A.** Since Lujayn does not replace the first coupon, the second event depends on the first event. The correct expression would be $\frac{18}{57} \times \frac{12}{56}$.

4. **D.** Since there are no shoe-store coupons in the packet, there is 0 probability of selecting a shoe-store coupon.

5. **D.** There are two favorable outcomes out of a possible twelve to have at least one 6 (answer D), 6 favorable outcomes out of a possible 36 to roll the same number on each cube (answer B), 6 favorable outcomes out of a possible 36 to roll a sum of 7 (answer C), and 1 favorable outcome out of a possible 36 that the sum of the numbers is 12 (answer A).

6. **D.** They are mutually exclusive choices, so add the probabilities together and convert to a percent: $\frac{2}{8} + \frac{3}{8} = \frac{5}{8} = 0.625 = 62.5\%$.

7. **B.** The second choice depends on the first choice, $\frac{1}{8} \times \frac{2}{7}$.

8. **A.** Multiply the probabilities together, $\frac{1}{8} \times \frac{1}{5} = \frac{1}{40}$.

9. **C.** $\frac{2}{8} \times \frac{3}{5} = \frac{6}{40} = \frac{3}{20} = 0.15 = 15\%$

10. **B.** Use the Counting Principle and find the factors that have a product of 90.

11. **C.** Use the Counting Principle. She wears either a pair of pants, blouse, and sweater ($7 \times 9 \times 4 = 252$) or a skirt, blouse, and sweater ($5 \times 9 \times 4 = 180$). $252 + 180 = 432$

12. **C.** *Wednesday will immediately follow Tuesday* is the only statement that is true 100% of the time.

Answer Key

CHAPTER 11 Measurement

Lesson 11.1

Key Concept, page 312
1. 23
2. 180
3. 480
4. 14
5. 5
6. 68,640

Think about Math, page 314
1. 48
2. 10
3. 5
4. 10,000
5. 74
6. 455
7. 2, 3
8. 3, 2

Vocabulary Review, page 314
1. length
2. weight
3. unit
4. capacity
5. time

Skill Review, page 315
1. pints
2. pounds
3. feet
4. gallons
5. minutes
6. miles
7. days
8. quarts
9. weeks
10. yards
11. 2 gallons
12. 86,400 seconds
13. Enrique
14. Sue

Skill Practice, page 315
1. **C.** One second is $\frac{1}{60}$ of a minute.
2. **C.** There are 4 quarts in one gallon, 2 pints in one quart, and 2 cups in one pint. So there are $4 \times 2 \times 2$ or 16 cups in one gallon. 20 cups − 16 cups = 4 cups left over.
3. **A.** There are 16 ounces in 1 pound so there are 8×16 or 128 ounces in 8 pounds.
4. **C.** Convert $60\frac{2}{3}$ yards to feet. There are 3 feet in 1 yard so
$60\frac{2}{3} \times 3 = \frac{182}{3} \times \frac{3}{1} = \frac{546}{3} = 182$ ft.

Lesson 11.2

Key Concept page 316
1. 280
2. 3,500
3. 5
4. 0.45
5. 90,000
6. 5,650

Think about Math, page 318
1. 0.001
2. 100
3. 1,000
4. 1,000
5. 100,000
6. 1,000,000
7. 0.00001
8. 0.01
9. 10
10. 0.000001
11. 0.001
12. 0.1

Think about Math, page 319
1. 3.2
2. 400
3. 5.5
4. 8.9
5. 1,400
6. 2
7. 0.24
8. 342
9. 6,500

Vocabulary Review, page 320
1. meter
2. liter
3. gram
4. power of 10

Skill Review, pages 320
1. 10 g
2. 125 g
3. 12 cm
4. 1,000 L
5. dm
6. B.
7. C.
8. A.

Skill Practice, page 321
1. **A.** There are 1,000 milligrams in 1 gram. Divide 70 mg by 1,000 mg per gram to get 0.07 gram.
2. **D.** Millimeter, centimeter, meter, kilometer is in order from smallest to largest unit of length.
3. **C.** Multiply 0.25 by 1,000 glasses to get 250 liters.
4. **A.** Divide 800 centimeters by 100 to get 8 meters.

Chapter 11 Review, pages 322–323

1. **D.** Multiply the number of yards by the number of inches in one yard. Multiply 12 by 36.

2. **C.** kilometer, meter, centimeter, millimeter

3. **B.** Betina has lived 5 months × 30 days per month in Tucson or 150 days. Mavis has lived there 136 days. Hector has lived there 18 weeks × 7 days per week or 126 days. Jumah has lived there 0.25 × 365 days or 91.25 days.

4. **B.** 2 cups equal 1 pint, and 2 pints equal one quart. There are 4 cups in one quart and 2 × 4 or 8 cups in 2 quarts.

5. **B.** Divide 75 inches by 12 inches per foot to get 6 feet 3 inches.

6. **C.** There are 1,000 mL in 1 liter. Divide 300 by 1,000 to get 0.3 liters.

7. **A.** Multiply 5,280 by 5 to get 26,400 feet.

8. Multiply 4 pounds times 16 ounces per pound to get 64.

9. **C.** There are 1,000 liters in 1 kiloliter, so 15 × 1,000 = 15,000.

10. **D.** Write both times in terms of minutes. Then compare the times.

11. **B.** $(12 \times 3) - 5 = 31$ feet.

12. **C.** *Kilo-* means 1,000, so 35 kilometers is equal to 35,000 meters.

13. **A.** 6,500 kilograms = 6,500 × 1,000 grams

14. **D.** Multiply 450 yards by 3 feet per yard to get 1,350 feet.

Answer Key

CHAPTER 12 Geometry

Lesson 12.1

Key Concept, page 326
1. square
2. circle
3. triangle
4. rectangle

Think about Math, page 327

Sample answers:

1.
P   Q

2.
M   N

3.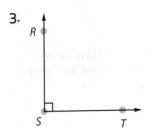
R
|
|
|
└─────►
S   T

Think about Math, page 328
1. acute equilateral triangle
2. obtuse scalene triangle

Vocabulary Review, page 330
A. segment
B. circle
C. quadrilateral
D. angle
E. rectangle
F. perpendicular lines
G. parallel lines
H. square

Skill Review, page 331
1. Archimedes found better approximations of pi by finding the perimeter of inscribed and circumscribed regular polygons with more and more sides. In other words, he found approximations of pi by finding the perimeter of regular polygons that were placed inside and outside of the circle, but only touched the circle. He then used more and more sides for these polygons to find better and better approximations of pi.

2. Rectangles and squares both have 4 sides and 4 angles. Both figures have all angles measuring 90 degrees. However, a square has all four sides measuring the same length, whereas a rectangle has opposite sides measuring the same length.

3. No; The sum of the measures of the three angles of a triangle is 180°. One right angle is 90°. One obtuse angle is greater than 90°, let us say 91°. 90 + 91 = 181. This measure for two angles is already greater than the total sum for three angles, so a triangle cannot have a right angle and an obtuse angle.

Skill Practice, page 331
1. D. The diameter is the distance across a circle through its center. Line segment ST is the diameter.

2. C. The measure of the angle is 45°, so it is acute.

3. C. One of the angles is equal 90°, and no sides are congruent. So, the triangle is a right scalene triangle.

4. A. The figure is a ray: a set of points continuing in one direction only, and it is named by the points MP.

Lesson 12.2

Key Concept, page 332
1. square
2. triangle

Think about Math, page 334
1. 100 yd.
2. 40 cm
3. 46 m

Think about Math, page 336
1. 62.8 m
2. 43.96 m or 44 m

Vocabulary Review, page 336
1. perimeter
2. radius
3. circumference
4. pi
5. diameter

Skill Review, page 336
1. There are two missing side lengths. Add 7 cm and 4 cm to find one missing side length. Subtract 10 cm from 15 cm to find the other side length.

   $15 + 7 + 5 + 4 + 10 + 11 = 52$

   The perimeter is 52 cm.
2. The radius given, and it is half of the diameter. To find the diameter, multiply the radius by 2. Then multiply the diameter by π to find the circumference. The circumference is 62.8 cm.

Skill Practice, page 337
1. **B.** To find the perimeter of the half circle, add the diameter, 1 m, to one-half of the circumference of the circle. $1 + (3.14 \div 2) = 2.57$ m
2. **B.** 8 sides $\times$ 8 mm = 64 mm
3. **C.** To find perimeter add the sides:

   $15 + 15 + 9 + 9 = 48$
4. **B.** 5

   $7 + 7 + x + x = 24$

   $14 + 2x = 24$

   $2x = 10$

   $x = 5$

Lesson 12.3

Think About Math, page 340
No, the answer does not make sense.

In the drawing, the table is 1.5 in. long. The scale is 1 in. = 4 ft. A dining room table would likely be longer than 0.375 feet. It may be necessary to revise the scale or the drawing.

Vocabulary Review, page 343
1. C.
2. B.
3. A.

Skill Review, page 343–344
1. 12
2. 22
3. 1.5 in.
4. 2 feet, 7 feet
5. A.
6. B.

Answer Key

(Lesson 12.3 cont.)

Skill Practice, page 345

1. **a.** height = 3 in.
   b. width = 4.5 in.
   c. diameter = $\frac{3}{4}$ in.
2. **A.**
3. 5 cm = 6 in.
4. In order to draw a triangle with side lengths 3 inches, 4 inches, 5 inches, first draw a line segment that measures the longest side (5 inches) using a ruler. Then from the one endpoint, draw a circle with radius 4 inches using a compass. From the other endpoint, draw a circle of radius 3 inches using a compass. The two circles created will intersect twice. Choose one intersection point to complete the triangle (the initial line segment is a side of the triangle). The drawn triangle will have side lengths 3 inches, 4 inches, and 5 inches.
5. In order for the new scale drawing to be to the same proportion, the ratio of the door lengths must be multiplied to every length. The ratio from new:old is 3:2 or 1.5:1. So multiply the old lengths be 1.5 to find the new lengths.

Lesson 12.4

Key Concept, page 346

1. 44 m
2. 12 cm
3. 18.8 mm
4. 44 ft

Think about Math, page 349

1. 60 cm^2
2. 154 ft^2
3. 153.9 in.^{2}
4. 64 mm^2

Vocabulary Review, page 350

1. height
2. length
3. area
4. width
5. base

Skill Review, page 351

1. 64 mm^2
2. 75 cm^2
3. 38.1 m^2

Skill Practice, page 351

1. **D.**
2. **B.** $96 \times 2 = 192$
   $72 \times 2 = 144$
   $192 + 144 = 336$
3. **D.** $25 \times 30 = 750$
   $750 - 100 = 530$
4. **C.** $30 = \frac{1}{2}(6)(h)$
   $10 = h$

Answer Key

Lesson 12.5

Think About Math, page 356
5

Vocabulary Review, page 357
1. G.
2. B.
3. A.
4. D.
5. E.
6. F.
7. C.

Skill Review, page 358–359
1. 6, –6
2. 11, –11
3. 3, –3
4. not a triple
5. Pythagorean triple
6. not a triple
7. $369 \approx 19.2$
8. $244 \approx 15.6$ ft.
9. 5 ft.

Skill Practice, page 359
1. $b^2 = 95, -95$
2. $a^2 = 45, -45$
3. $x^2 = 24, -24$
4. a Pythagorean triple
5. not a triple
6. a Pythagorean triple
7. $\sqrt{48} \approx 6.9$
8. $\sqrt{161} \approx 12.7$
9. $\sqrt{40} \approx 6.3$
10. 10
11. $\sqrt{544} \approx 23.3$
12. $\sqrt{20} \approx 4.5$

Lesson 12.6

Key Concept, page 360
1. right triangle
2. rectangle
3. circle

Think about Math, page 363
1. 288 m³
2. 512 cm³

Vocabulary Review, page 364
1. C.
2. B.
3. E.
4. D.
5. F.
6. G.
7. H.
8. A.

Skill Review, page 364
1. A square and a cube are similar in that they are both figures whose sides have equal lengths. The angles of both figures are right angles. They are different in that a square is a two-dimensional figure, and a cube is a three-dimensional figure. You square the length of side ($s \times s$) to find the area of a square, while you cube the length of a side ($s \times s \times s$) to find the volume of a cube.
2. A cube is a type of rectangular solid in which all of the edges are equal in length.

Answer Key

(Lesson 12.6 cont.)

Skill Practice, page 365

1. **D.** $V = l \times w \times h$
   $126 = 7 \times 3 \times h$
   $126 = 21 \times h$
   $\frac{126}{21} = \frac{21 \times h}{21}$
   $6 = h$

2. **A.** $V = s \times s \times s$
   so $V = 14 \times 14 \times 14$
   and $V = 2{,}744$ in.^{3}

3. **C.** $1{,}728 = s^3$
   $\sqrt[3]{1{,}728} = s$
   $12 = s$

4. **D.** $V = 18 \times 15 \times 8 = 2{,}160$ in.^{3}

Lesson 12.7

Think About Math, page 367
The package is a cylinder. $V = \pi r^2 h$
$r = \frac{d}{2} = \frac{6}{2} = 3$
$V = (3.14)\,(3^2)\,(12)$
The cylinder's volume is ≈ 339 in.^{3}

Think About Math, page 368
The disk is shaped like a cone. $V = \frac{1}{3}\pi r^2 h$
$r = \frac{d}{2} = \frac{7.5}{2} = 3.75$
$V = \frac{1}{3}(3.14)\,(3.75^2)\,(2.25)$
The disc's volume is ≈ 33 in.^{3}

Vocabulary Review, page 370

1. A 3-D shape with one end that ends in a point is a <u>cone</u>.
2. A <u>cylinder</u> had two equally sized circular ends.
3. The <u>base</u> of a cone is a circle.
4. A <u>frustum</u> is a cone with its top cut off.
5. The point of a cone is the <u>apex</u>.
6. All of the points on the outside surface of a <u>sphere</u> are the same distance from the shape's center.

Skill Review, page 370

1. Only choice C uses cubic units of measurement.

2. $V = \pi r^2 h$
   $= (3.14)\,(16)\,(12)$
   ≈ 603 cm^3

3. ≈ 5.96 in.^{3}

4. $V = \pi r^2 h$
   $r = \frac{d}{2} = \frac{32}{2} = 16$
   $= (3.14)\,(16^2)\,(81)$
   $\approx 65{,}111$ m^3

5. $V = \frac{1}{3}\pi r^2 h$
   $r = \frac{d}{2} = \frac{5.8}{2} = 2.9$
   $= (3.14)\,(2.9^2)\,(12.1)$
   ≈ 107 m^3

Skill Practice, page 371

1. $V = $ volume of half-sphere $+$ volume of cylinder
   $V = \frac{1}{2}\left(\frac{4}{3}\pi r^3\right) + (\pi r^2 h) = 0.002$ m^3 $+ 0.016$ m^3 $= 0.018$ m^3

2. $V = $ volume of two half-spheres $+$ volume of cylinder
   $V = \left(\frac{4}{3}\pi r^3\right) + (\pi r^3 h) = 268$ mm^3 $+ 452$ mm^3 $= 720$ mm^3

3. $V = $ (volume of cone 1) $-$ (volume of cone 2)
   $V = \left(\frac{1}{3}\pi r_1^2 h_1\right) - \left(\frac{1}{3}\pi r_2^2 h_2\right) = 105$ in.^{3} $- 44$ in.^{3} $= 61$ in.^{3}

Chapter 11 Review, pages 372–373

1. **A.** Multiply the length and width of the side of the truck to find the area of that side.
   $14 \times 13 = 182$ square feet

2. 1,456 Multiply the length and the width and the height to find the volume, or capacity, of the truck.
   $14 \times 13 \times 6 = 1,102$ cubic feet

3. **C.** All squares are quadrilaterals with only right angles, so they are rectangles. However, there are rectangles with two pairs of congruent sides that are not all the same length. No triangles in the plane have more than one right angle, and there are triangles that are not equilateral.

4. The formula for the perimeter of a rectangle is $2l + 2w$. $15 = 2l + 2(2.5)$, so $2l = 10$, and $l = 5$ feet.

5. **D.** The formula for the area of a circle is πr^2.
   $\pi(12)^2 = 144\pi$, or about 452.16 square inches.

6. **B.** The perimeter is the sum of the length of the sides. $6 + 24 + 18 + 6 + 4 + 6 + 4 + 6 + 12 + 6 = 92$ feet

7. **C.** One way to find the area is to find the area of the large 24 by 18 rectangle, then subtract the smaller rectangles that are not included.
   $24 \times 18 - (6 \times 12) - (6 \times 4) =$
   $432 - 72 - 24 = 336$

8. **B.** There are 3 circles in the rectangle, each with a diameter of 18 inches. The length of the rectangle is $18 \times 3 = 54$ inches.

9. **A.** The length of the rectangle is $18 \times 3 = 54$ inches, and the width is 18 inches. The area of the rectangle is $54 \times 18 = 972$ square inches.

Glossary

A

abbreviate (uh BREE vee ayt) to use a short form of a word, often followed by a period, in order to save space

abbreviation (uh bree vee AY shuhn) a shortened form of a word or phrase

absolute value (ab suh LOOT VAL yoo) the distance between a number and zero; it is always a positive number

acute angle (uh KYOOT ANG guhl) an angle with measure greater than 0° and less than 90°

addends (AD endz) the numbers being added in an addition problem

addition (uh DISH uhn) the combining of two or more numbers

addition method (uh DISH uhn METH uhd) the principle stating that adding the same value to each side of an equation does not change the equality of the relationship

adjust (uh JUHST) to change

affect (uh FEKT) to have an impact on

algebraic expression (al juh BRAY ik ek SPRESH uhn) a combination of numbers, one or more variables, and operations

align (uh LYNE) to line up

analyze (AN uh lyze) to examine something carefully

angle (ANG guhl) a figure of two rays extending from the same point

annex zeros (AN eks ZEER ohz) to add zeros

annexed (AN eksd) added

apex (AY peks) the top, peak, or highest point of an object

apply (uh PLEYE) to use information you know in a different way or new situation

approximate 1. (uh PRAHKS uh mit) *adjective*, close to; 2. (uh PRAHKS uh mayt) *verb*, to estimate

approximation (uh prahks uh MAY shuhnz) a reasonable guess

area (AIR ee uh) the amount of surface covered by a figure

arithmetic expression (a RITH met ik ek SPRESH uhn) an expression that has a number value

arithmetic sequence (a RITH met ik SEE kwuhnss) a sequence with a common difference

aspect (AS pekt) part or feature

B

bar graph (bahr graf) a graph made up of rectangular bars that extend horizontally or vertically; the height of each bar corresponds to one number in the data

base (bayss) 1. the number to be multiplied in a power; 2. the side in a triangle that is perpendicular to the height

breaking apart numbers (BRAY king uh PART NUHM burs) a mental math strategy for working with large numbers into two or more small numbers in order to facilitate calculations; for example, $153 + 62 = 100 + (50 + 60) + (3 + 2) = 100 + 110 + 5 = 215$

C

calculate (KAL kyoo layt) to find the answer using a mathematical process

capacity (kuh PASS i tee) the measure of the amount that can be held or contained

cause (kawz) why something happens

cell (sel) a place in a table or spreadsheet where a column and a row intersect

cent (sent) one hundredth of a dollar; a penny

center (SENT ur) the middle of something

certain event (SUR tn i VENT) probability of 1; it always happens

chart (chahrt) an arrangement of numbers or other information; a diagram that shows information

circle (SUR kuhl) a curved flat figure every point of which is the same distance from the center

circle graph (SUR kuhl grahf) a circle divided into parts; all parts add up to 100%

circumference (sur KUHM fur uhnss) the distance around a circle

clarify (KLAIR uh feye) to make meaning understood

classify (KLASS uh feye) to sort things into groups by their characteristics

cluster (KLUS ter) a grouping

coefficient (koh uh FISH uhnt) a number that multiplies a variable

column (KAHL uhm) cells of information in a table arranged vertically

combination (kahm buh NAY shuhn) an arrangement of items in which order does not matter

common denominator (KAHM uhn di NAHM uh nay tur) a common multiple of the denominators of two fractions

common difference (KAHM uhn DIF ur uhnss) the difference between any two consecutive terms in an arithmetic sequence

common multiple (KAHM uhn MUHL tuh puhl) a number that is a multiple of two or more different integers; 12 is a common multiple of 2, 3, and 4

Commutative Property of Multiplication (kom MYOOT uh tive PRAH per tee uhv muhl ti pli KAY shuhn) the product of two numbers is unaffected by the order in which they are multiplied

compare (kuhm PAYR) for numbers, to decide which has the greater value

comparison (kuhm PAYR uh suhn) examining similarities, or how two or more people, things, or ideas are alike

compatible numbers (kuhm PAT uh buhl NUHM burz) numbers that are close to the original numbers, used to make an estimation easier or quicker to find

compensation (kahm pen SAY shuhn) a mental math strategy in which numbers in a sum or difference are changed so it is easier to perform the operation

complex (KAHM pleks) complicated

complex shape (KAHM pleks shayp) a figure composed of two or more shapes

compliance (kom PLY uns) the ratio of movement to the force applied to a spring; the opposite of the spring's stiffness

compound event (KAHM pound i VENT) the result of two or more events

compound interest (KAHM pound IN trist) the total amount of money earned on an investment when the earned interest is added to the principal

concept (KAHN sept) idea

conclusion (kuhn KLOO zhuhn) a decision made about information

cone (kohn) a circular base with a single curved side that meets at a point, or apex, opposite the circle

confirm (kuhn FURM) to validate a point

congruent (kuhn GROO uhnt) having the same measure (angles) or length (sides)

congruent (kon GREW uhnt) identical shapes or angles

connection (kuh NEK shuhn) a relationship

cons (kahnz) negative outcomes

consecutive (kuhn SEK yuh tiv) following in order one after the other

constant of proportionality (KON stahnt uhv pruh PAWR shuhn AL uh tee) the constant value of the ratio between two variables, represented by k in the equation $y = kx$

constant term (KAHN stuhnt turm) a number that is added or subtracted in a variable expression

context (KAHN tekst) the setting, events, or ideas surrounding something

context clue (KAHN tekst kloo) a word or words that give meaning to the words around them

contrast 1. (kuhn TRAST) *verb,* to examine differences between people, things, or ideas; 2. (KAHN trast) *noun,* the differences between people, things, or ideas

convert (kuhn VURT) to change

coordinate plane (koh AWR duhn it playn) formed by two number lines perpendicular to each other

correlation (kor uh LAY shun) a connection or relationship between two or more events or occurrences

Counting Principle (KOUNT ing PRIN suh puhl) using multiplication to get the total number of possible outcomes in a compound event

cross multiplication (krahss muhl ti pli KAY shuhn) finding the product of the numerator of one ratio multiplied by the denominator of another ratio

cube (kyoob) a rectangular solid in which the length, width, and height have the same measure

cube root (kyoob root) the cube root of a number is the number that, multiplied three times, will give the original number

currency (KUR en see) money amount

cylinder (SIL ihn dur) a three-dimensional shape with two circular ends and one curved side

D

data (DAY tuh) information that is collected and analyzed

decimal (DES uh muhl) based on a whole being split into ten equal parts one or more times

decimal point (DES uh muhl point) a period that separates whole numbers from decimal numbers

definition (def uh NISH uhn) the meaning of a word

denominator (di NAHM uh nay tur) the number on the bottom of a fraction; shows how many equal parts the number has been broken into

dependent events (di PEN duhnt i VENTSS) events that affect each other

dependent variable (di PEN duhnt VAYR ee uh buhl) an unknown number represented by a letter or symbol whose value depends on another number

details (DEE taylz) information that supports the main idea

diagram (DEYE uh gram) an illustration or picture that shows mathematical or other types of information

diameter (deye AM i tur) a line segment that crosses a circle through its center from one side to the other

difference (DIF ur uhnss) the answer to a subtraction problem

digit (DIJ it) the ten number symbols: 0, 1, 2, 3, 4, 5, 6, 7, 8, 9

distinguish (di STING gwish) to tell things apart

Distributive Property of Multiplication (dis TRIB yoo tiv PRAH per tee uhv muhl ti pli KAY shuhn) multiplying a sum by a number results in the same product as multiplying each addend by the number and adding the products; for example, $a \times (b + c) = (a \times b) + (a \times c)$

dividend (DIV i dend) the number that is divided in a division problem

division (di VIZH uhn) the operation that is used to separate a quantity into parts

divisor (di VEYE zur) the number that is dividing in a division problem

E

edge (ej) a line segment where two faces of a solid intersect

effect (i FEKT) what happens as a result of a cause

eliminate (ee LIM uh nayt) to completely remove or get rid of

ellipsis (i LIP siss) the symbol "..."; indicates that something has been omitted for space; at the end of a list, indicates that the list goes on forever

equal sign (EE kwuhl syne) (=) a symbol that means the two expressions on either side of it have the same value

equation (i KWAY zhuhn) a mathematical statement that two expressions have the same value

equilateral triangle (ee kwuhl LAH tur uhl TREYE ang guhl) a triangle with three sides of the same length

equilibrium point (EE kwuhl LIB ree uhm poynt) the point where two equations, when plotted on a graph, intersect and are equal

equivalent (i KWIV uh luhnt) representing the same value

equivalent equations (i KWIV uh luhnt i KWAY zhuhnz) equations that have the same solution

equivalent fractions (i KWIV uh luhnt FRAK shuhnz) fractions that have the same value

error (ER ur) a mistake

estimate 1. (ES tuh mayt) *verb,* to find an approximate answer; 2. (ES tuh mit) *noun,* an approximate answer

evaluate (i VAL yoo ayt) 1. to find the value of an expression; 2. to study closely

event (i VENT) one or more outcomes of an experiment

experimental probability (ek spair uh MEN tul prahb uh BIL i tee) the ratio of the number of success to the total number of trials

explain (ek SPLAYN) to tell

exponent (ek SPO nuhnt) the number that indicates how many times to multiply the base in a power

exponential curve (ek SPO nehn tshul kuhrv) a line produced by a formula in which the variable is not in the base of the function but in the exponent

expression (eks PRESH uhn) numbers, variables, and operators grouped together to show a value

extremes (ek STREEMZ) a and d in any proportion $\frac{a}{b} = \frac{c}{d}$

F

face (fayss) a flat surface of a solid

factor (FAK tur) a number that is multiplied

favorable outcomes (FAY vur uh buhl OUT kuhmz) the outcomes that are specified by a problem

formula (FAWR myoo luh) an equation that shows a relationship among its parts

fraction (FRAK shuhn) a way to represent parts of a whole

front-end digits (fruhnt end DIJ itss) the far-left digits in a number, used in front-end estimation

frustum (FRUHS tuhm) a three-dimensional shape created by removing the apex of a cone or pyramid

function (FUNK shun) a mathematical equation that has two variables

G

generalize (JEN ur uh lyz) to make a general statement inferred from particular information

geometric mean (jee oh MET rik meen) the square root of the product of two numbers; the geometric mean of x and y is $\sqrt{xy}$

gradient (GRAY dee uhnt) the rate of change of a line or plane; also slope

gram (gram) the basic unit of metric measure for mass

graph (graf) a diagram that gives a visual picture of data

greatest common factor (GRAY test KAHM uhn FAK tur) the largest number that divides exactly, without remainder into two or more numbers

grid (grid) a network of evenly spaced horizontal and vertical lines

grouping symbol (GROOP ing SIM buhl) something that groups numbers and variables together; for example, parentheses, brackets, fraction bars, and radical signs

H

height (hyte) the length of the segment perpendicular to the base of a triangle and extending to the "top" of the triangle

horizontal (hor i ZAHN tuhl) left to right

horizontal axis (hor i ZAHN tuhl AK siss) the left-to-right axis of a graph

hundredth (HUHN dridth) $\frac{1}{100}$ of a whole

hypotenuse (hi POT uh noos) the longest side of a right triangle, opposite the right angle

I

identify (eye DEN tuh feye) to find

illustrate (IL uh strayt) to use pictures to explain

impossible event (im PAWSS uh buhl i VENT) an outcome of an experiment with a probability of 0; it can never happen

improper fraction (im PRAWP ur FRAK shuhn) a fraction with a numerator greater than or equal to its denominator

independent events (in di PEN duhnt i VENTSS) events that do not affect each other

independent variable (in di PEN duhnt VAYR ee uh buhl) an unknown number represented by a letter or symbol whose value does not depend on another number

inequality (in i KWAWL i tee) a statement in which an inequality symbol is placed between two expressions

infinite (IN fuh nit) extends without end; endless

input (IN puht) something put in or given to a system

input variable (IN put VAIR ee uh buhl) a variable to which a rule is applied

integer (IN ti jur) a number that has both distance from zero and a direction (positive or negative), made up of the whole numbers and their opposites

interest (IN trist) money earned by an investment or paid when money is borrowed

intersect (IN tur sect) meet

inverse (IN vurss) opposite

inverse operations (IN vurss awp uh RAY shuhnz) operations that are opposite of each other and undo each other's results; for example, addition and subtraction are inverse operations

invert (in VURT) to flip a fraction, or to change the places of the numerator and denominator

irregular polygon (i REG yuh lur PAHL ee gahn) a polygon in which some or all of the side lengths and angles are not congruent

irrelevant (i REL uh vuhnt) unnecessary

isolate (EYE suh layt) to get the variable by itself on one side of an equation

isosceles triangle (eye SAHS uh leez TREYE ang guhl) a triangle with two sides of the same length

K

key (kee) a legend on a map or chart; explains what items in the map or chart stand for

key words (kee wurdz) the most important words in a sentence

L

label (LAY buhl) words or numbers written on a diagram to identify what the images are or what they represent

leaf (leef) a number in the right column in a stem-and-leaf plot; always has only 1 digit

leg (lehg) the two sides of a right triangle that are on either side of the right angle

length (length) the measure of the distance from one point to another; in a rectangle, length is the longer dimension

line (lyne) a set of points continuing in opposite directions

linear correlation (LIN ee uhr kor uh LAY shun) data that has a relationship producing a line when graphed

linear equation (LI nee uhr i KWAY zhuhn) an equation with two variables that produces a straight line when plotted on a graph

linear function (LIN ee uhr FUNK shuhn) an equation with two variables that produces a straight line on a graph; $y = mx + b$

linear relationship (LI nee uhr re LAY shun ship) a relationship that can be plotted as a straight line on a graph; as the independent variable increases, the dependent variable increases or decreases proportionally

like denominators (lyke di NAHM uh nay turz) denominators in two or more fractions that are the same

line graph (lyne graf) a graph made up of points that are connected by line segments; often used to display data over a period of time

line plot (lyne plawt) a data display that uses a number line with Xs or other marks to show how often each data value occurs

line segment (lyne SEG ment) 1. a line drawn to connect two points on a graph; 2. a set of points forming the shortest path between two points

liter (LEE tur) the basic unit of metric measure for liquid capacity

lowest terms (LOH ist turmz) a fraction in which the numerator and denominator cannot be divided evenly by the same number

M

main idea (mayn eye DEE uh) what a paragraph, article, or lesson is about

mathematical expression (math uh MAT i kuhl ek SPRESH uhn) any combination of symbols, numbers, and operations

mean (meen) the average value of a data set

means (meenz) b and c in any proportion $\frac{a}{b} = \frac{c}{d}$

measures of central tendency (MEZH urz uhv SEN truhl TEN duhn see) measures that describe the center of a data set

median (MEE dee uhn) the middle value of a data set listed in order from least to greatest

mental math (MEN tl math) applying certain strategies to find an answer without writing

meter (MEE tur) the basic unit of metric measure for length

mislead (miss LEED) to lead the reader to make a wrong conclusion

mixed number (mikst NUHM bur) the sum of a whole number and a fraction

mnemonic device (ni MAHN ik di VYSSE) a memory aid

mode (mohd) the item(s) that occurs most often in a data set

monetary units (MAW nuh tair ee YOO nitss) money amount

multiplication (muhl ti pli KAY shuhn) repeated addition

multiplicative inverse (muhl ti PLIK uh tiv IN vurss) another name for the reciprocal

mutually exclusive events (MYOO choo uhl ee ik SKLOO siv i VENTSS) events that cannot happen at the same time

N

negative (NEG uh tiv) (–) numbers to the left of 0 on a number line

negative correlation (NEG uh tihv kor uh LAY shun) two variables that have a relationship where, if one increases, the other decreases

no correlation (noh kor uh LAY shun) when data points do not have any definite relationship; on a scatter plot, the data points do not follow any linear or nonlinear trend

nonlinear function (NAHN lin ee uhr FUNK shuhn) a function that does not have the form $y = mx + b$ and that does not produce a straight line on a graph

number line (NUHM bur lyne) a list of numbers arranged in order from left to right on a line

numerator (NOO mur ay tur) the number on top of a fraction; shows how many of the equal parts are being counted

numerical pattern (NOO mer i kul PAT urn) a set of numbers related by a rule

O

objective (uhb JEK tiv) a goal that is set out to be achieved

obtuse angle (uhb TOOSS ANG guhl) an angle with measure greater than 90° and less than 180°

operation (ahp uh RAY shuhn) one of the mathematical processes: addition, subtraction, multiplication, or division

opposite (AHP uh zit) the opposite of a number is a number that is the same distance from zero on a number line as the original number but in the other direction

order of operations (AWR dur uhv ahp uh RAY shuhnz) the order in which operations should be performed

ordered pair (AWR durd PAYR) two numbers that tell exactly where a point lies in a plane; written (x, y)

organize (AWR guh nyze) to write a problem in a way that makes it easy to understand

origin (AWR uh jin) 1. the point to which the number 0 is assigned on a number line; 2. the point (0, 0) where the x-axis and y-axis intersect in a coordinate plane

outcome (OUT kuhm) the result of an experiment, such as flipping a coin and having it land on tails

outlier (OUT leye ur) a data value that falls well outside the range of other values in a set

output (OUT puht) something produced by a system

output variable (OUT put VAIR ee uh buhl) the result when a rule is applied to an input variable

overlapping events (o vur LAP ing i VENTSS) events that can occur at the same time

P

parallel lines (PAYR uh lel lynes) lines that run in the same direction that never cross or intersect

parallelogram (PAYR uh LEL uh gram) a quadrilateral with both pairs of opposite sides parallel and congruent

paraphrase (PAYR uh frayz) to use your own words to restate information

part (pahrt) a specified fraction of a whole

passage (PAS ij) a piece of writing

pattern (PAT urn) a repeated set of characteristics

percent (pur SENT) a way of expressing a number as part of a whole; means "for each 100"

perfect cube (PUR fikt kyoob) a number whose cube root is an integer

perfect square (PUR fikt skwair) a number whose square root is an integer

perimeter (puh RIM i tur) the distance around a figure, such as a triangle, rectangle, or square

periods (PEER ee uhdz) groups of three digits starting from the right of a number

permutation (pur myoo TAY shuhn) an arrangement of items or events in which order is important

perpendicular (pur puhn DIK yuh lur) when two lines intersect to form right, or 90°, angles

persuade (pur SWAYD) to convince someone of something

pi (peye) (π) the ratio of the circumference of a circle to its diameter; the value of pi is 3.1415926...

place value (playss VAL yoo) tens, ones, tenths, hundredths, thousandths

point (point) a mark made on a graph to represent the position of a data value; a specified place

point-slope form (poynt slohp fohrm) an equation that allows points on a line to be calculated if one point and the slope is known; $y - y_1 = m(x - x_1)$, where (x_1, y_1) is a known point, m is the slope, and (x, y) is any other point on the line

polygon (PAHL ee gahn) a closed flat figure made up of three or more line segments that are joined together

portion (PAWR shuhn) a part

positive (PAHZ i tiv) (+) numbers to the right of 0 on a number line

positive correlation (PAH zuh tihv kor uh LAY shun) two variables that have a relationship where, if one increases, the other increases as well

power (POU ur) a number with a base and an exponent where the base is multiplied by itself the number of times of the exponent

power of ten (POU ur uhv ten) 10 raised to a power; $10^1 = 10$, $10^2 = 100$, $10^3 = 1,000$, $10^4 = 10,000$, and so on

predict (pri DIKT) to make a logical guess

prediction (pri DIK shuhn) an attempt to answer the question, "What will happen next?"

prefix (PREE fikss) one or more syllables that are added to the beginning of a word

principal (PRIN suh puhl) the amount of money invested or borrowed

probability (prahb uh BIL i tee) the chance of an event occurring

product (PRAWD uhkt) the answer to a multiplication problem

proof (pruhf) a logical progression of true statements showing that a statement is factual

proper fraction (PRAWP ur FRAK shuhn) a fraction with a numerator less than its denominator

proportion (pruh PAWR shuhn) an equation with two equivalent ratios on opposite sides of the equal sign

proportional relationship (pruh PAWR shuhn ahl re LAY shuhn ship) sets of value pairs that have the same ratio

pros (prohz) positive outcomes

Pythagorean theorem (pi thag ohr EE uhn THEER uhm) the sum of the squares of the legs equals the sum of the square of the hypotenuse; $a^2 + b^2 = c^2$

Q

quadratic equation (kwah DRAT ik ih KWAY zhuhn) an equation containing a variable to the power of 2, but no higher powers, typically $ax^2 + bx + c = 0$, where $a \neq 0$

quadrilateral (kwahd ruh LAT ur uhl) a polygon with four sides

quotient (KWOH shuhnt) the answer to a division problem

R

radical sign (RAD i kuhl syne) $\sqrt{\ }$; indicates the square root

radius (RAY dee uhss) the distance from the center of a circle to any point on the curve of the circle; *plural* radii

range (raynj) the difference between the greatest and least items of a data set

rate (rayt) a relationship between two quantities measured in different units; the annual interest rate, usually given as a percent

ray (ray) a set of points continuing in one direction only

ratio (RAY shee oh) a comparison of two numbers

reasonable (REE zuh nuh buhl) sensible

reasoning (REE zuh ning) thinking

reciprocal (ri SIP ruh kuhl) the reciprocal of a number is the number that has a product of 1 when multiplied by the original number

rectangle (REK tang guhl) a parallelogram with four right angles

rectangular prism (rek TANG gyoo luhr PRIZ um) a three-dimensional solid that is made up of rectangles

rectangular solid (rek TANG gyoo luhr SAHL lid) another name for a rectangular prism

reduce (ri DOOSS) to simplify

regular polygon (REG yuh lur PAHL ee gahn) a polygon in which all the side lengths and angles are congruent

rename (ree NAYM) to find equivalent fractions

repeated (ri PEE tid) done again and again

repeating decimal (ri PEE ting DES uh muhl) a decimal with digits that repeat over and over

replacement (ri PLAYSS muhnt) drawing an object from a container and then returning it before drawing again

represent (rep ri ZENT) to stand for

restate (ree STAYT) to say again

restating (ree STAYT ing) putting an explanation into one's own words

reverse (ri VURSS) to change the direction

revise (ri VYZE) to rework

right angle (ryte ANG guhl) an angle with measure equal to 90°

rise (ryez) the vertical distance between points

rounding (ROUN ding) a common estimation strategy in which a number is increased or reduced to a place value that makes an estimation easier

row (roh) cells of information in a table arranged horizontally

rule (rool) an operation or operations applied to an input variable to form a pattern

run (ruhn) the horizontal distance between points

S

sample space (SAM puhl spayss) a list of possible outcomes

scale drawing (skayl DRAHW ing) a drawing of a place or object that is smaller or larger than the real place or object but that has the same ratios of length, width, and other measurements

scale factor (skayl FAK tur) the ratio found on a map key; for example 1 inch might represent 20 miles

scalene triangle (skay LEEN TREYE ang guhl) a triangle where no two sides are equal

scatter plot (SKAT uhr plaht) a graph with numerous points that do not all lie on the same line

scientific notation (seye uhn TIF ik no TAY shuhn) a way to write very large or very small numbers using multiplication and powers of ten

segment (SEHG mehnt) a finite portion of a line connecting two points, called end points

sequence (SEE kwuhnss) 1. the order in which events happen or things are arranged; 2. actions that happen in a certain order; 3. a set of numbers in a specific order

sign (syne) the symbol that indicates a number is positive or negative

similarity (sim uh LAYR i tee) things in common

simple interest (SIM puhl IN trist) the most basic type of interest; it depends on principal, rate, and time

simplify (SIM pluh feye) to rename to lowest terms

simultaneous (sy mull TAYN ee us) happening or occurring at the same time

slope (slohp) the rate of change of a line or plane; also steepness, grade, or gradient

slope-intercept form (slohp IN tur sept fohrm) an equation that produces a straight line using the slope and the y-intercept of the line

solid (SAHL id) a figure that has length, width, and height; a three-dimensional figure that occupies space

solution (suh LOO shuhn) 1. the value of the variable in an equation that makes the equation a true statement; 2. an answer

solution of an inequality (suh LOO shuhn uhv an in i KWAWL i tee) the set of all numbers that make the inequality true

solve (sahlv) to find the solution

sphere (sfeer) a round three-dimensional object like a ball or globe

square (skwair) a parallelogram with four congruent sides and four right angles

square root (skwair root) the square root of a number is the number that, multiplied by itself, will yield the original number

squared (skwaird) a number times itself

standard notation (STAN durd noh TAY shuhn) the most common way of writing numbers

stem (stem) the left column in a stem-and-leaf plot; can have one or more digits

stem-and-leaf plot (stem and leef plaht) a graph that has two columns; used to show distribution of data

straight angle (strayt ANG guhl) an angle with measure equal to 180°

strategy (STRAT uh jee) a plan

strong correlation (strahng kor uh LAY shun) a very close relationship between two variables, as seen on a scatter plot where the data points align very tightly along a trend line

subscript (SUHB skript) a letter or number written smaller and lower than the other text; 1 in y_1 or a in y_a

substitution (sub stih TOO shun) a mental math strategy in which numbers in a sum or difference are changed so it is easier to perform the operation

substitution method (sub stih TOO shun METH uhd) a direct way to solve a pair of linear equations without graphing by solving one equation and then substituting the solution of one of the variables in the second equation

subtraction (suhb TRAK shun) deducting, or taking away, an amount from another amount

success (suhk SESS) a favorable outcome

sum (suhm) the total; the answer to an addition problem

summarize (SUHM uh ryze) to restate the most important information from a passage in your own words

support (suh PAWRT) to give weight or credibility to something

symbolic expression (sim BAHL ik ik SPRESH uhn) an expression that uses variables, numbers, and symbols for operations

synthesize (SIN thuh syze) to combine two or more ideas to create a new, more complex idea

system of simultaneous linear equations (SIS tum uhv sy mull TAYN ee us LIN ee uhr i KWAY zhuhns) a collection of equations that produce lines when graphed and that all intersect at some point

T

table (TAY buhl) a way to organize information in rows and columns

tenth (tenth) $\frac{1}{10}$ of a whole

term (turm) a number in a sequence

theoretical probability (thee uh RET i kuhl prahb uh BIL i tee) the ratio of the number of favorable outcomes to the total number of possible outcomes; it is based on outcomes that are equally likely

three-dimensional figure (three duh MEN shuh nuhl FIG yur) a solid with length, width, and height; it occupies space

theorem (THEER uhm) a mathematical statement that is not self-evident, but shown to be true through a proof

tic mark (tik mahrk) a division on a number line

time (tyme) the measure of a period during which something exists; in the interest formula, it is the length of time in years the money is invested or borrowed

title (TYTE uhl) the name of book, passage, chart, or graph

topic (TAH pik) the main idea of a passage

tree diagram (tree DEYE uh gram) a diagram that shows sample space and the number of possible outcomes

trend (trend) a general direction in which data tends to move

trend line (trehnd leyn) a line that can be drawn through nearly linear points on a scatter plot

trials (TREYE uhlz) the number of times an experiment is repeated

triangle (TREYE ang guhl) a polygon that has three sides

two-point form (too poynt fohrm) a method used to generate the equation for a straight line using two given points

two-step equation (too step i KWAY zhuhn) an equation that contains two different operations

two-variable equation (too VAIR ee uh buhl i KWAY zhuhn) an equation that contains two different variables

U

unit (YOO nit) an amount used to measure length, capacity, weight, and time

unit price (YOO nit pryse) price for one unit of a quantity

unit rate (YOO nit rayt) the rate for one unit of a quantity

unlike denominators (uhn LYKE di NAHM uh nay turz) denominators in two or more fractions that are different

V

value (VAL yoo) how much a digit represents

variable (VAYR ee uh buhl) a symbol that stands for an unknown number or value

Venn diagram (ven DEYE uh gram) a graphic organizer used for showing membership in sets

verbal expression (VER buhl ik SPRESH uhn) an expression written in words and numbers

vertex (VUR tekss) a point where edges of a solid come together; *plural* vertices

vertically (VUR ti kuhl le) up and down

vertical line test (VER tih kal leyn tehst) a method to determine if an equation is a function; if a vertical line crosses a graphed line at more than one point, the equation is not a function

vertical axis (VUR ti kuhl AK siss) the up-and-down axis of a graph

volume (VAHL yoom) the measure of the space inside a three-dimensional figure

W

weak correlation (week kor uh LAY shun) a loose relationship between two variables, as seen on a scatter plot, where the data points are not tightly aligned to a trend line

weight (wayt) the measure of the heaviness of something

whole (hohl) all the parts of something taken together

whole number (hohl NUHM bur) the number system beginning with 0, 1, 2, 3, and so on

width (width) the shorter dimension in a rectangle

X

x-axis (ekss AK siss) the horizontal line in a coordinate plane

x-coordinate (ekss koh AWR duhn it) the first number in an ordered pair

Y

y-axis (weye AK siss) the vertical line in coordinate plane

y-coordinate (weye koh AWR duhn it) the second number in an ordered pair

y-intercept (why IN tur sept) the value at which a line crosses the *y* axis

Index

Distances:
    between points on coordinate graph, 356
    on scale drawings, 339–340
Distributive Property, 29
Distributive Property of Multiplication, 29
Dividend:
    definition of, 23
    in dividing decimals, 64
Division:
    choosing operations, 42
    of decimals, 64–67
    definition of, 23
    of fractions, 89–90
    of integers, 120–121
    of mixed numbers, 95
    by power of 10, 319
    to solve inequalities, 154
    of whole numbers, 23–24
Division equations:
    to solve inequalities, 152–153
    solving, 142
Division Property of Equality, 140, 151
Divisor:
    definition of, 23
    in dividing decimals, 64
Draw a picture strategy, 43
Draw Evidence from Text (reading skill), 24
Drawing conclusions, 33

E

Edge:
    of a cube, 363
    definition of, 360
    of rectangular solid, 360
Ellipsis (...), 104, 233
Equal sign (=), 140
Equation(s):
    definition of, 140
    one-step, 140–142
    quadratic, 354
    two-step, 144–147
Equilateral triangle, 328
Equilibrium point, 186
Equivalent equations, 140
Equivalent fractions, 76–78
    definition of, 76
    and proportions, 224–225
Equivalent ratios, 224, 226
Estimate (term), 32
Estimation, 34–35
Ethics, probability and, 297
Evaluate:
    arithmetic expressions with exponents, 251–252
    definition of, 137
    exponents, 250–251
    expressions, 137–138

Evaluate Arguments (reading skill), 94
Evaluate Expressions (core skill), 136, 147, 251, 313
Evaluate Reasoning (core skill), 65, 153, 220, 238, 255, 298, 328
Evaluation (term), 29
Event(s):
    certain, 296
    compound, 292, 302–305
    definition of, 292
    dependent, 305
    impossible, 296
    independent, 304
    mutually exclusive, 302–303
    overlapping, 303
Experimental probability, 299
Exponent(s), 250–252
Exponential curve, 194
Exponential model, 194
Expression(s), 134–138
    definition of, 29
    evaluating, 137–138
    representing patterns, 156–159
    verbal, 136–137
    verbal and symbolic representations of, 134–137
Extremes, 238

F

Faces, 360
Factoring, 26–30
Factors:
    comparing sets of, 28
    definition of, 22, 26
    finding, 26–27
Favorable outcomes, 296
Fewer than (<), 150
Find Reverse Operations (core skill), 22, 23
Five-step approach (problem solving), 40–41
FOIL method, 354
Fractions, 74–79
    add and subtract, 82–85
    compare and order, 78–79
    as decimals, 233–234
    decimals as, 233
    definition of, 75
    divide, 89–90
    equivalent, 76–78, 224–225
    improper, 92
    multiply, 88–89
    as percents, 233–234
    percents as, 237
    proper, 92
    writing percents as, 232
    writing ratios as, 214
Frameworks, 136
Front-end digits, 34
Front-end estimation (*See* Whole numbers)
Frustum, 368, 371

Mathematical Formulas

Area of a:	square	$Area = side^2$
	rectangle	$Area = length \times width$
	triangle	$Area = \frac{1}{2} \times base \times height$
	parallelogram	$Area = base \times height$
	trapezoid	$Area = \frac{1}{2} \times (base_1 + base_2) \times height$
	circle	$Area = \pi \times radius^2$; π is approximately equal to 3.14

Perimeter of a:	square	$Perimeter = 4 \times side$
	rectangle	$Perimeter = 2 \times length + 2 \times width$
	triangle	$Perimeter = side + side + side$

Circumference of a circle: $Circumference = \pi \times diameter$: π is approximately equal to 3.14

Volume of a:	cube	$Volume = edge^3$
	rectangular solid	$Volume = length \times width \times height$
	square pyramid	$Volume = \frac{1}{3} \times (base\ edge)^2 \times height$
	cylinder	$Volume = \pi \times radius^2 \times height$; π is approximately equal to 3.14
	cone	$Volume = \frac{1}{3} \times \pi \times radius^2 \times height$; π is approximately equal to 3.14

Coordinate Geometry

distance between points $(x_2 = x_1)^2 + (y_2 - y_1)^2$; (x_1, y_1) and (x_2, y_2) are two points in a plane

slope of a line $= \dfrac{y_2 - y_1}{x_2 - x_1}$; (x_1, y_1) and (x_2, y_2) are two points on the line

Pythagorean Relationship

$a^2 + b^2 = c^2$; in a right triangle, a and b are legs, and c is the hypotenuse

Trigonometric Ratios

$\sin = \dfrac{opposite}{hypotenuse}$ $\cos = \dfrac{adjacent}{hypotenuse}$ $\tan = \dfrac{opposite}{adjacent}$

Measures of Central Tendency

mean $= \dfrac{x_1 + x_2 + \ldots + x_n}{n}$, where the x's are the values for which a mean is desired, and n is the total number of values for x

median $=$ the middle value of an odd number of ordered scores, and halfway between the two middle values of an even number of ordered scores

Simple Interest

$interest = principal \times rate \times time$

Distance

$distance = rate \times time$